Welding Print Reading

by

John R. Walker

South Holland, Illinois
THE GOODHEART-WILLCOX COMPANY, INC.
Publishers

Library of Congress Catalog Card Number 91-18309
International Standard Book Number 0-87006-890-3

234567890-91-098765432

Library of Congress Cataloging in Publication Data

Walker, John R.
Welding print reading / by John R. Walker.

p. cm.
Includes index.
ISBN 0-87006-890-3
1. Blueprints. 2. Welding--Drawings. I.Title.
T379.W28 1991
671.5'2-dc20 91-18309
CIP

IMPORTANT SAFETY NOTICE

The procedures and practices described in this textbook are effective methods of performing given tasks. However, note that this information is general and applies to most situations. You must make sure that all weld requirements are being fulfilled and that the resulting weldment will meet design specifications. You must also make sure that you are following all safety rules!

This book contains the most complete and accurate information that could be obtained from authoritative sources at the time of publication. The Goodheart-Willcox Company, Inc. cannot assume responsibility for changes, errors, or omissions.

INTRODUCTION

WELDING PRINT READING provides instruction on interpreting and using the type of engineering drawings or prints found in the welding trade. It is a write-in text or text-workbook that starts out with the basics and progresses to more specialized coverage of welding symbols and notations.

Welding symbols and notations allow a large amount of data about the weld to be condensed into a small amount of space on a print. They simplify communications between the designer/engineer and the welder and also between other workers associated with the production of a weldment. Symbols and notations help assure that welds meet design requirements.

Just about every manufactured product uses welding, either directly or indirectly. Over the years, a complex system has been developed to convey exact weld specifications. A welder, or anyone else (technician, engineer, drafter, etc.) working with welding prints, must know how to use these symbols and notations. This text is designed to help you grasp this information as quickly and easily as possible.

When using this book, you will find that the example prints, symbols, and problems provide a realistic up-to-date coverage of welding prints. The information in this text follows the recently revised standards set up by the American Welding Society (AWS) and the American National Standards Institute (ANSI). This will prepare you to work with the prints now being used in industry.

WELDING PRINT READING is intended for students in vocational/technical schools, for apprentices, and for workers on the job. The text may also be used as a self-study course for those unable to attend print reading classes.

John R. Walker

CONTENTS

Unit 1

PRINTS—THE LANGUAGE OF INDUSTRY

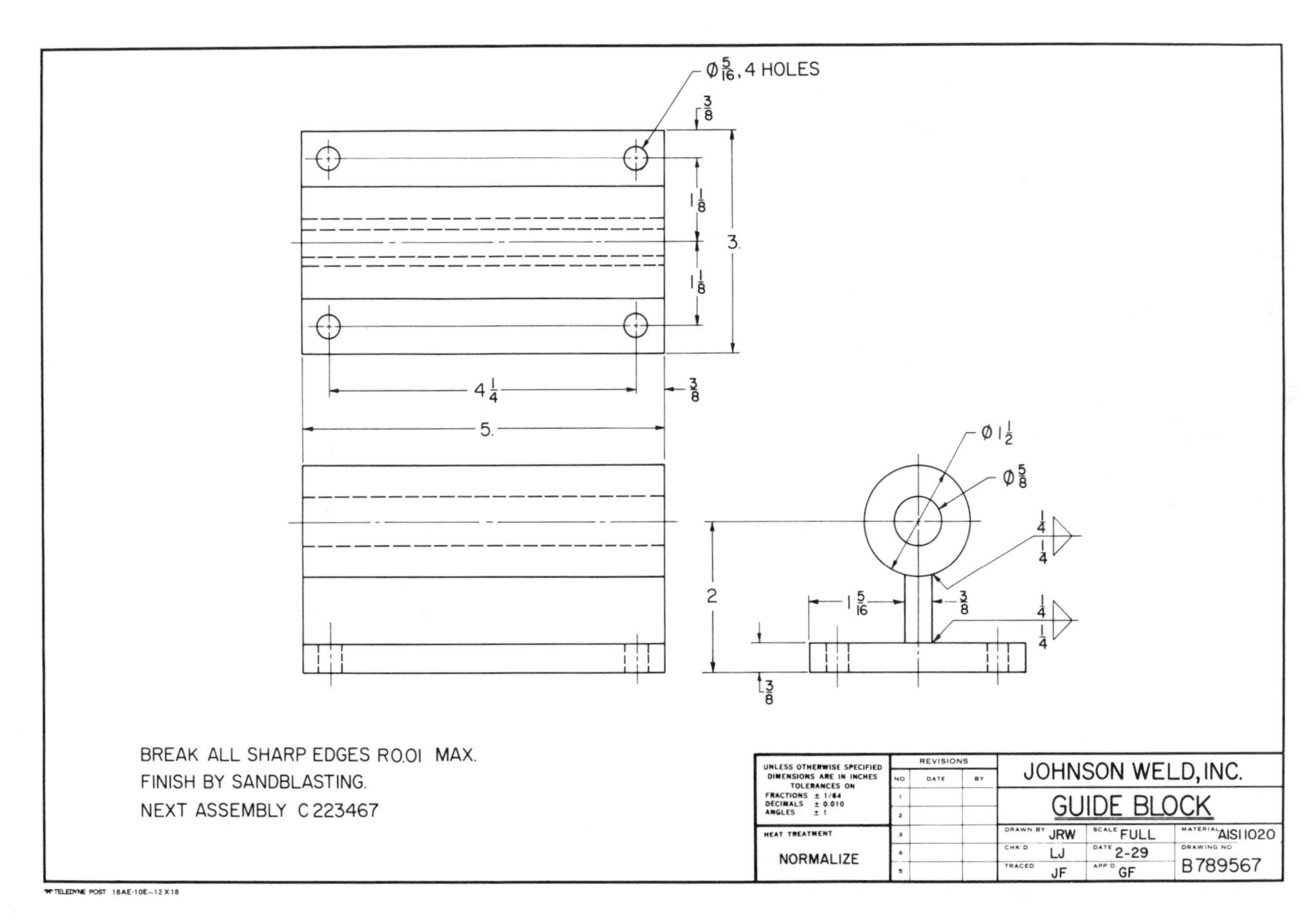

Fig. 1-1. Drawings or prints are utilized to commmunicate ideas in graphic or picture form.

Drawings are used by the welding industry to communicate ideas in graphic or picture form, Fig. 1-1. They are often the only means of showing the construction of a complex product. Refer to Fig. 1-2.

Drawings are called the "LANGUAGE OF INDUSTRY." It is a precise language that utilizes symbols.

Symbols (lines and figures) have specific meanings to accurately describe the shape, size, type of material, finish, and fabrication of an object. The symbols have been standardized and are used as a universal language over most of the free world. This standardization makes it possible to interpret and understand drawings made in other countries.

WHAT IS A PRINT?

A *print* is a duplicate of an original drawing. Basically, a print consists of:

1. Lines (show part surfaces and points of machining).
2. Dimensions (give sizes of part).
3. Notes (provide information not given by lines and dimensions).

4. Specifications (special notes for standards, type material, or specific process to be used).
5. Views (normally front, side, and top of part).

Fig. 1-2. Drawings are best way of showing how to construct something as complex as this 265,000 ton tanker. It is difficult to imagine trying to explain how to fabricate this ship using only words. (Bethlehem Steel Corp.)

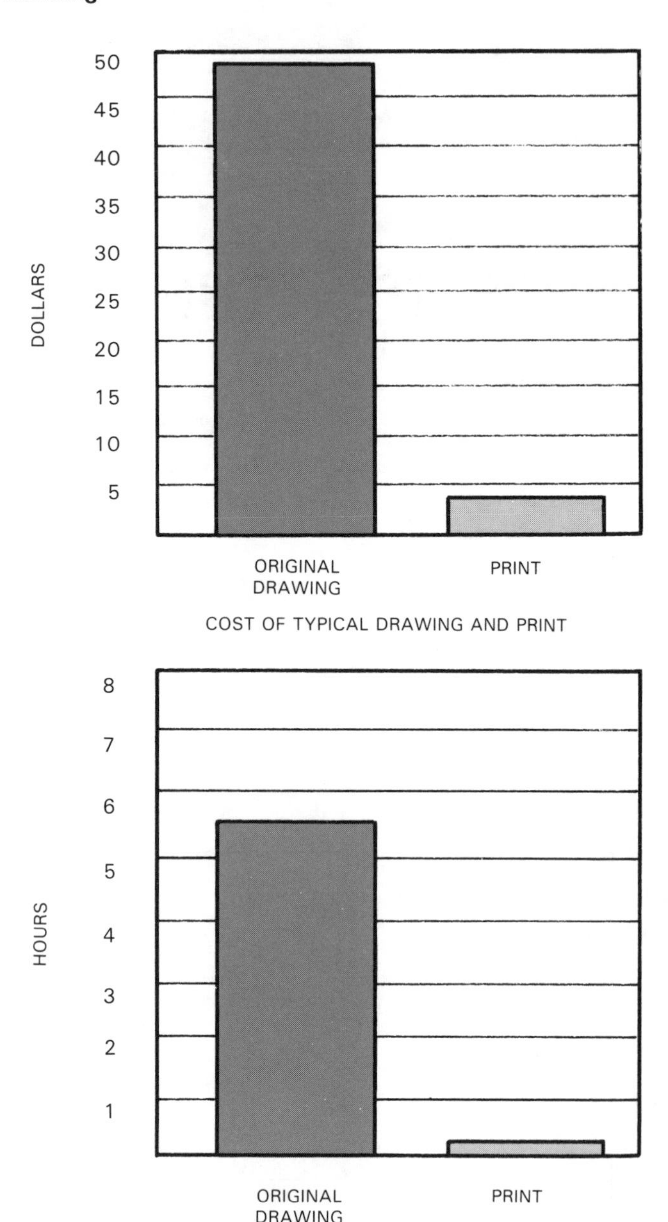

Fig. 1-3. Compare cost and time of using original drawings and prints. It would be impractical to draw a set of plans for each worker and/or to replace original drawings when they became worn out, damaged, or destroyed.

WHY PRINTS ARE USED

Original engineering drawings are seldom, if ever, used on a job. If they were, they would soon become worn, soiled, and difficult to read. Also, several sets of identical drawings are usually needed at the same time because welders at different locations are working on the same product or structure.

It would be very expensive and impractical to draw a set of plans for each worker who needed them. It is also impractical to replace original drawings when they wore out or became damaged in normal shop use. Fig. 1-3 compares the cost of and time to produce original drawings and prints.

Instead, a reproduction technique is used to make accurate copies of the original drawings quickly and inexpensively. These copies are usually *white prints* (dark lines on a light background). In the shop, they are known as *prints, drawings,* and *blueprints.* Fig. 1-4 illustrates a common print and a less common blueprint.

The original drawings are NOT damaged or destroyed by the printmaking process. This permits them to be stored for reference purposes and/or future copying.

Prints are usually discarded or destroyed when they are no longer needed.

HOW PRINTS ARE MADE

Original engineering drawings are made on a semitransparent (translucent) material such as tracing paper, cloth, or plastic film. The drawings are frequent-

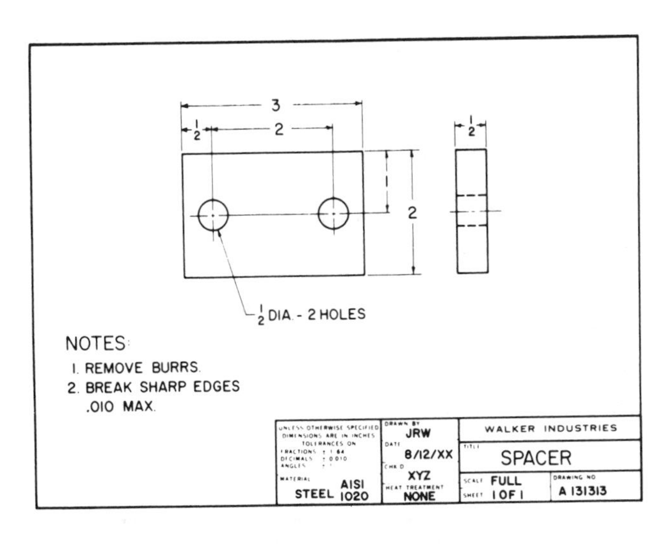

POSITIVE PRINT

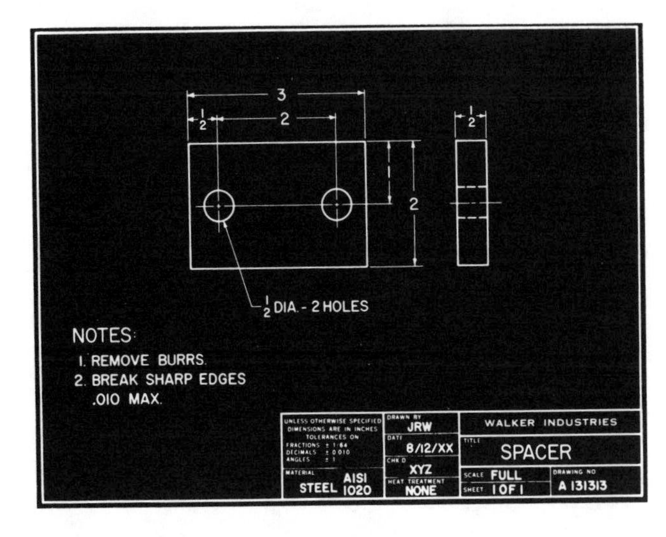

BLUEPRINT

Fig. 1-4. True blueprint is a print with white lines on a blue background. The term "blueprint" is commonly used when referring to all types of prints regardless of color of lines or background.

ly called *tracings*. They are prepared by DRAFTERS. Refer to Fig. 1-5.

Blueprints

The *blueprint process* is the oldest of the techniques employed to duplicate drawings. Blueprints are seldom used today.

To make a print by the blueprint method, the tracing is placed in contact with a sensitized paper (unexposed blueprint paper). The two sheets are exposed to a bright light. The light cannot penetrate the opaque lines of the tracing and exposure takes place only where the light strikes the sensitive coating on the print paper.

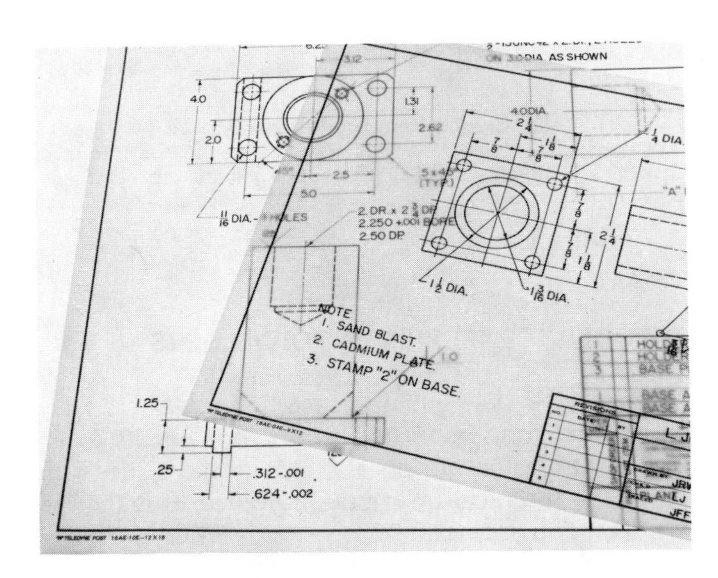

Fig. 1-5. Original engineering drawings are done on a semitransparent material such as tracing paper, cloth, or plastic film. These tracings are used to make prints.

The exposed print is developed by washing in water. The lines are "fixed" by washing in a solution of potassium dichromate crystals dissolved in water to prevent the lines from fading. After a final washing, the print is dried and pressed.

Diazo process

The *diazo process* is a common and versatile copying technique for making direct positive copies or white prints (dark lines on light background). It can be used to reproduce anything that is drawn, written, or typed on semitransparent paper, cloth, or film. The process produces the copies quickly and inexpensively.

The print is made by placing the tracing in contact with sensitized copying material (paper, cloth, or film) and exposing it to light. After the tracing has been removed, the exposed copy is developed by passing it through ammonia vapors. See Fig. 1-6.

With most modern diazo copying machines, Fig. 1-7, the print is made in one continuous process. In addition to turning out the finished print in seconds, it is clean, odorless, and quiet.

Using this process, black-on-white, black-on-color, color-on-white, or color-on-color prints can be produced.

Microfilm process

The *microfilm process* was originally designed to reduce storage facilities and to protect prints from loss. With this process, the drawing is reduced by photographic means. Finished negatives can be stored in roll form, or on cards adaptable to computerized storage and retrieval.

To produce a working print, the microfilm image

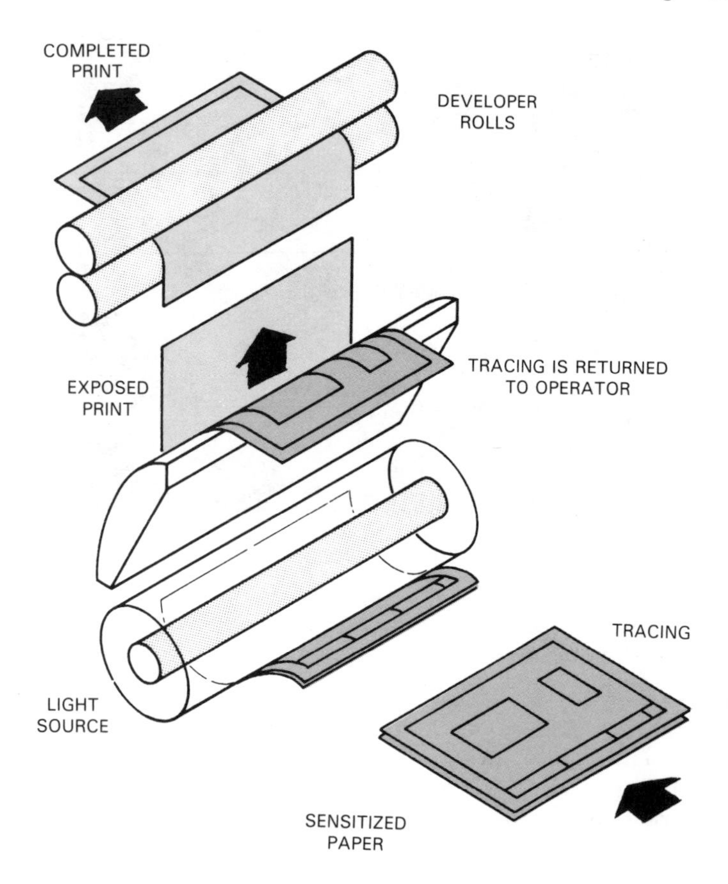

Fig. 1-6. Study diazo process. On most machines, print is made in one continuous operation.

is retrieved from the files and enlarged (called BLOW-BACKS) on photographic paper. The print is discarded or destroyed when it is no longer needed. Microfilms can also be viewed on a reader to check details.

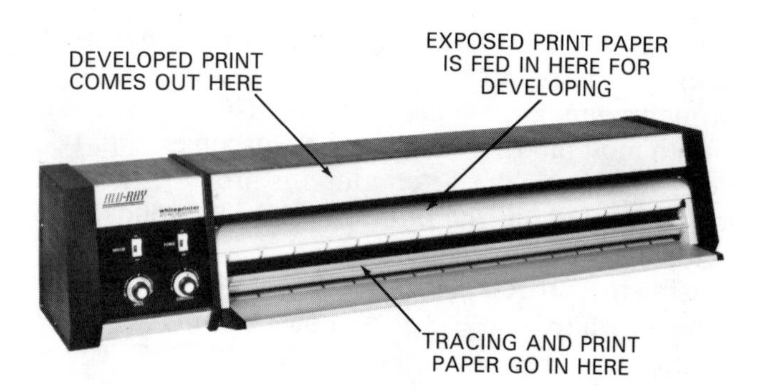

Fig. 1-7. A combination printer and developer type diazo copying machine. (Blue-Ray, Inc.)

Xerography process

Xerography (pronounced ze-rog'-ra-fee) prints from dry materials. No fluids are used. The process uses an electrostatic charge (stationary electric charge) to duplicate an original written, drawn, or printed image. Look at Fig. 1-8.

Xerography, also called *electrostatic process,* is based on the principle that like electrical charges repel and unlike charges attract.

The xerographic process makes a copy exactly duplicating the orginal. It is clean and dry. The copy can be enlarged or decreased in size if desired. Full color copies can be made on color copying xerographic machines.

The main use of xerographic printing is to make inexpensive duplicate copies.

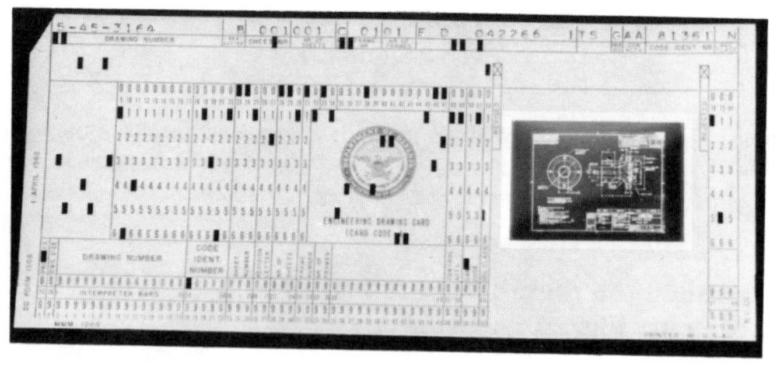

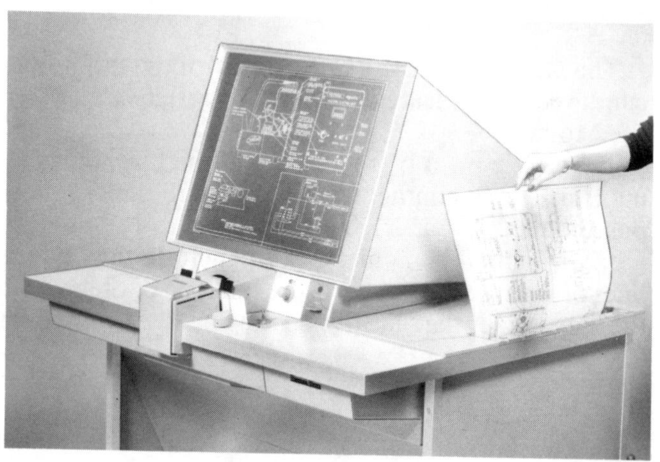

Fig. 1-8. A—Small negative on microfilm aperture card is enlarged by a photographic process to desired print size. Card is easier to store than full size prints and can be retrieved by computerized techniques. B—A microfilm reader. Enlarged print can be studied on a view screen in this unit and then made into a print of required size. (RECORDAK)

IMPORTANCE OF PRINTS TO WELDERS

Prints are of special importance to the welder. They show where and how the various components to be welded fit together. Prints also provide the welder with all of the technical information needed to make the weld(s). This includes:

1. Welding process to be used, Fig. 1-9.
2. Weld size.
3. Weld type.
4. Kind of filler metal.

5. How the weld is to be finished.
6. Other pertinent data needed to make welds meet design specifications.

In addition to having welding skills, welders must know how to read and interpret prints. Otherwise, there is little chance the welds will be made to specifications,

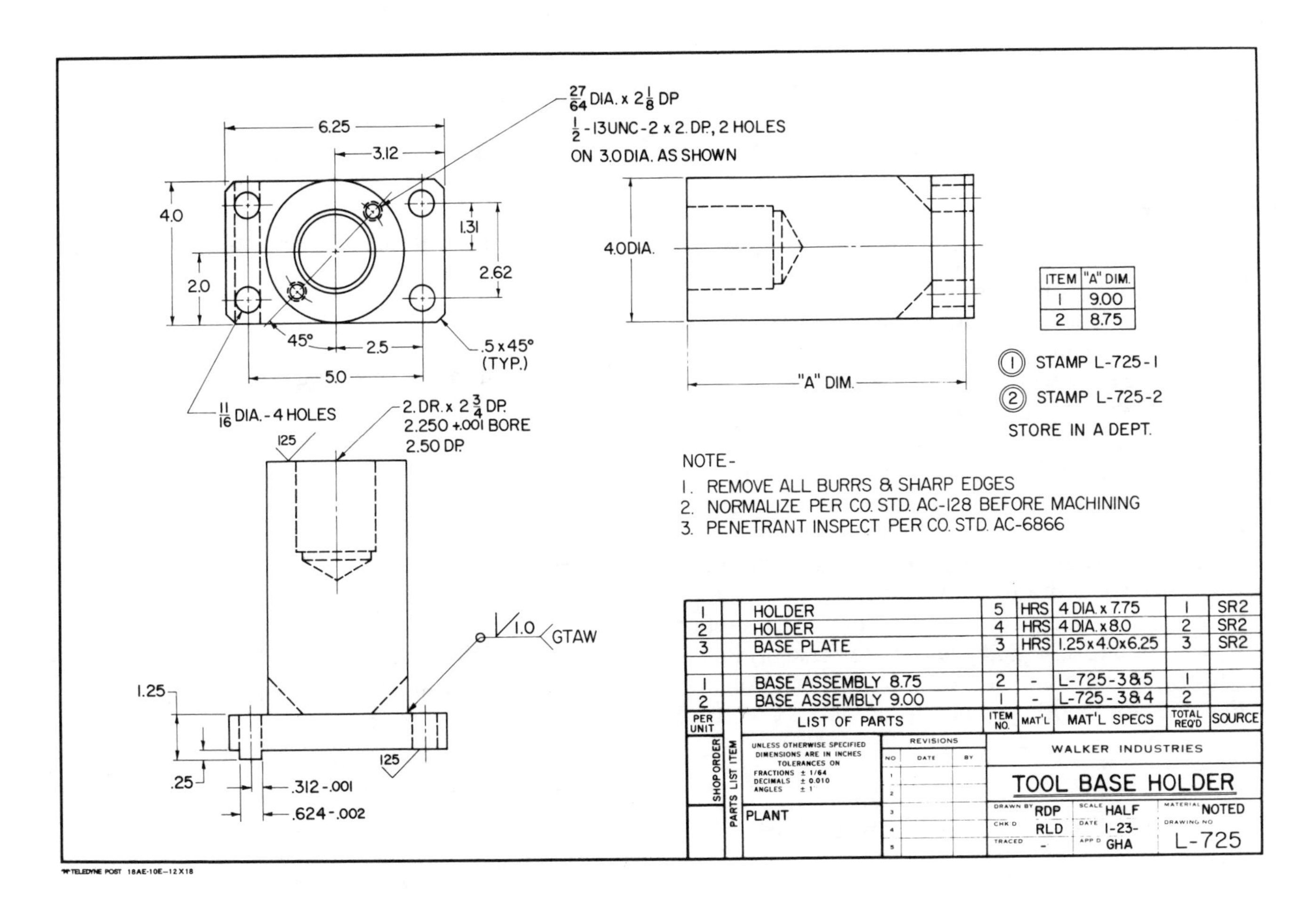

Fig. 1-9. In addition to having welding skills, welders must know how to read and interpret prints.

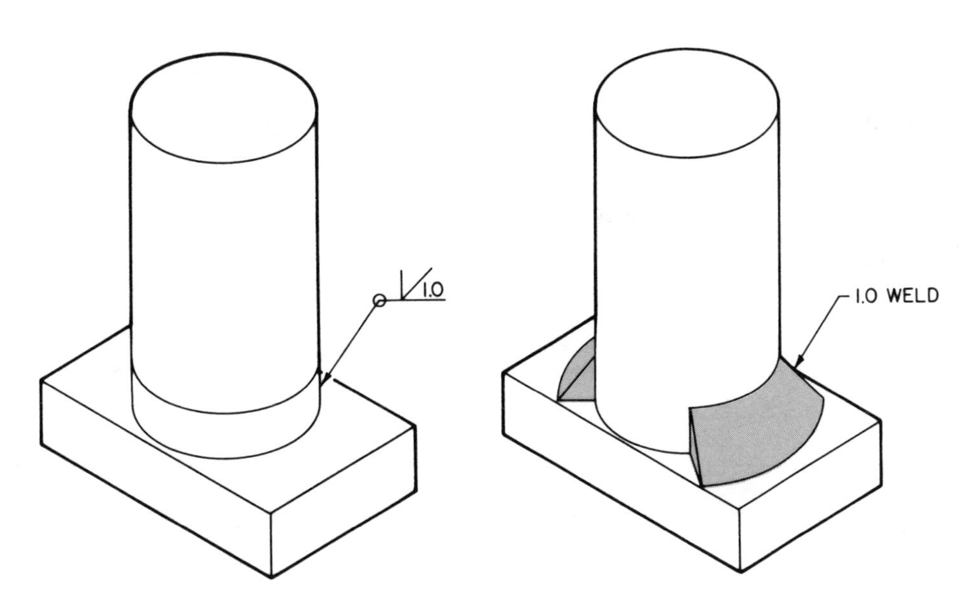

Fig. 1-10. Without a print describing welding process, weld size and type, kind of filler metal, method of finish, and other pertinent data, there is no way welder can be sure of making a weld that will meet design specifications.

Refer to Fig. 1-10.

Not being able to read and interpret prints could result in a welded product that was:

1. Very costly because more welding was done than necessary.
2. Unsafe or dangerous because the welds were NOT strong enough.
3. Rejected because it does not meet design specifications.

CARE OF PRINTS

If given proper care, a print will have a long, useful life before it must be discarded because of wear.

1. Keep prints clean. Dirty prints are difficult to read. This can cause errors when welding.
2. Avoid tearing prints. Fold and unfold prints carefully when you use them.
3. Avoid laying tools and other objects on prints.
4. Do NOT use a print as a "table cloth" when eating lunch.
5. Do NOT make revisions (changes) on a print unless you have written authority to do so.
6. Some prints are concerned with work of a secret or classified nature. Handle and store these prints according to company or agency security policies. Penalties for ignoring security precautions can prove costly to you.

UNIT 1—TEST YOUR KNOWLEDGE

1. How are drawings used by industry? ________

____

____

____.

2. Drawings are often called the ____________

____.

3. List the reasons why it is impractical to use the original engineering drawings in the shop.

   a. ________________________________

   ____

   b. ________________________________

   ____

   c. ________________________________

   ____

   d. ________________________________

   ____

4. What is used in the shop in place of the original engineering drawings? ________________

____.

5 The above (question 4) are called __________

____.

6. What is a white print?________________

____

____

7. Prints provide the welder with ____________

____

____.

8. Why is it important for a welder to be able to read prints?________________________

____

____

____

____

9. Given proper care, a print will have a long useful life. What does this care involve? __________

____

____

____

____

____

____

____

____

10. Summarize four methods of making prints. __

____

____

____

____

____

____

____

____

____

____

____.

Unit 2

REVIEW OF MEASUREMENT

Precise measurement is extremely important in welding. Mistakes made in measuring are very costly—in both time and material. A welder must know how to make measurements accurately and quickly. He or she must also be familiar with the proper way to use the measuring tools of the welding trade, Fig. 2-1.

Reading a fractional inch graduated rule

The *inch rule* can read as small as 1/64 in. (one-sixty fourth of an inch). For general welding work, you should be able to read a rule to 1/16 (one-sixteenth) of an inch. Precise weldments require more accuracy.

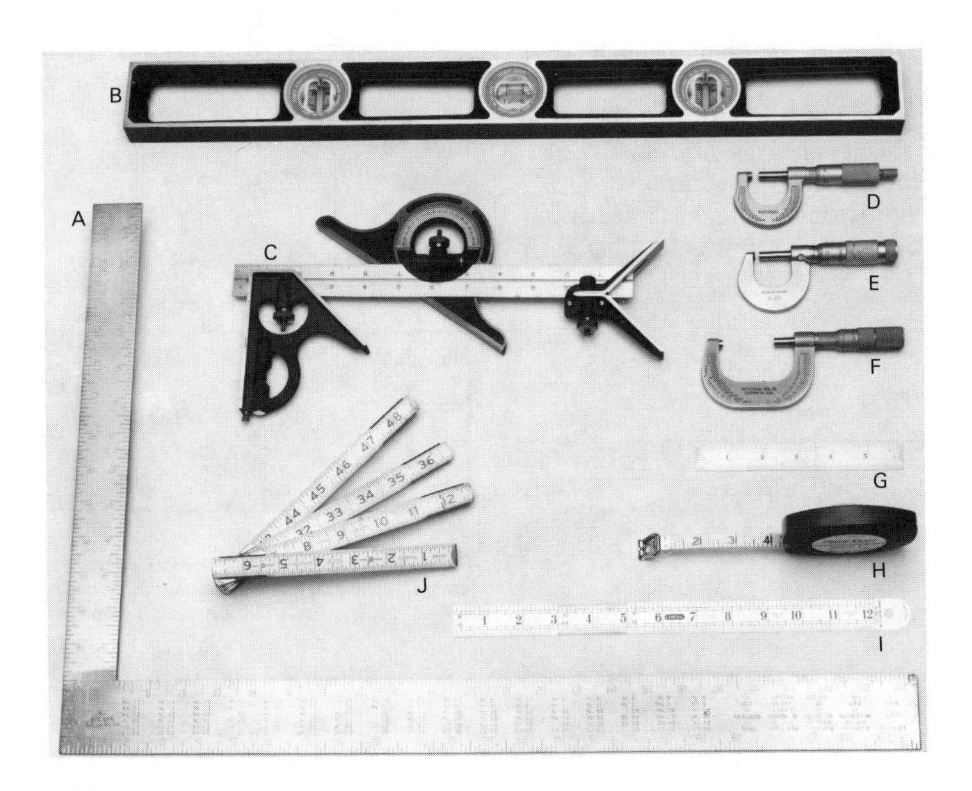

Fig. 2-1. Note some of the basic measuring tools used by welders. How many of them can you use correctly? A—Square. B—Level. C—Combination square. D, E, and F—Micrometers. G—Six-inch rule. H—Tape measure. I—One foot rule. J—Folding rule.

TYPES OF MEASUREMENT

A welder must be able to read rules and tapes graduated in *inches* (fractions and decimals) and *metrics* (millimeters). Fig. 2-2 shows inch, metric, and decimal inch rules.

Secure a rule or tape measure and examine the edges used for measuring. Each mark or division on the edge of the rule is called a *graduation.*

Look carefully at the edge used to measure 1/16 in. Many rules have 16 or 1/16 stamped or engraved on this edge. If your rule does NOT, count the

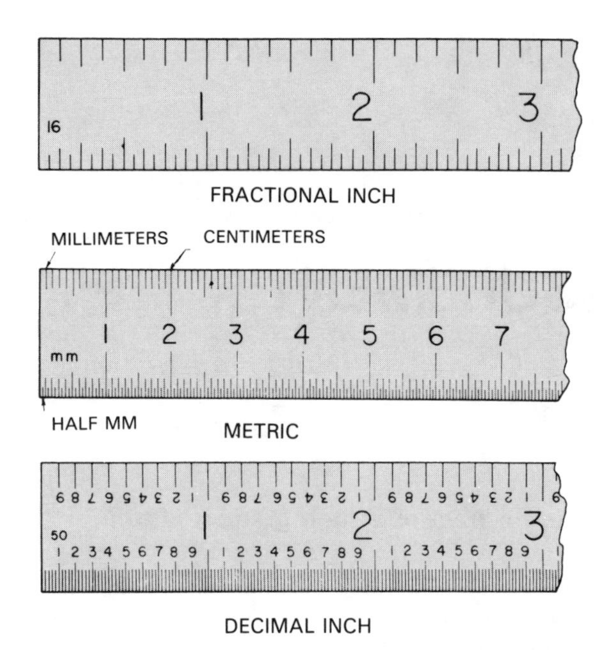

Fig. 2-2. Study these rules. Compare the divisions of each carefully.

graduations. There will be 16 of them per inch. This is shown in Fig. 2-3.

Note that the 1 in. graduation is the longest. The 1/2 in. graduation is the next longest, and so on down to the 1/16 graduation which is the shortest.

Until you are familiar with the rule, visualize or imagine that each 1/16 in. graduation is numbered as shown in Fig. 2-4.

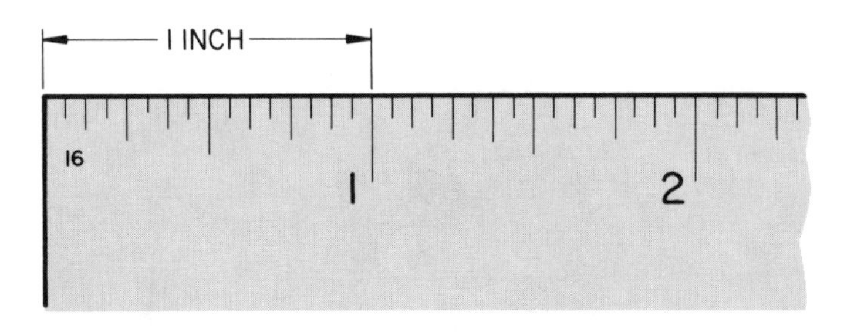

Fig. 2-3. There are 16 equal divisions in each inch of this rule. Each division or graduation is equal to 1/16 in.

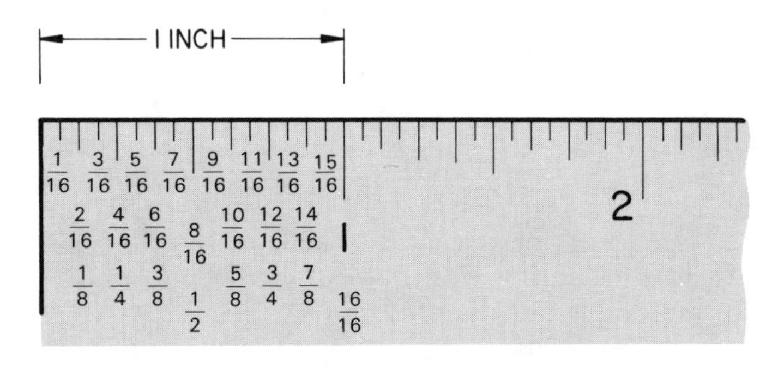

Fig. 2-4. Until you are familiar with the rule, memorize the value of each division. Study this scale closely.

You may find it easier to count the graduations or divisions when you first start to use the rule. After some practice, however, this should not be necessary.

One thing to remember, fractional measurements are always reduced to the *lowest terms* (numerator and denominator have no common factor). For example, a measurement of 8/16 (divide top and bottom by eight) would be read as 1/2, or 2/16 (divide by two) would read as 1/8.

Practice making measurements until you become proficient enough to read the scale accurately and quickly.

Reading a decimal inch graduated rule

The more common *decimal rule* has 10 divisions per inch (one division equals 0.1 inch) on one edge and 50 divisions (one division equals 0.02 inch) or 100 divisions (one division equals 0.01 inch) on a second edge. Fig. 2-5 illustrates a decimal inch rule.

General welding seldom requires measurements as close as 1/50 (0.2) or 1/100 (0.01) inch. Only special welding jobs need this much accuracy.

The same learning sequence is used to make accurate decimal measurements as was used with the fractional inch rule.

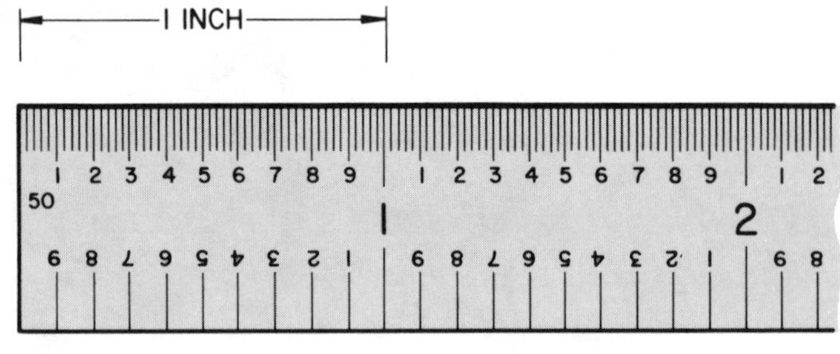

Fig. 2-5. More common decimal rule has 10 divisions per inch, with each equaling 0.1 in., on one edge and 50 divisions, each equaling 0.02 in., on second edge.

Reading a metric graduated rule

A *metric rule* is graduated in millimeters, abbreviated mm. In addition to using fractional and decimal inch measurements, the welder must also be familiar with metric measure.

You may find the metric unit rule easier to read than a fractional inch rule. There are no fractions to complicate your calculations. Numbers in the metric system are added, subtracted, multiplied, and divided as with decimal inch numbers.

A metric rule or tape is graduated in millimeters (mm) on one edge and with one-half millimeter (0.5 mm) graduations on the second edge. Every ten millimeters (each centimeter) is numbered. This is shown in Fig. 2-2.

The United States is in the process of adapting the INTERNATIONAL SYSTEM OF UNITS (the metric system). The measurement units of the International System (abbreviated SI) are shown in Fig. 2-6.

MEASUREMENTS ON DRAWINGS

Many large firms that use prints are presently converting to the metric system. During the transition period that will take many years, welders and other craftworkers will use drawings that are:

PREFIXES, EXPONENTS, AND SYMBOLS

DECIMAL FORM	EXPONENT OR POWER	PREFIX	PRONUNCIATION	SYMBOL	MEANING
1 000 000 000 000 000 000	$= 10^{18}$	exa	ex'a	E	quintillion
1 000 000 000 000 000	$= 10^{15}$	peta	pet'a	P	quadrillion
1 000 000 000 000	$= 10^{12}$	tera	tĕr'á	T	trillion
1 000 000 000	$= 10^{9}$	giga	ǰi'gá	G	billion
1 000 000	$= 10^{6}$	mega	mĕg'á	M	million
1 000	$= 10^{3}$	kilo	kĭl'ō	k	thousand
100	$= 10^{2}$	hecto	hĕk'to	h	hundred
10	$= 10^{1}$	deka	dĕk'a	da	ten
1					base unit
0.1	$= 10^{-1}$	deci	dĕs'ī	d	tenth
0.01	$= 10^{-2}$	centi	sĕn'tǐ	c	hundredth
0.001	$= 10^{-3}$	milli	mĭl'ǐ	m	thousandths
0.000 001	$= 10^{-6}$	micro	mi'krō	μ	millionth
0.000 000 001	$= 10^{-9}$	nano	năn'ō	n	billionth
0.000 000 000 001	$= 10^{-12}$	pico	péc ō	p	trillionth
0.000 000 000 000 001	$= 10^{-15}$	femto	fĕm'tō	f	quadrillionth
0.000 000 000 000 000 001	$= 10^{-18}$	atto	ăt'tō	a	quintillionth

Most commonly used

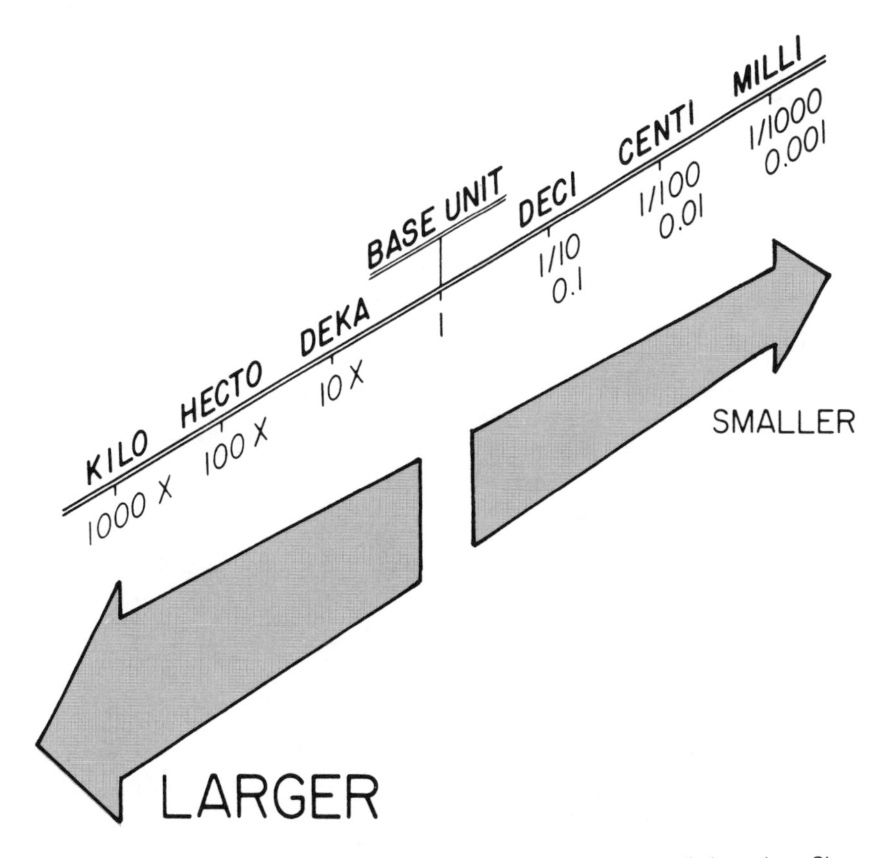

Fig. 2-6. The United States is in the process of adopting the International System of Units (also called metric or SI system). Study basic units of this system.

1. Dimensioned with inch units (fractional and/or decimal), Fig. 2-7.
2. *Dual dimensioned,* Fig. 2-8. Dimensions are given in both inches (usually decimal inches) AND in the metric system (usually in millimeters).
3. Dimensioned in SI or metric units, Fig. 2-9.

To understand the metric system, welders have to THINK in SI units. In addition, they must remember that the symbols employed in the metric system are all important. Inaccuracies will result if symbols are used carelessly.

For example, lower-case mm means millimeter (0.001 meter) while Mm means megameter (1,000,000 meters), a considerable difference. See Fig. 2-10.

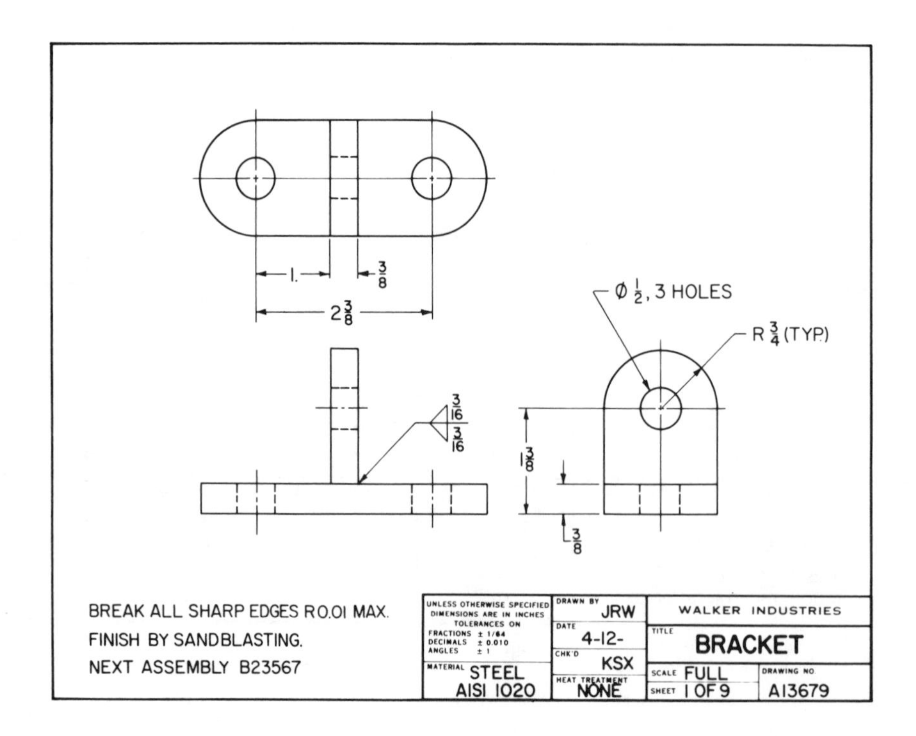

Fig. 2-7. This drawing is dimensioned with inch units.

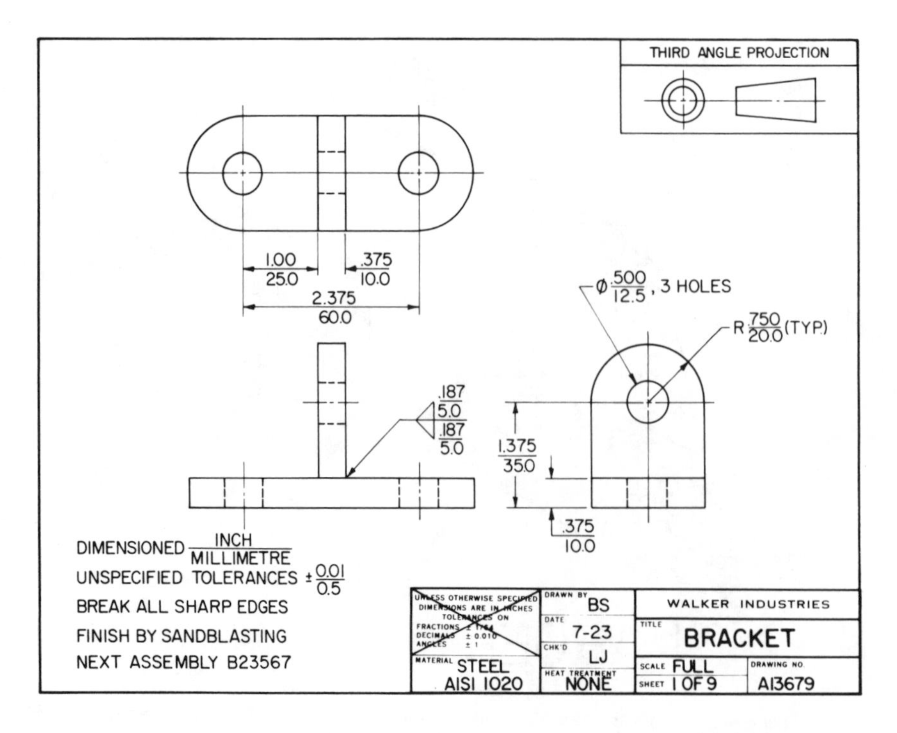

Fig. 2-8. This is a dual dimensioned drawing. Inches are on top and millimeters are on bottom.

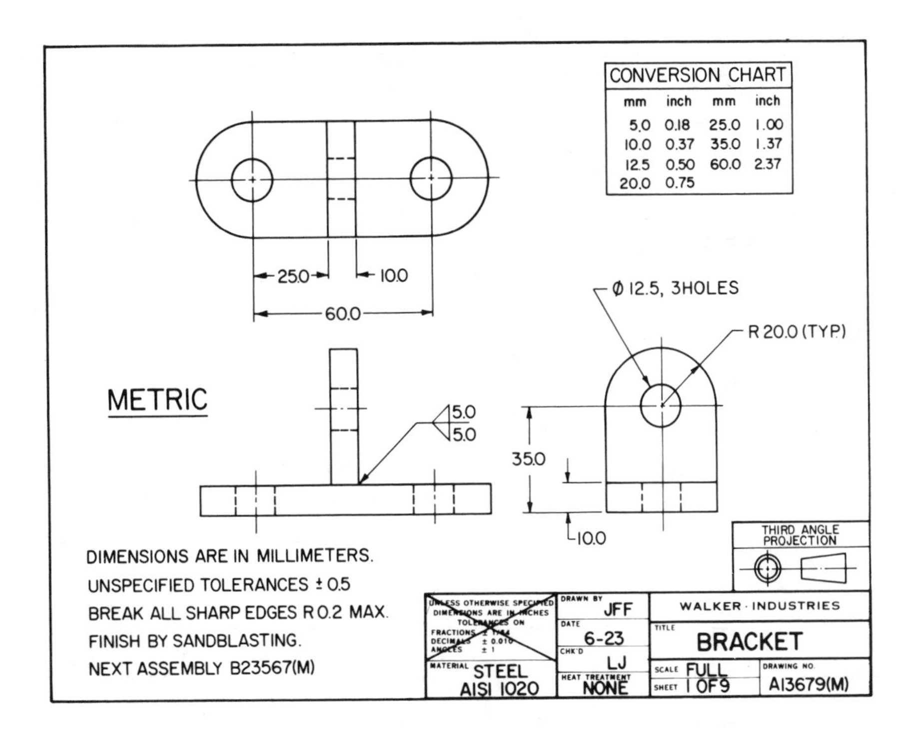

Fig. 2-9. Note how this drawing is dimensioned in metric units.

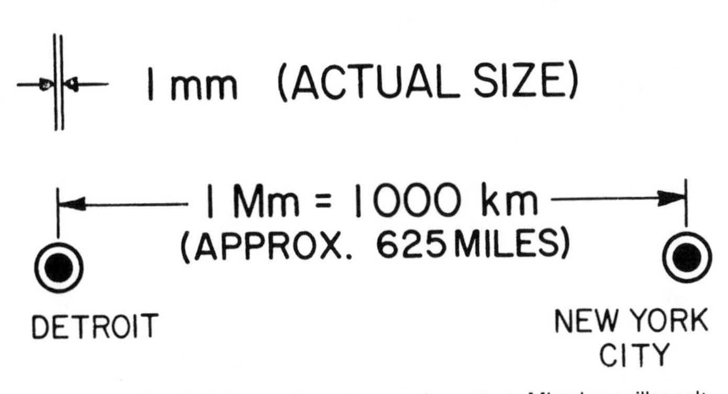

Fig. 2-10. Symbols in metric system are important. Mistakes will result if they are used improperly. Note difference made by upper and lower case letters.

The millimeter (mm) is the basic unit for linear dimensions on drawings used by MANUFACTURING INDUSTRIES. In BUILDING CONSTRUCTION, the meter (m) is normally used.

Two methods have been designated to display dual dimensions on a drawing, Fig. 2-11. The POSITION or BRACKET method may be used on a drawing.

Drawings employing the *position method,* and used primarily in the United States, will display the inch dimension above or to the left of the millimeter dimension.

The *bracket method* encloses the dimension in square brackets. Drawings used primarily in Europe will have

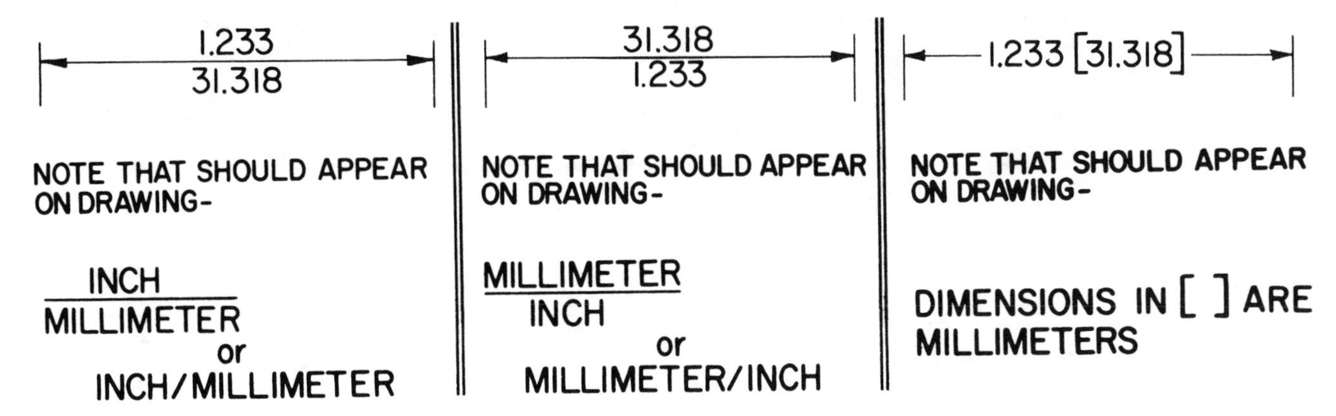

A—Method used when drawing is to be used in United States.

B—Method used when drawing is to be used in a metric country and United States.

C—Brackets are sometimes used to indicate metric equivalent on a drawing to be used in United States.

Fig. 2-11. Study how inch and metric dimensions are indicated on a dual dimensioned drawing.

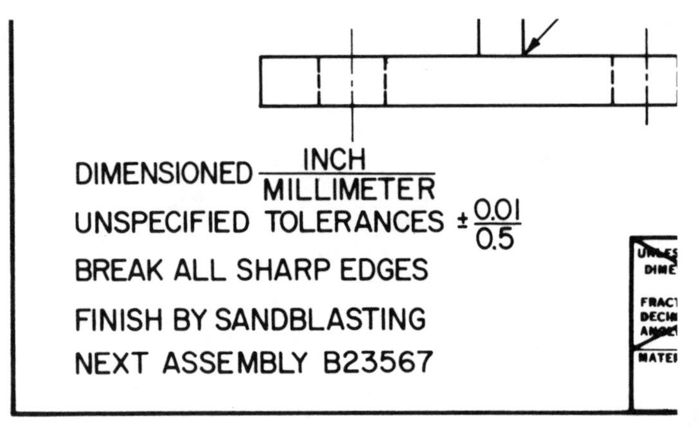

Fig. 2-12. A note will be placed on drawing to show how inch and millimeter dimensions can be identified.

the inch dimension enclosed in square brackets.

A *note* will be placed on the drawing to show how the inch and millimeter dimension can be identified. Refer to Fig. 2-12.

Hole diameter or *shaft diameter* is indicated on SI unit drawings with the number and symbol ϕ.

UNIT 2—TEST YOUR KNOWLEDGE

PART I

Place the correct reading for each inch measurement in the blank space provided. Reduce fractions to their lowest terms; for example: 10/16 = 5/8.

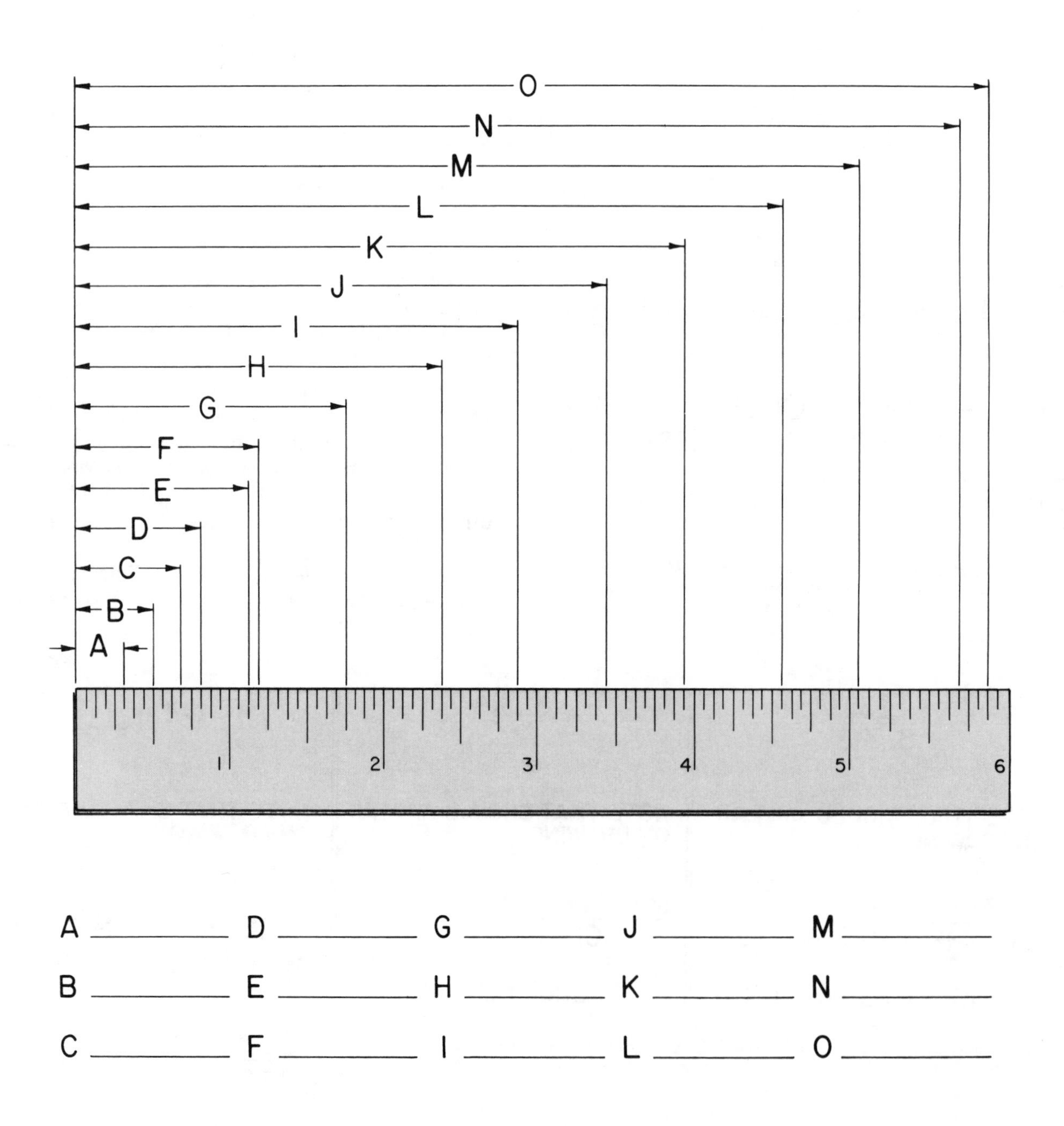

A ________ D ________ G ________ J ________ M ________

B ________ E ________ H ________ K ________ N ________

C ________ F ________ I ________ L ________ O ________

PART II

Place the correct decimal for each measurement in the blank space provided. Be careful to place the decimal point in the proper location.

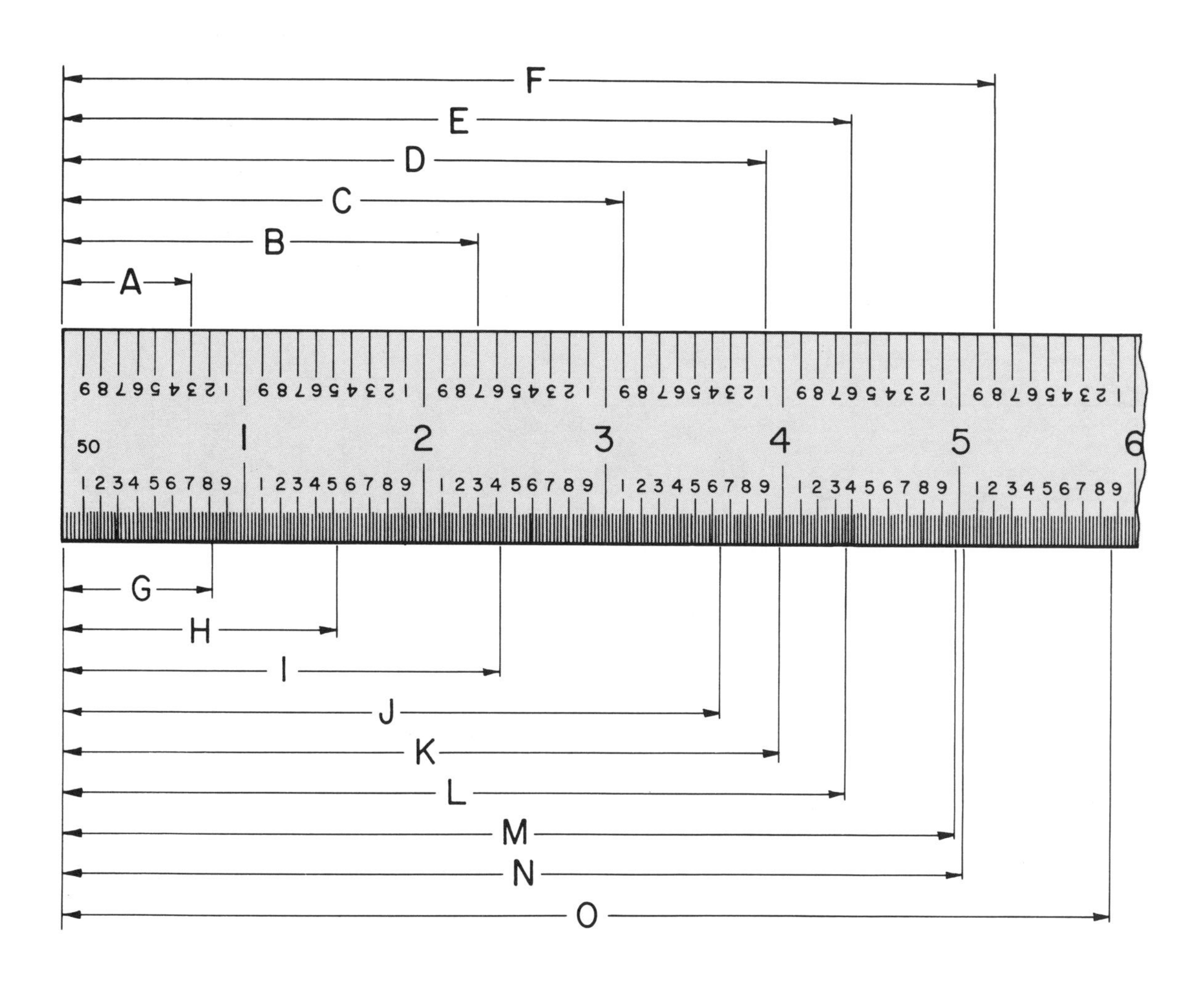

A 0.7 in	D ________	G 0.82 in.	J ________	M ________
B ________	E ________	H ________	K ________	N ________
C ________	F ________	I ________	L ________	O ________

PART III

Place the correct metric reading for each measurement in the blank space provided.

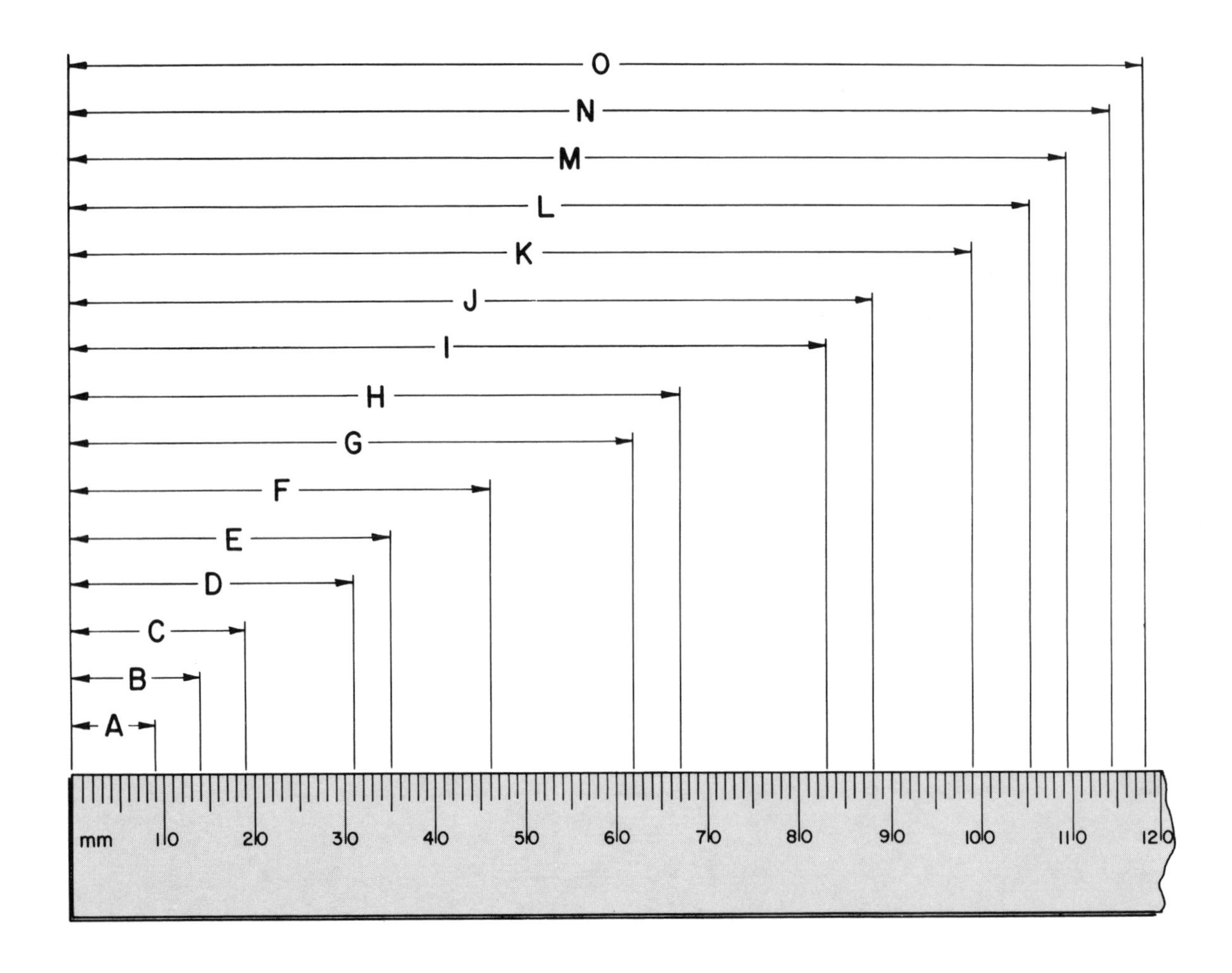

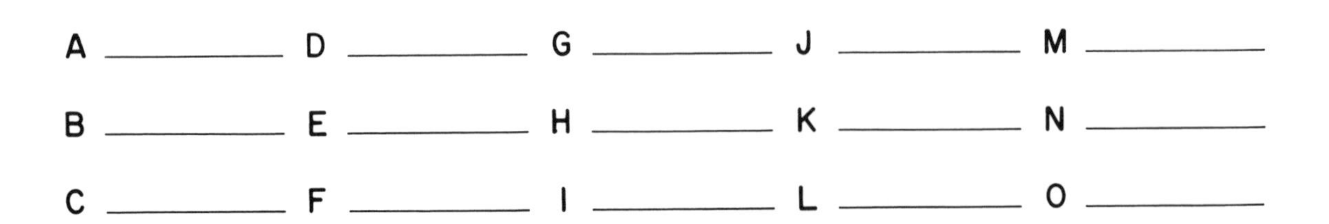

PART IV

Secure a rule with 1/16 in. graduations and measure the length of each line. Place your answer in the space to the left of each line. Reduce fractions to their lowest common denominator; for example: 6/8 = 3/4.

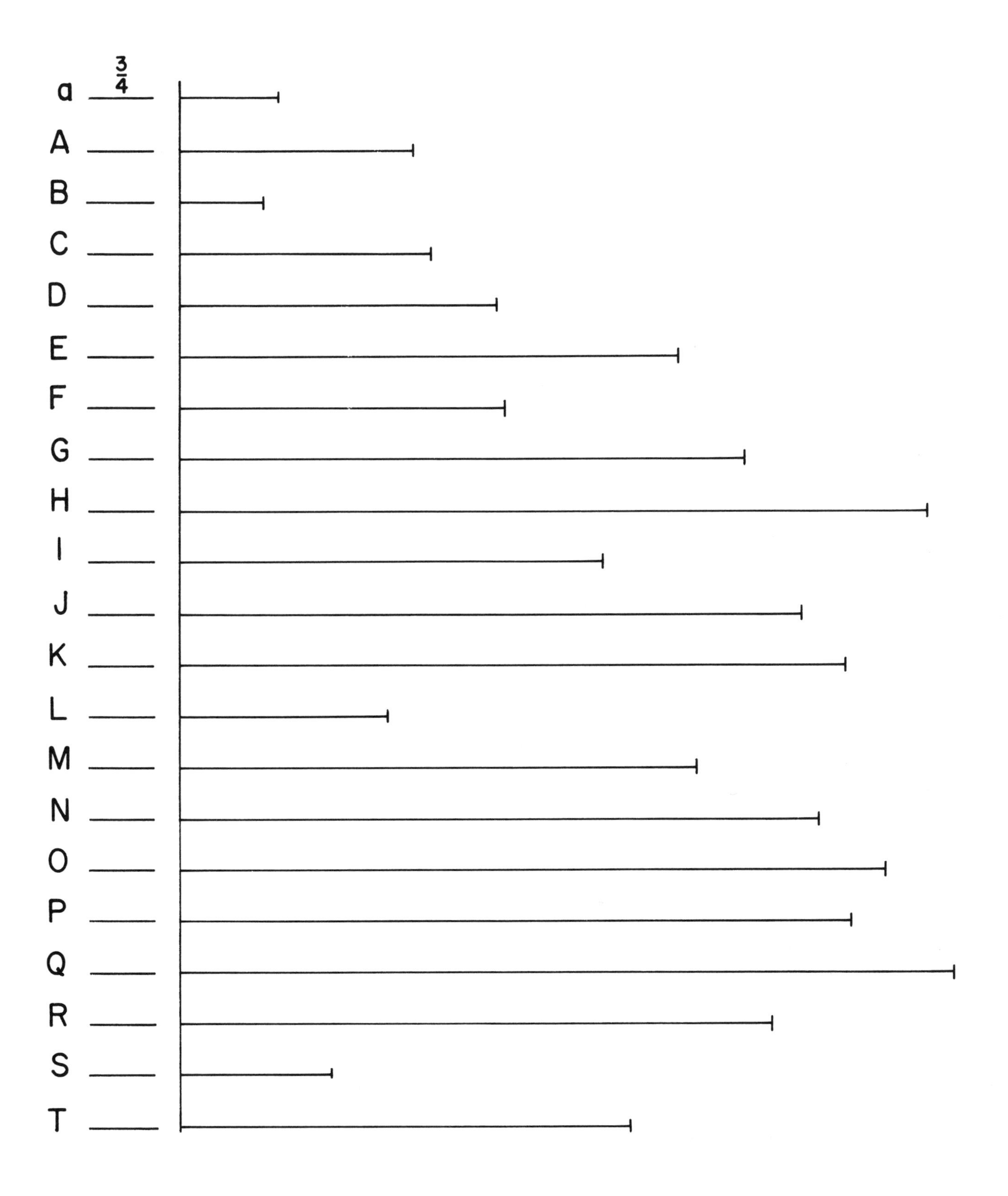

PART V

Secure a rule with metric graduations and measure the length of each line. Place your answer in the space to the left of each line.

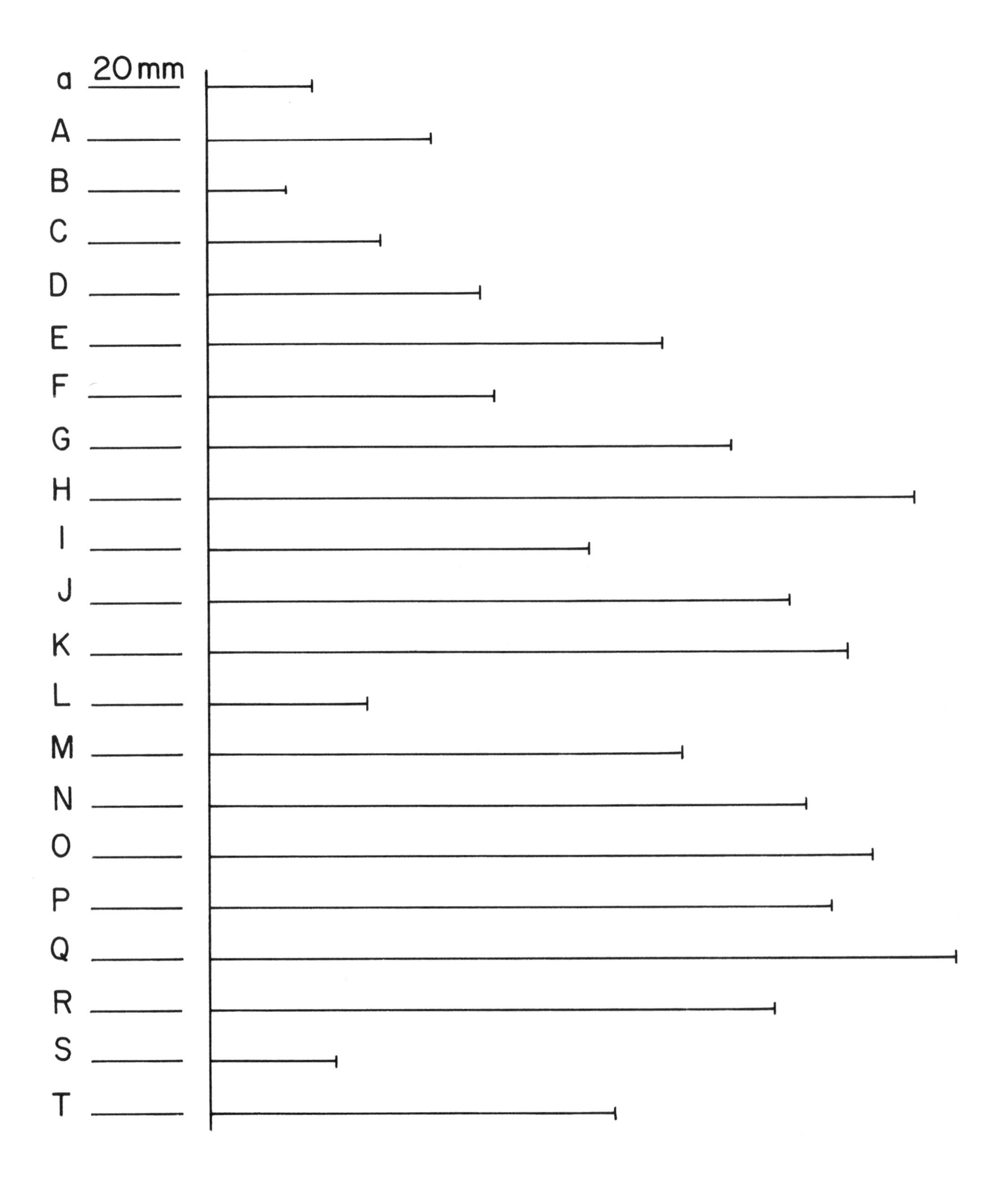

Unit 3

A REVIEW OF FRACTIONS AND DECIMALS

Anyone using welding prints on the job must be able to make calculations using fractions and decimals. For this reason, this unit will provide a quick review of basic mathematical skills.

FRACTIONS

In math, a *fraction* is a number that tells what part of something is taken.

A *common fraction* has one number placed above the other. A bar or short line is placed between the numbers. The fraction thirteen sixteenths, for example, would be written $\frac{13}{16}$.

The number BELOW the bar, 16 in this example, is called the *demoninator*. The denominator tells into how many parts a whole unit has been divided, Fig. 3-1.

The number on TOP of the bar, 13 in this example, is called the *numerator*. The numerator indicates how many parts of the whole unit are taken.

Equal fractions are fractions that have the same value, but may have different forms. For instance, $\frac{1}{2}$ equals $\frac{2}{4}$ equals $\frac{4}{8}$.

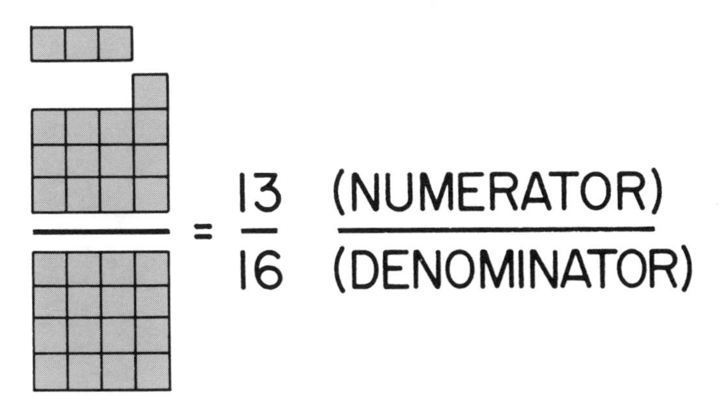

Fig. 3-1. Common fraction has one number placed above other. A bar or short line is placed between numbers. Number below bar is called denominator; it tells into how many parts a whole unit has been divided. Number on top of bar is called numerator; it indicates how many parts of whole unit are taken.

A *proper fraction* is a fraction whose numerator is smaller than the denominator. Two examples are $\frac{3}{8}$ and $\frac{7}{9}$.

An *improper fraction* is a fraction whose numerator is equal to, or greater than, its denominator. The fractions $\frac{3}{3}$ and $\frac{8}{7}$ are examples.

A *mixed number* is the sum of a whole number and a common fraction, as: $1\frac{1}{3}$ and $2\frac{3}{4}$.

RULES FOR USING FRACTIONS

1. *Whole numbers* are changed to fractions by multiplying the numerator and denominator by the same number. For example, change 5 into eighths.
   a. Each unit will contain 8 eights; 5 units will contain 5 times 8 eighths or 40 eights.
   $\frac{5}{1} \times \frac{8}{8} = \frac{40}{8}$
2. *Mixed numbers* are changed to fractions by converting the whole number to a fraction with the same denominator as the fractional part of the mixed number and then adding the two fractions. For example, change $4\frac{3}{4}$ to a fraction.
   a. Change the whole number 4 to a fraction with the same denominator as the fraction $\frac{3}{4}$. Since each unit will contain 4 fourths, 4 units will contain 4 times 4 fourths or 16 fourths.
   $\frac{4}{4} \times \frac{4}{1} = \frac{16}{4}$
   b. Adding the $\frac{3}{4}$ part of the mixed number to $\frac{16}{4}$ will give $\frac{19}{4}$.
   $\frac{3}{4} + \frac{16}{4} = \frac{19}{4}$
3. *Improper fractions* are reduced to a whole or mixed number by dividing the numerator by the

denominator. For example, convert $\frac{35}{16}$ to a whole or mixed number.

a. $\frac{35}{16} = 35 \div 16 = 2\ \frac{3}{16}$

4. Fractions are reduced to their *lowest form* by dividing the numerator and the denominator by the same number. The value of a fraction is not changed if the numerator and denominator are divided by the same number. For example, reduce $\frac{12}{16}$ to its lowest form.

a. $\frac{12 \div 4}{16 \div 4} = \frac{3}{4}$

5. Fractions can be changed to a *higher form* (equal fraction) by multiplying the numerator and denominator by the same number. The value of a fraction is not changed if the numerator and denominator are multiplied by the same number. For example, convert $\frac{3}{4}$ to sixteenths.

a. $\frac{3 \times 4}{4 \times 4} = \frac{12}{16}$

ADDING FRACTIONS

Fractions can be added only if the denominators are the same. Then add the numerators. For example, adding $\frac{2}{5} + \frac{1}{2} + \frac{1}{3} = ?$

1. The fractions must be converted to fractions that have the same denominator. The common denominator is found by multiplying their present denominators ($5 \times 2 \times 3 = 30$). Next, each fraction must be converted to an equal fraction with a denominator of 30.

$$\frac{2}{5} = \frac{?}{30} = \frac{2 \times 6}{5 \times 6} = \frac{12}{30}$$

$$\frac{1}{2} = \frac{?}{30} = \frac{1 \times 15}{2 \times 15} = \frac{15}{30}$$

$$\frac{1}{3} = \frac{?}{30} = \frac{1 \times 10}{3 \times 10} = \frac{10}{30}$$

2. Add only the numerators. The denominator remains the same.

$$\frac{12}{30} + \frac{15}{30} + \frac{10}{30} = \frac{37}{30} = 37 \div 30 = 1\ \frac{7}{30}$$

Part I—Addition problems (fractions)

Solve the following problems in the addition of fractions. Reduce your answers to their lowest form.

1. $\frac{1}{2} + \frac{2}{3} + \frac{3}{4} =$ ________
2. $\frac{3}{8} + \frac{15}{16} + \frac{5}{8} =$ ________
3. $\frac{7}{16} + \frac{7}{8} + \frac{5}{6} =$ ________
4. $\frac{5}{32} + \frac{1}{8} + \frac{9}{16} =$ ________
5. $\frac{9}{64} + \frac{3}{32} + \frac{5}{16} + \frac{3}{4} =$ ________
6. $\frac{3}{20} + \frac{7}{10} + \frac{3}{5} =$ ________
7. $3\frac{3}{8} + \frac{7}{10} + 1\frac{5}{16} =$ ________
8. $5\frac{7}{8} + 3\frac{1}{4} + \frac{3}{8} + 3 =$ ________
9. $\frac{3}{64} + \frac{15}{32} + \frac{1}{2} + 4 =$ ________
10. $1\frac{3}{4} + 3\frac{1}{8} + 1\frac{1}{4} + \frac{3}{2} =$ ________
11. $\frac{5}{16} + \frac{5}{32} + \frac{5}{64} + \frac{5}{16} =$ ________
12. $\frac{3}{8} + \frac{6}{16} + \frac{12}{32} + \frac{1}{8} =$ ________
13. $2\frac{3}{8} + 1\frac{1}{8} + 5\frac{1}{16} =$ ________
14. $1\frac{2}{3} + \frac{3}{5} + \frac{4}{5} + 2\frac{3}{10} + 1\frac{5}{6} =$ ________
15. $\frac{1}{16} + \frac{3}{32} + \frac{5}{64} + \frac{17}{16} + 1\frac{1}{8} =$ ________

SUBTRACTING FRACTIONS

Fractions cannot be subtracted unless the denominators are the same. Then subtract the numerators. For example, subtract $\frac{3}{8}$ from $\frac{3}{4}$.

1. The fractions must be converted so that they have the same denominator. In this case, the common denominator would be 8.

$$\frac{3}{4} = \frac{?}{8} = \frac{3 \times 2}{4 \times 2} = \frac{6}{8}$$

$$\frac{3}{8} = \frac{?}{8} = \frac{3 \times 1}{8 \times 1} = \frac{3}{8}$$

2. Subtract only the numerators. The denominator remains the same.

$$\frac{6}{8} - \frac{3}{8} = \frac{3}{8}$$

Part II—Subtraction problems (fractions)

Solve the following problems in the subtraction of fractions. Reduce your answers to their lowest form.

1. $\frac{5}{8} - \frac{1}{4} =$ ________

2. $\frac{1}{3} - \frac{1}{4} =$ ________

3. $\frac{4}{5} - \frac{1}{6} =$ ________

4. $\frac{11}{16} - \frac{15}{32} =$ ________

5. $1\frac{3}{4} - \frac{15}{16} =$ ________

6. $12\frac{3}{8} - 12\frac{3}{16} =$ ________

7. $9\frac{13}{16} - 5\frac{5}{32} =$ ________

8. $4\frac{5}{8} - 1\frac{1}{3} =$ ________

9. $17\frac{1}{16} - \frac{15}{32} =$ ________

10. $11\frac{3}{64} - \frac{7}{8} =$ ________

11. $13\frac{7}{32} - 3\frac{7}{16} =$ ________

12. $5\frac{1}{10} - \frac{1}{8} =$ ________

13. $1\frac{5}{8} - \frac{11}{16} =$ ________

14. $12\frac{1}{2} - \frac{15}{32} =$ ________

15. $20\frac{9}{16} - 7\frac{3}{32} =$ ________

MULTIPLYING FRACTIONS

Fractions can be multiplied if the following are done:

1. All mixed numbers are changed to improper fractions.
2. Multiply the numerators to get the numerator part of the answer.
3. Multiply the denominators to get the denominator part of the answer.
4. Reduce the fraction answer to lowest form. For example, what does $1\frac{1}{2}$ times $2\frac{1}{8}$ times 2 equal?

$$\frac{3}{2} \times \frac{17}{8} \times \frac{2}{1} = \frac{102}{16}$$

$$\frac{102}{16} = 6\frac{6}{16} = 6\frac{3}{8}$$

Part III—Multiplication problems (fractions)

Solve the following problems in the multiplication of fractions. Reduce your answers to lowest form.

1. $\frac{1}{2} \times \frac{1}{2} \times \frac{1}{2} \times \frac{1}{2} =$ ________

2. $3\frac{1}{4} \times \frac{1}{2} \times 2\frac{3}{4} =$ ________

3. $15 \times \frac{7}{8} =$ ________

4. $10 \times \frac{3}{5} \times 5\frac{1}{2} =$ ________

5. $2\frac{3}{4} \times \frac{3}{4} \times 1\frac{1}{3} =$ ________

6. $4\frac{3}{8} \times 5\frac{3}{4} \times \frac{1}{2} =$ ________

7. $2 \times 198 \times \frac{3}{3} =$ ________

8. $9\frac{3}{5} \times \frac{1}{5} \times \frac{4}{5} \times \frac{2}{5} =$ ________

9. $\frac{1}{3} \times 1\frac{2}{3} \times 1\frac{5}{6} =$ ________

10. $15 \times \frac{3}{5} =$ ________

DIVIDING FRACTIONS

Fractions can be divided as follows:

1. All mixed numbers must be changed to improper fractions.
2. The number being divided is called the *dividend.* The number by which the dividend is divided is the *divisor.*
3. To divide by a divisor that is a fraction, invert (turn upside down) the fraction and multiply. For example, divide $4\frac{3}{4}$ by $\frac{3}{4}$.

$$\frac{19}{4} \div \frac{3}{4} =$$

$$\frac{19}{4} \div \frac{4}{3} = \frac{76}{12} = 6\frac{4}{12} = 6\frac{1}{3}$$

Part IV—Division problems (fractions)

Solve the following problems in the division of fractions. Reduce your answers to lowest form.

1. $2\frac{1}{2} \div \frac{1}{4} =$ ________

2. $16 \div \frac{3}{8} =$ ________

3. $8\frac{1}{2} \div 4\frac{1}{16} =$ ________

4. $5\frac{1}{8} \div 5\frac{1}{16} =$ ________

5. $15 \div 5\frac{1}{5} =$ ________

DECIMAL FRACTIONS

Decimal fractions are fractions using the base 10 for the denominator ($\frac{1}{10}$, $\frac{1}{100}$, $\frac{1}{1000}$, etc.). Decimal fractions, however, are always written WITHOUT a denominator.

A *decimal point* (.) is placed to the left of a decimal fraction.

Decimal place is the position of a digit to the right of a decimal point. It indicates the value of the digit. See Fig. 3-2.

For example: $\frac{1}{10}$ is written .1 (one tenth)

$\frac{1}{100}$ is written .01 (one hundredth)

$\frac{1}{1000}$ is written .001 (one thousandth)

$\frac{7}{10}$ is written .7 (seven tenths)

$\frac{93}{100}$ is written .93 (ninety three hundredths)

$\frac{625}{1000}$ is written .625 (six hundred twenty five thousandths)

$3\frac{753}{1000}$ is written 3.753 (three and seven hundred fifty three thousandths)

Note that a zero (s) is used as a place holder (.01 = one hundredth, .001 = one thousandth, etc.).

Part V—Writing problems (decimal fractions)

Write the decimal fractions for the following:

1. $\frac{3}{10}$ = ________

DECIMAL PLACE

DECIMAL PLACE	
9	billionths
8	hundred-millionths
7	ten-millionths
6	millionths
5	hundred-thousandths
4	ten-thousandths
3	thousandths
2	hundredths
1	tenths

.1 2 3 4 5 6 7 8 9

Fig. 3-2. Decimal place is position of digit to right of a decimal point and indicates value of digit.

2. $\frac{5}{100}$ = ________
3. $\frac{9}{1000}$ = ________
4. $2\frac{53}{100}$ = ________
5. $7\frac{625}{1000}$ = ________

ADDING AND SUBTRACTING DECIMAL FRACTIONS

Decimal fractions are added and subtracted the same way as whole numbers. The decimal points, however, must line up vertically.

For example, what does 1.317 plus 5.81 plus 3.1 plus 7.01 equal?

```
  1.317
  5.81
  3.1
 +7.01
 17.237
```

Another example, 10.001 minus 9.3 equals what?

```
 10.001
- 9.300
   .701
```

The decimal point in the answer is placed directly below its position in the columns of numbers being added or subtracted.

Part VI—Addition problems (decimal fractions)

Solve the following problems in the addition of decimal fractions. Double-check your answers and make sure you align the decimal points.

1. 2.123 + 3.234 + 4.345 = ________

2. 5 + 1.125 + 3.250 + 5.500 = ________

3. 10.534 + 7.3 + 5.152 + 3.304 = ________

4. 1 + .005 + .825 + .0234 = ______

5. 3.217 + .01 + 9.5 + 4.05 + 2.005 = ______

Part VII—Subtraction problems (decimal fractions)

Solve the following problems in the subtraction of decimal fractions.

1. 9.750 − 5.650 = ______

2. 7.897 − 4.925 = ______

3. 5.650 − 3.750 = ______

4. 15.345 − 2.456 = ______

5. 30.7 − 19.875 = ______

MULTIPLYING DECIMAL FRACTIONS

Decimal fractions are multiplied in the same way as whole numbers. Do not be concerned with the placement of the decimal point until multiplication is completed. Then the number of decimal places in the product (answer) is the sum of the decimal places in the two numbers multiplied. The decimal point in the answer is set off from the right.

For example, what does 23.53 times 2.1 equal?

```
   23.53        2 decimal places
 ×   2.1       +1 decimal place
 -------
   2353
  4706
 -------
  49.413        3 decimal places
```

Part VIII—Multiplication problems (decimal fractions)

Solve the following problems. Check your answers carefully! Make sure the decimal point is correctly located.

1. 15.1 × .25 = ______
2. 6.3756 × 4 = ______
3. 18.312 × 2.001 = ______
4. 6.250 × .123 = ______
5. 3.1416 × 8.2 = ______
6. 9 × .125 = ______
7. 8.125 × .25 = ______
8. 9.373 × 10.2 = ______
9. 93.73 × 1.02 = ______
10. 17.33 × .333 = ______

DIVIDING DECIMAL FRACTIONS

Divide decimal fractions in the same way you divide whole numbers. The decimal point, however, must be considered and properly placed in the *quotient* (answer).

To aid in the proper placement of the decimal point, a device called a *caret mark* (^) is used.

The decimal point in the quotient will be located directly above the caret mark in the *dividend* (number being divided).

To locate the position of the decimal point in the answer:

1. Count the number of decimal places in the *divisor* (dividing number). Indicate the last position with a caret mark.
2. The number of places in the divisor is counted to the right of the decimal point in the dividend and indicated with a caret mark.
3. The decimal point in the answer (quotient) is placed directly above the caret mark in the dividend.

For example: What does 4.21875 divided by 3.75 equal? (Work to 3 decimal places in answer.)

```
            1.125
 3.75^)4.21^875
       375
       ---
        468
        375
        ---
         937
         750
         ---
         1875
         1875
```

Part IX—Division problems (decimal fractions)

Solve the following problems to three decimal places.

1. 8 ÷ 2.5 = ________
2. 4.25 ÷ .875 = ________
3. 5.123 ÷ 5.1 = ________
4. 567.765 ÷ 12.37 = ________
5. 24 ÷ .8 = ________
6. 123.786 ÷ 12.3786 = ________
7. 8.875 ÷ .125 = ________
8. 7.50 ÷ 1.25 = ________
9. 750.5 ÷ 125.375 = ________
10. 19.875 ÷ .025 = ________

ADDITIONAL INFORMATION ON DECIMAL FRACTIONS

1. To change a common fraction to a decimal fraction, divide the numerator by the denominator. For example: change $\frac{3}{4}$ to a decimal fraction.

$$\begin{array}{r} .75 \\ 4\,\overline{)3.00} \\ \underline{2\,8} \\ 20 \\ \underline{20} \end{array}$$

2. To change a decimal fraction to a common fraction, remove the decimal point and write in the proper denominator. For example: change .875 to a common fraction.
   .875 = eight hundred seventy five thousandths
   $\frac{875}{1000} = \frac{7}{8}$
3. Decimal fractions must often be ROUNDED OFF. A number less than halfway between two values is rounded off to the LESSER value. For example: 3.812 would round off to 3.81.
   A number halfway between two values is rounded off to the next GREATER value. For example: 3.815 would round off to 3.82.
4. Some common fractions when converted to a decimal fraction produce *repeating decimal fractions.* For practical purposes, this means you cannot complete the division. In this situation, use as many decimal places as necessary for the particular problem and round off. For example:

$\frac{1}{3}$ = .333333333 . . . round off to .333

$\frac{2}{3}$ = .666666666 . . . round off to .667

METRICS

In welding, as in other occupations using drawings and prints, metric dimensioning may be encountered. Fig. 3-3 is a conversion chart which can be used to change from U.S. to metric values. It also gives approximate equivalents.

The basic unit of the metric system is the *meter* (m). Units that are multiples or fractional parts of the meter are designated as such by prefixes to the word "meter."

For example:

1 millimeter (mm) = 0.001 meter or 1/1000 meter
1 centimeter (cm) = 0.01 meter or 1/100 meter
1 decimeter (dm) = 0.1 meter or 1/10 meter
1 meter (m) = 1.0 meter or 10/10 meter
1 decameter (dkm) = 10 meters
1 hectometer (hm) = 100 meters
1 kilometer (km) = 1000 meters

These prefixes may be applied to any unit of length, weight, volume, etc. The meter is adopted as the basic unit of length, the gram for mass, and the liter for volume.

In the metric system, area is measured in square millimeters (mm^2), square centimeters (cm^2), etc. Volume is commonly measured in cubic millimeters, centimeters, etc. One liter (L) is equal to 1000 centimeters.

The following are some examples of length equivalents.

10 millimeters = 1 centimeter
10 centimeters = 1 decimeter
10 decimeters = 1 meter
1000 meters = 1 kilometer

When working with metrics in welding, you should NOT convert metric dimensions to inches, feet, yards, etc. Calculations involving metrics are done in much the same manner as calculations made with decimal fractions. The major difference being that ALL UNITS must be the same. Values must be either in millimeters, centimeters, meters, etc., before they can be added, subtracted, multiplied, or divided.

For example, add the following:

25.8 mm + 0.82 m + 15.7 mm + 30.52 mm =

Before the problem can be worked, the 0.82 m must be converted to millimeters (m). Then all units will be the same.

Since there are 1000.0 mm in each m, multiply 0.82 x 1000.0 to get 820.0 mm. Now addition can be performed.

$$\begin{array}{r} 25.80 \\ 820.00 \\ 15.70 \\ \underline{+\ 30.52} \\ 892.02 \text{ mm} \end{array}$$

Part X—Metric addition problems

Solve the following problems. Double-check your answers. Remember to convert to equal units.

SI UNITS & CONVERSIONS

PROPERTY	UNIT	SYMBOL	EXACT CONVERSION			APPROXIMATE EQUIVALENCY
			FROM	TO	MULTIPLY BY	
length	meter centimeter millimeter	m cm mm	inch inch foot	mm cm mm	2.540×10 2.540 3.048×10^{-4}	25mm = 1 in. 300mm = 1 ft.
mass	kilogram gram tonne (megagram)	kg g t	ounce pound ton (2000 lb)	g kg kg	2.835×10 4.536×10^{-1} 9.072×10^{2}	2.8g - 1 oz. kg = 2.2 lbs. = 35 oz. 1t = 2200 lbs.
density	kilogram per cub. meter	kg/m^3	pounds per cu. ft.	kg/m^3	1.602×10	$16 kg/M^3$ = 1 $lb./ft^3$
temperature	deg. Celsius	°C	deg. Fahr.	°C	(°F−32) x 5/9	0°C = 32°F 100°C = 212°F
area	square meter square millimeter	m^2 mm^2	sq. inch sq. ft.	mm^2 m^2	6.452×10^{2} 9.290×10^{-2}	$645 mm^2$ = 1 $in.^2$ $1 m^2$ = 11 $ft.^2$
volume	cubic meter cubic centimeter cubic millimeter	m^3 cm^3 mm^3	cu. in. cu. ft. cu. yd.	mm^3 m^3 m^3	1.639×10^{4} 2.832×10^{-2} 7.645×10^{-1}	$16400 mm^3$ = 1 $in.^3$ $1 m^3$ = 35 $ft.^3$ $1 m^3$ = 1.3 $yd.^3$
force	newton kilonewton meganewton	N kN MN	ounce (Force) pound (Force) Kip	N kN MN	2.780×10^{-1} 4.448×10^{-3} 4.448	1N = 3.6 oz. 4.4N = 1 lb. 1kN = 225 lb.
stress	megapascal	MPa	$pound/in^2$ (psi) Kip/in^2 (ksi)	MPa MPa	6.895×10^{-3} 6.895	1MPa = 145 psi 7MPa = 1 ksi
torque	newton-meters	N.m	in-ounce in.pound ft pound	N.m N.m N.m	7.062×10^{3} 1.130×10^{-1} 1.356	1N.m = 140 in.oz. 1N.m = 9 in.lb. 1N.m =.75 ft.lb. 1.4N.m = 1 ft.lb.

Fig. 3-3. Study SI units conversion chart. Multiply by given factor to convert from U.S. to metric value.

1. 12.5 mm + 17.5 mm + 0.85 mm + 0.15mm = __________
2. 1.25 mm + 1.25 m + 1.25 cm = __________
3. 987.98 cm + 456.7 m + 456.7 cm + 987.98 mm = __________
4. 0.98 cm + 0.89 cm + 89.0 cm + 10.0 mm = __________
5. 1.000 mm + 1000.0 mm + 1.0 cm + 10.10 cm = __________

Part XI—Metric subtraction problems

Solve the following metric subtraction problems. Convert to equal values before subtracting.

1. 12.5 cm − 12.5 mm = __________
2. 125.0 mm − 12.5 cm = __________
3. 87.5 mm − 37.5 mm = __________
4. 625.0 mm − 2.5 cm = __________
5. 375.5 mm + 37.55 cm − 562.5 mm = __________

Part XII—Metric multiplication problems

Solve the following metric multiplication problems.

1. 12.5 mm × 12.5 mm = __________
2. 56.3 mm × 5.63 cm = __________
3. 875.25 mm × 12.3 cm = __________
4. 438.2 cm × 2.15 cm = __________
5. 75.0 mm × 6.5 cm = __________

Part XIII—Metric division problems

Solve the following division problems to two (2) decimal places.

1. 8.0 mm ÷ 2.5 mm = __________
2. 8.0 cm ÷ 2.5 mm = __________
3. 88.75 mm ÷ 0.25 cm = __________
4. 19.99 mm ÷ 0.25 cm = __________
5. 25.0 mm ÷ 2.5 cm = __________

Unit 4

BASIC METALWORKING PROCESSES

Welding is but one of the techniques employed in the manufacture of the products we see and use, Fig. 4-1. Most welded products are fabricated from parts produced by one or more of the basic metalworking processes (casting, forging, extruding, machining, grinding, forming, etc.).

A good welder must have an understanding of these processes. Then, he or she will be fully prepared to do complex weldments.

Fig. 4-1. Many products require welding when being made or repaired. (Santa Fe Railway)

CASTINGS

Casting is a metalworking process of making objects by pouring molten (melted) metal into a mold. Almost any metal that can be heated to the molten state can be cast. See Fig. 4-2.

The *mold* is a cavity made in a material suitable for holding the molten metal until it cools and solidifies. The mold cavity is almost the same shape and size of the object to be cast, Fig. 4-3.

NOTE! The mold cavity must be made slightly larger than the object being cast because molten metal shrinks or contracts as it cools and becomes solid. The type of metal being cast determines how much larger the cavity must be made because each metal shrinks a different amount per foot (meter) of casting length.

Permanent mold casting

Permanent mold castings are made in metal molds that are NOT destroyed when the casting is removed. Fig. 4-4 shows a permanent mold casting.

Fig. 4-2. Casting is a metalworking process of making an object by pouring molten metal into a mold.

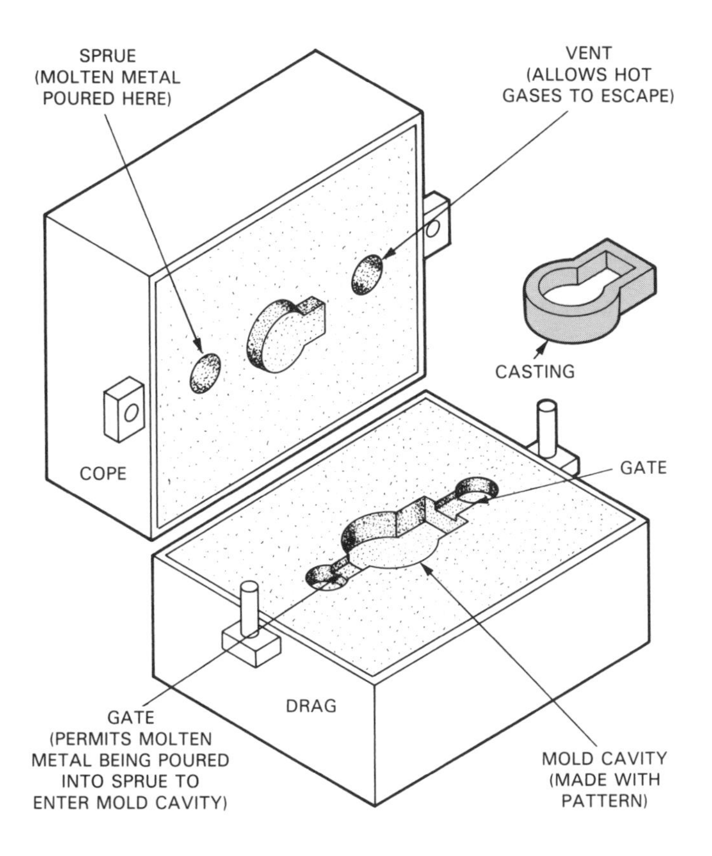

Fig. 4-3. Study example of simple sand mold. Mold cavity is about same shape and size as object to be cast.

Fig. 4-4. Worker is removing initial core from large permanent mold casting prior to taking iron mold from around casting. Permanent molds and cores are reusable thousands of times. (Aluminum Company of America)

Centrifugal casting is a form of permanent mold casting. Molten metal is poured into a SPINNING circular mold. Centrifugal force holds the molten metal against the wall of the mold until it solidifies, Fig. 4-5.

Cylinderical objects, like large pipes, brake drums, and cylinderical objects, can be cast by this technique. Wall thickness of the object being cast is determined by the amount of molten metal poured into the mold.

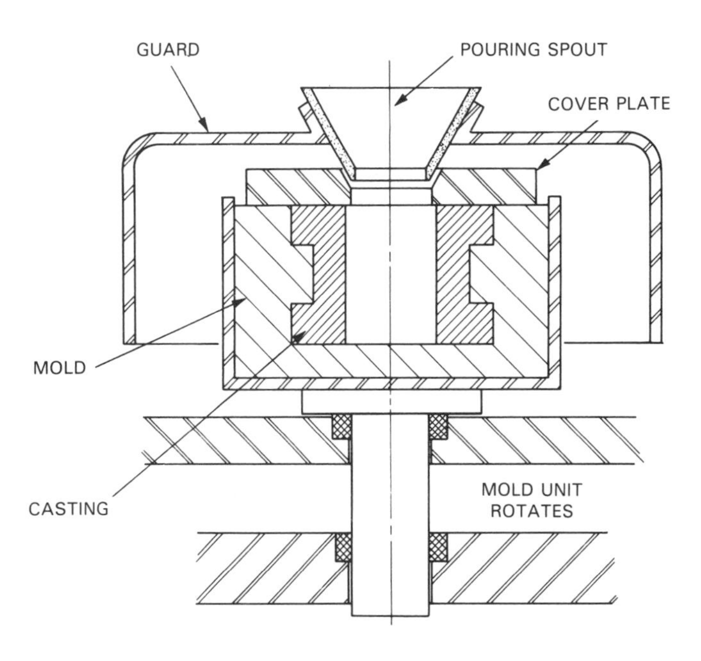

Fig. 4-5. Centrifugal force from spinning action holds molten metal against wall of mold until it solidifies.

Die casting

Die casting is a variation of permanent mold casting. Molten metal is forced into a metal mold under pressure, as in Fig. 4-6. After solidification, ejector pins force the completed casting from the mold or die.

Castings produced by this technique have smoother surface finishes, finer details, and greater accuracy than objects cast by other processes. Fig. 4-7 shows one example of a die cast object.

Investment or lost wax casting

Investment or *lost wax casting* is a foundry process used where a very accurate casting of a complex shape or intricate design must be produced. Fig. 4-8 illustrates this process.

Patterns of wax or plastic are *invested* (placed) in a *refractory mold* (mold that will withstand high

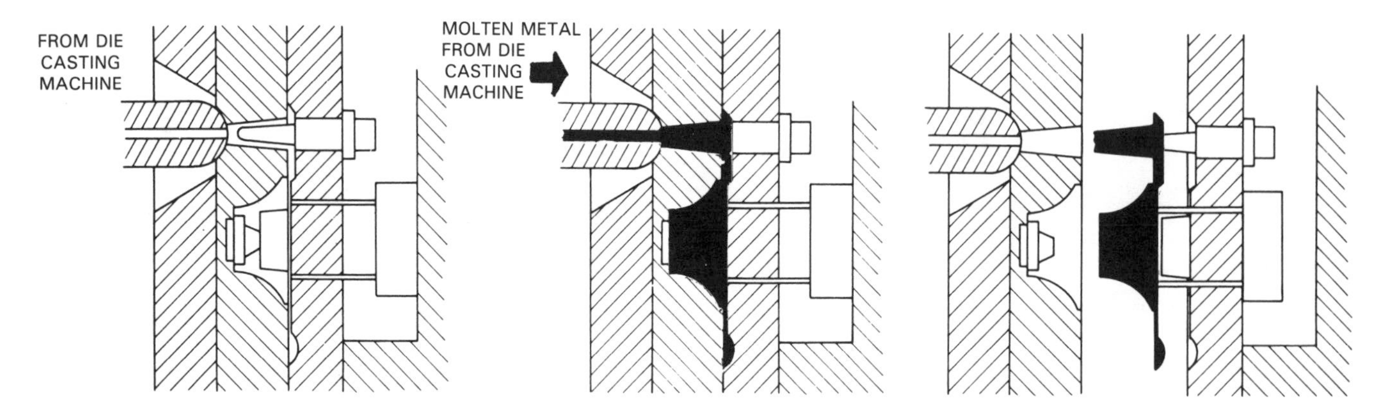

Fig. 4-6. Left. Here are two sections of a die, closed and locked to receive "shot" of molten metal to form casting. Center. Cavity of mold is now completely filled. Note metal in overflow well at bottom of cavity which provides an outlet for air entrapped in die cavity. Right. Die is opened to permit ejection of casting. Note two pins which free casting from die.

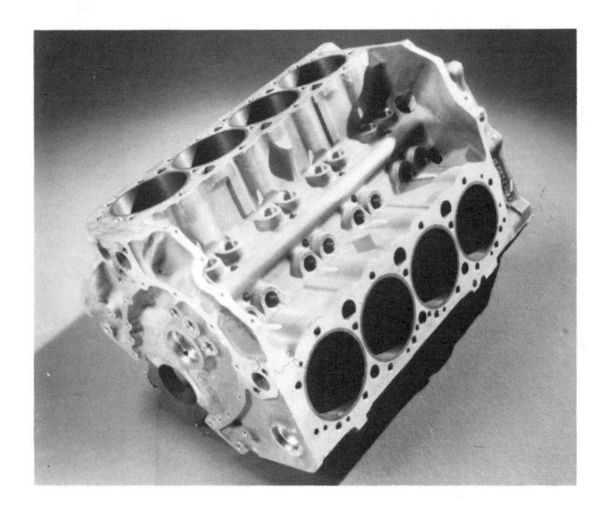

Fig. 4-7. This is a die cast aluminum engine block. Block weighs only 68 lbs., including 14 lbs. of cast iron cylinder liners.

temperatures). When the mold has hardened, it is placed in an oven and heated until the wax or plastic pattern is burned out (lost). A cavity the shape of the pattern is left in the mold. Molten metal is forced into the mold and allowed to cool. The mold is then broken apart to remove the casting.

Fig. 4-9 shows a part made by this process.

Shell molding

Shell molding makes use of a mold that is a thin sand shell. One is shown in A of Fig. 4-10. A metal pattern attached to a steel plate is fitted into a molding machine. After the pattern is heated to the required temperature, a measured amount of thermosetting resin (heat causes it to take a permanent shape) and sand is deposited on it. Both halves of the mold can be made at the same time. The thin, soft shell is cured until the desired hardness is obtained. After cooling, the pattern is removed.

The shell halves are bonded together using a special

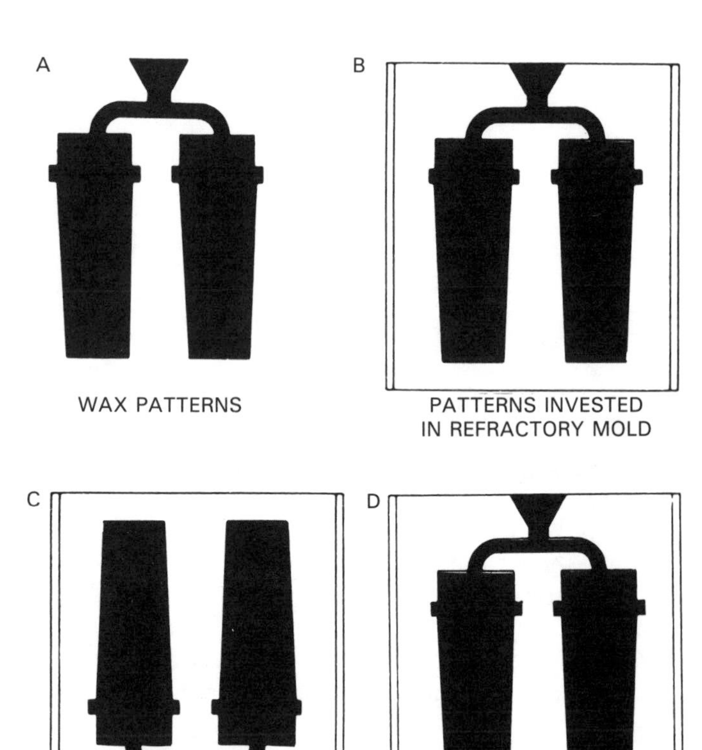

Fig. 4-8. Study basic lost wax or investment casting process.

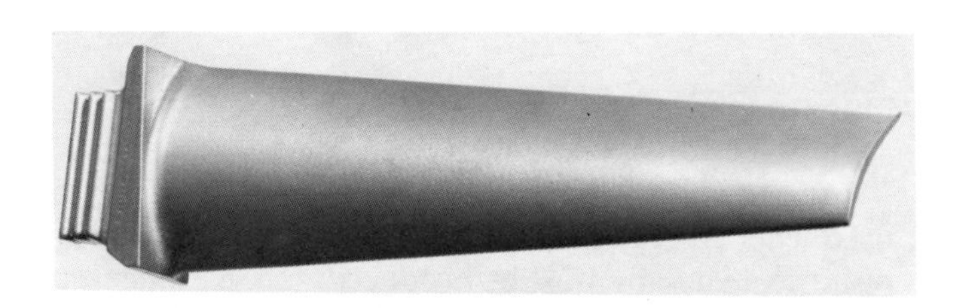

Fig. 4-9. Note complex shape of this turbine blade produced by lost wax or investment casting process.

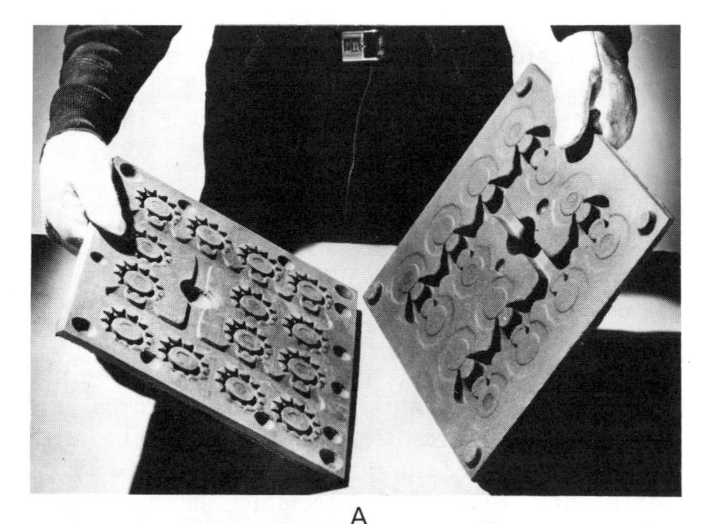

A

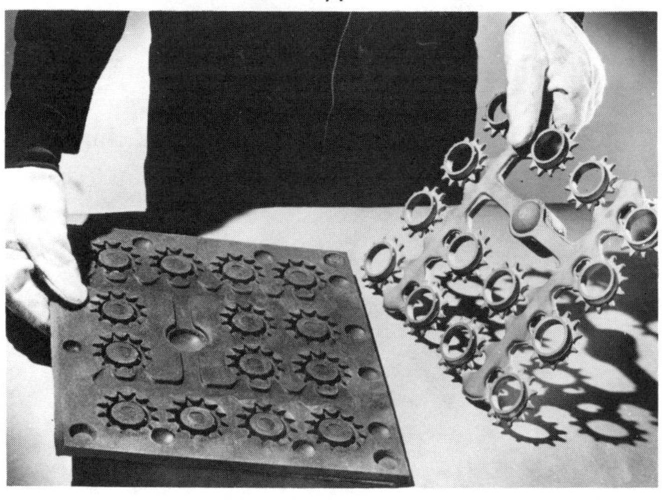

B

Fig. 4-10. A—This is a shell mold. B—Shell molding is well suited to quantity production of quality castings. (Link-Belt)

Fig. 4-11. Patterns used for sand castings are made of wood or metal.

adhesive. Molten metal is poured into the mold. The shell mold is broken away from the casting after the metal solidifies, Fig. 4-10B.

Sand casting

Sand casting uses a mold composed of a mixture of sand, clay, and a plastic and/or oil binder. Patterns are wood or metal (usually aluminum). Fig. 4-11 pictures one example.

A typical sand mold is made by packing sand around the pattern that has been positioned in a wood or metal box, called a *flask*.

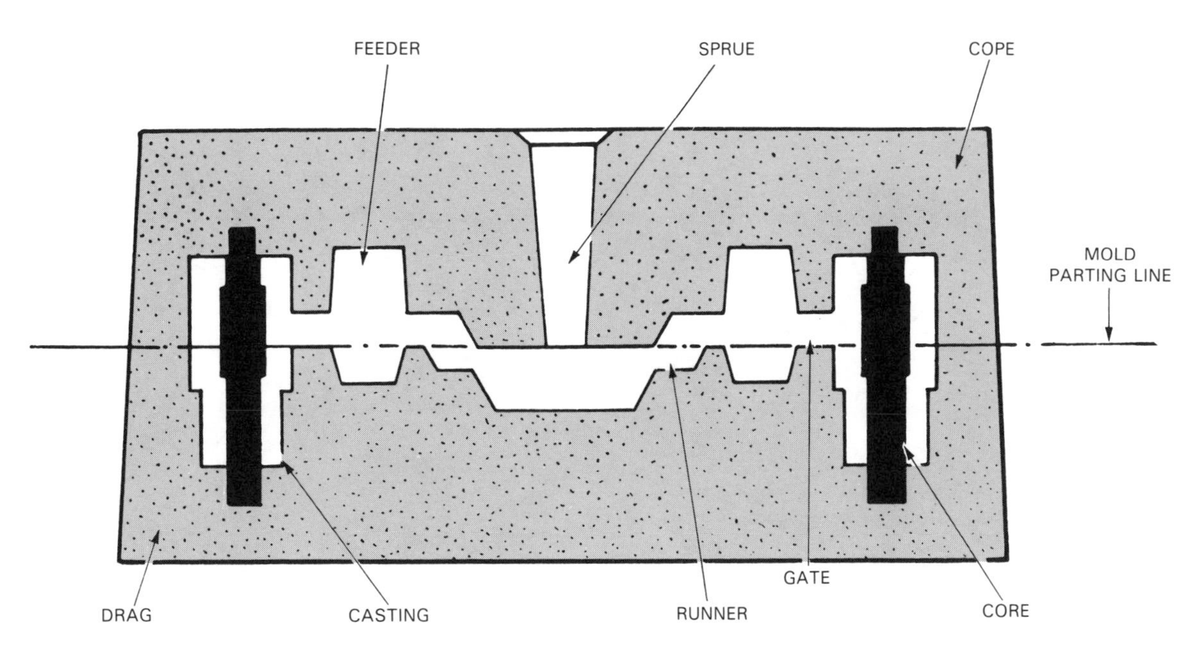

Fig. 4-12. To allow sand cast pattern to be removed, mold is made in two parts.

To allow the pattern to be removed, the mold is made in two parts, Fig. 4-12. The top half of the mold is called the *cope,* the lower half the *drag.* The sand mold is destroyed when the casting is shaken out (removed).

Openings (holes or cavities) are often required in a casting. These openings are made by inserts of sand called *cores.* The cores are positioned in the mold cavity before the mold is closed. See Fig. 4-13.

FORGING

Forging is the process of using pressure to shape metal. The pressure is usually applied in a hammering action. The metal is heated almost to melting point to improve its plasticity before pressure is applied.

Forging improves the physical characteristics of most metals. That is, a forged piece is stronger than an identical piece machined from a solid bar of stock or made by casting. Fig. 4-14 illustrates improved grain structure in forged metal.

Dies are used with the pressure to shape the metal. Pressure may be applied hydraulically, by dropping weights (drop forge), with explosives, expanding gases, electrical discharge, and by hand. The various forging techniques are shown in Figs. 4-15 through 4-17.

EXTRUSION

The *extrusion process* is used to manufacture irregular shapes, both solid and hollow, that cannot be made economically by any other process. Several extruded shapes are given in Fig. 4-15.

During extrusion, the metal is heated to a plastic, but not molten, state and inserted in the extrusion press. Tremendous pressures are exerted on the metal to literally squeeze it through a die, Fig. 4-19. After forming, the extruded section is cut to length and straightened.

MACHINE TOOLS

A *machine tool* is a power driven machine, used to

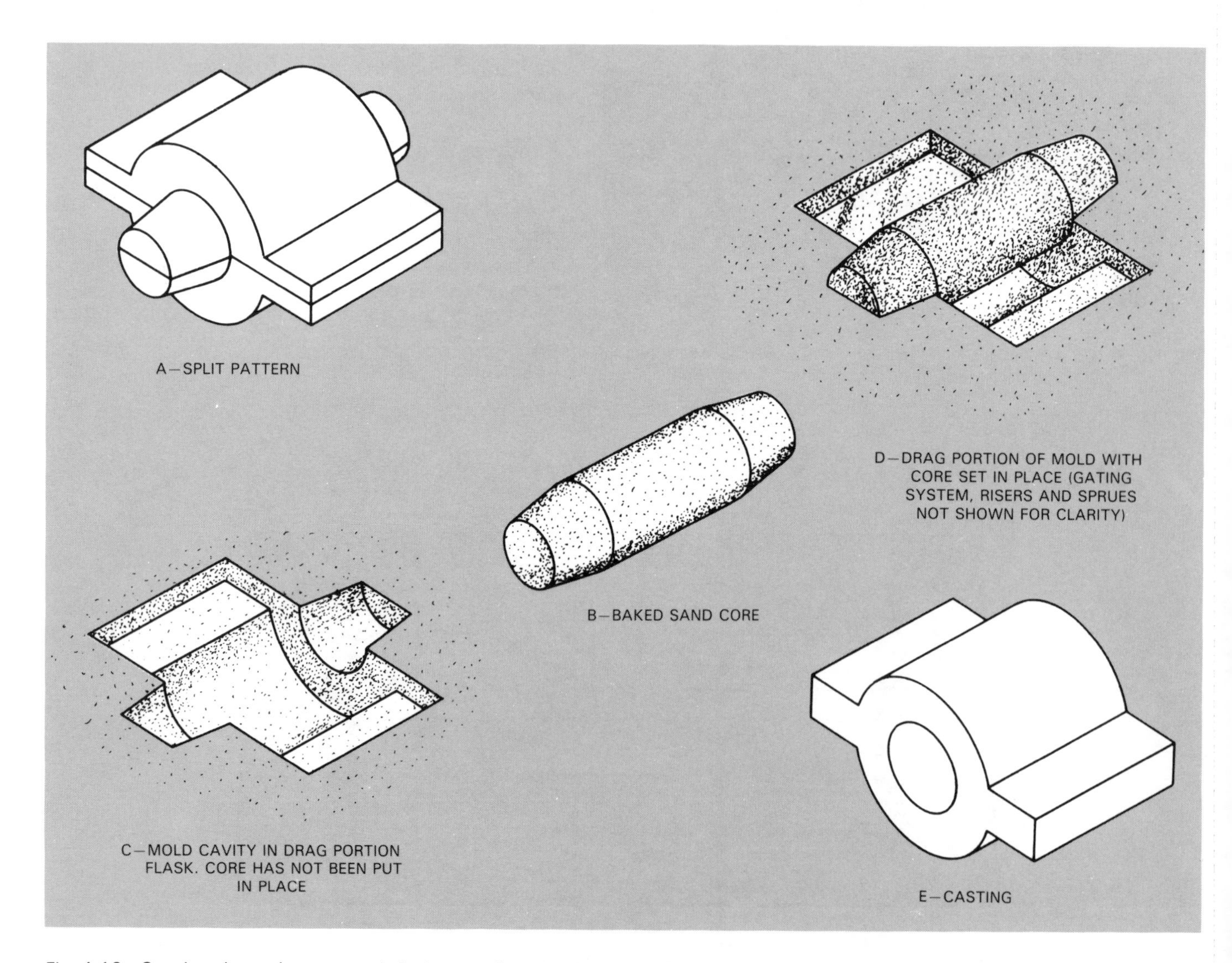

Fig. 4-13. Openings in castings are made by inserts of sand called cores. Cores are positioned in mold cavity before mold is closed. While a round core is shown, cores can be almost any shape.

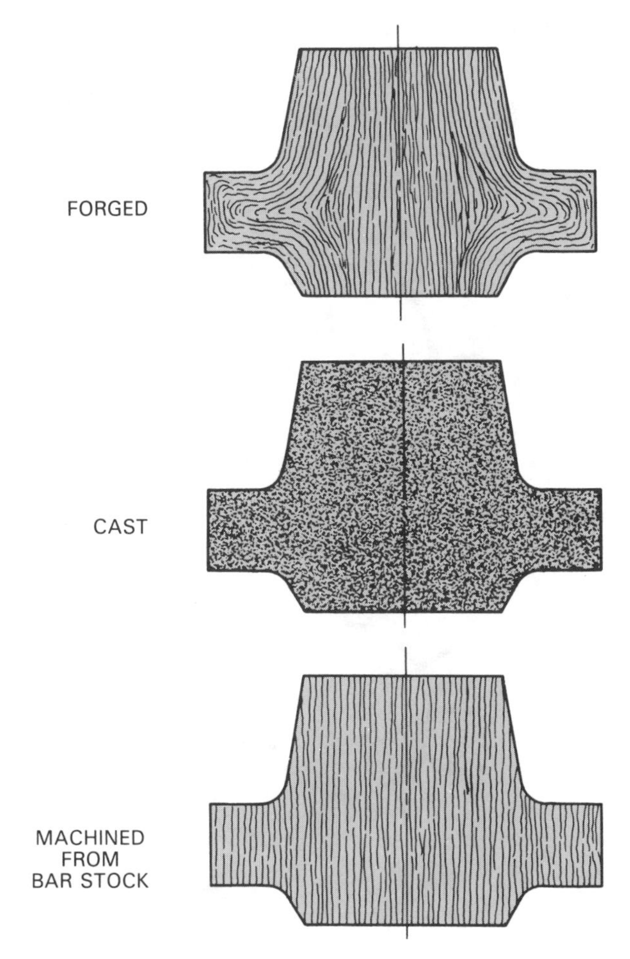

Fig. 4-14. Forged piece is stronger than an identical piece machined from a solid bar of stock or made by casting. Note grain structure in forged piece.

Fig. 4-15. This 35,000 lb. steam hammer is used to forge aircraft parts and conventional forgings. Here, an aircraft engine crankcase is being made.

Fig. 4-16. These are two giant forging presses for making forgings. A 50,000 ton press is in foreground and a 35,000 ton press is in background. (Aluminum Co. of America)

shape metal by a cutting process. It is not hand portable. Machine tools are manufactured in a large range of styles and sizes.

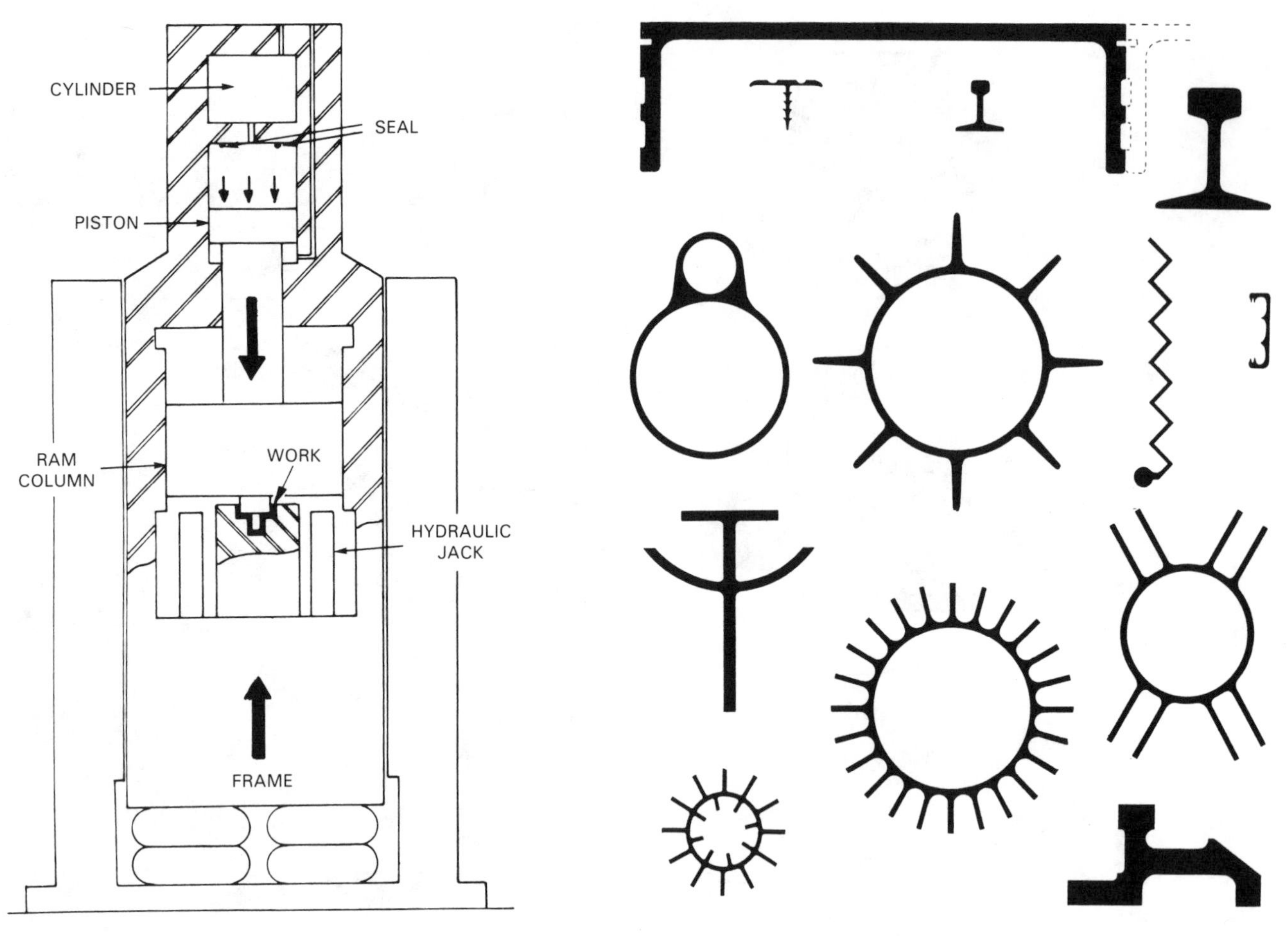

Fig. 4-17. Cross section drawing of a pneumatic-mechanical forming press. High pressure gases are used to drive ram.

Fig. 4-18. Wide scope of extrusion process is shown by this group of aluminum sections. (Reynolds Metals Co.)

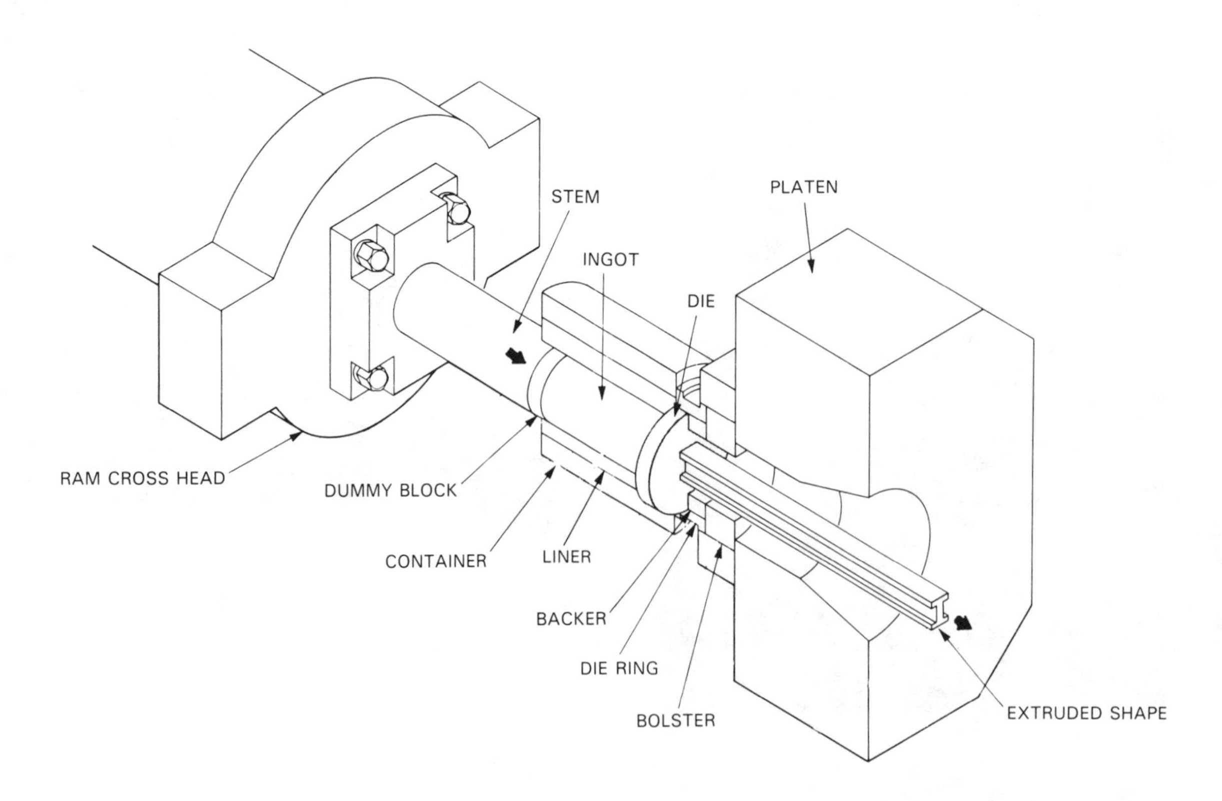

Fig. 4-19. A cross sectional view of an extrusion press showing metal (billet) being extruded in shape of a channel.

Lathe

The *lathe,* Fig. 4-20, operates on the principle of the work being rotated against the edge of a cutting tool. This is shown in Fig. 4-21. The cutting tool can be controlled and can be moved lengthwise and across the face of the material being turned (machined).

Drill press

A *drill press,* Fig. 4-22, uses a cutting tool called a *twist drill.* The drill press rotates the twist drill against the work with sufficient pressure to cut through the material. Fig. 4-23 illustrates drill press action.

Other operations which can be performed on the drill press include:

1. REAMING finishes a drilled hole to close tolerances, Fig. 4-24.

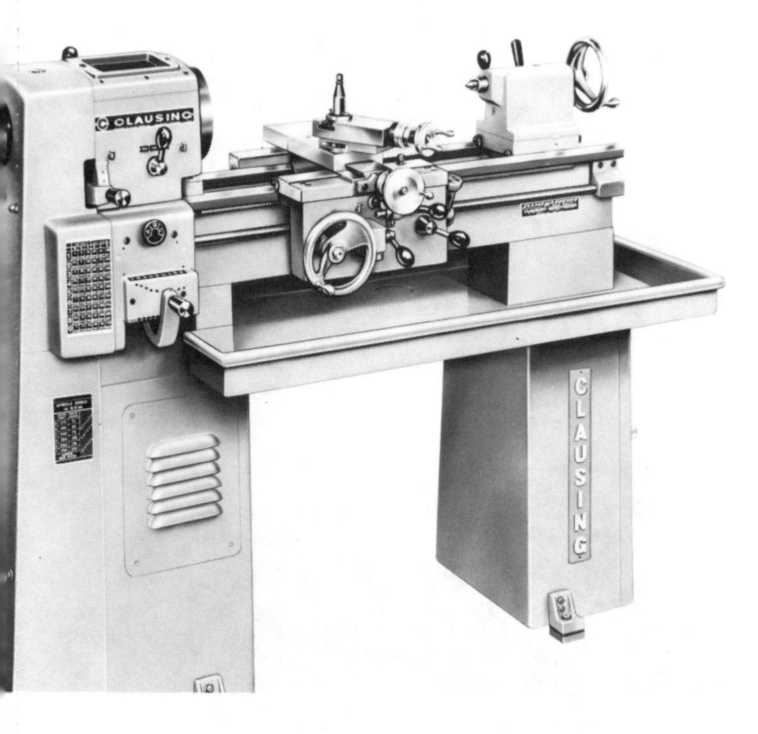

Fig. 4-20. Metal cutting lathe is available in a great number of sizes. (Clausing)

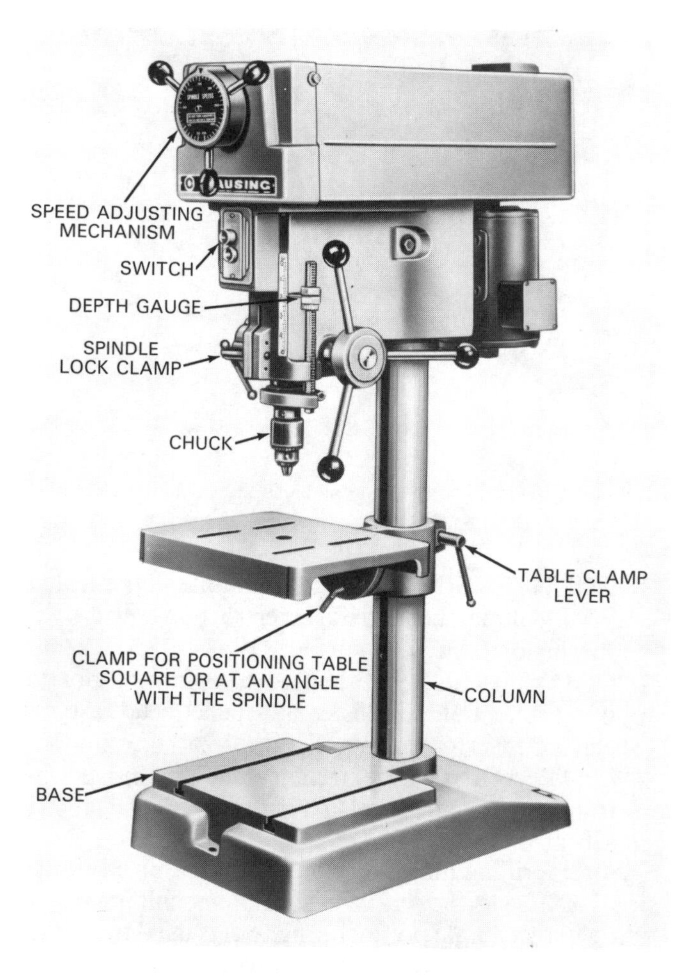

Fig. 4-22. Study basic parts of bench type drilling machine. It is commonly found in welding shops. (Clausing)

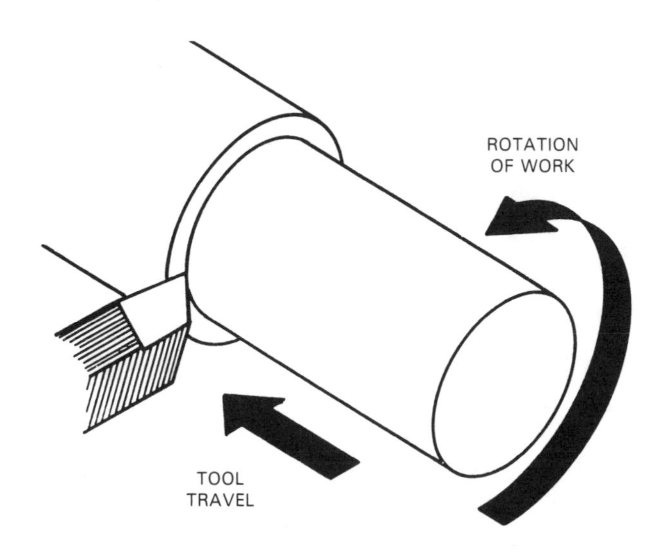

Fig. 4-21. Note operating principle of lathe.

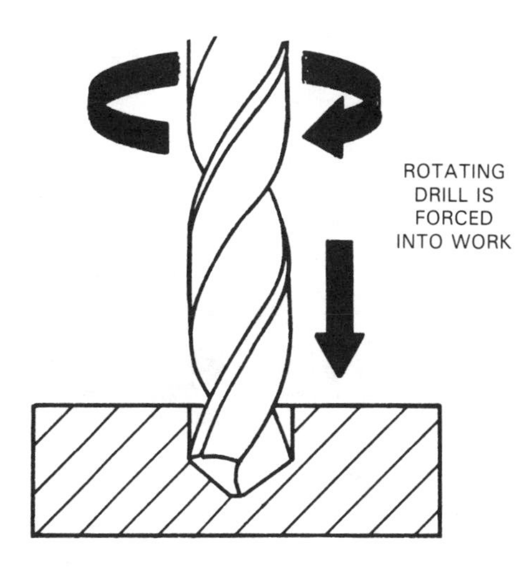

Fig. 4-23. Note how drill press works.

Fig. 4-24. Reaming on drill press after drilling hole.

2. COUNTERSINKING cuts a chamfer on a hole to permit a flat headed fastener to be inserted with the head flush with surface, Fig. 4-25.
3. COUNTERBORING is used to enlarge a portion of a drilled hole so fillister and socket head fasteners can be inserted properly, Fig. 4-26.
4. SPOTFACING is a circular spot machined on a rough surface to furnish a true bearing surface for a bolt or nut, Fig. 4-27.

Note! Drilling must NOT be mislabeled or misunderstood as boring. *Boring* is an internal machining operation where a single point cutting tool is used to enlarge a hole, Fig. 4-28. Boring can be done on the lathe, drill press, or milling machine.

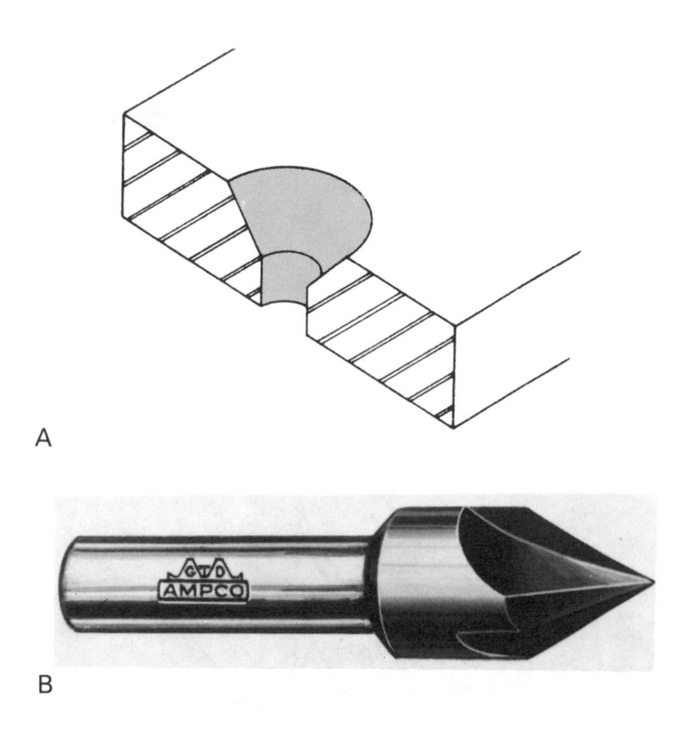

Fig. 4-25. A—Cross section of a hole that has been countersunk. B—Countersink chucks in drill in same way as twist drill. (Greenfield Tap & Die)

A

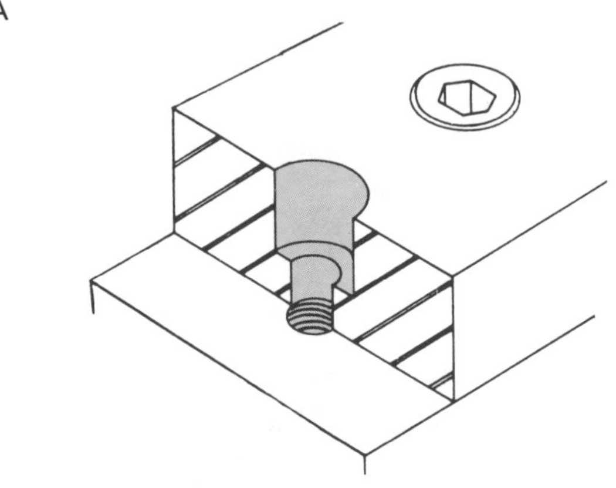

B

Fig. 4-26. A—Counterboring can be used to prepare a hole to receive a fillister or socket head screw. B—A sectional view of a hole that has been drilled and counterbored.

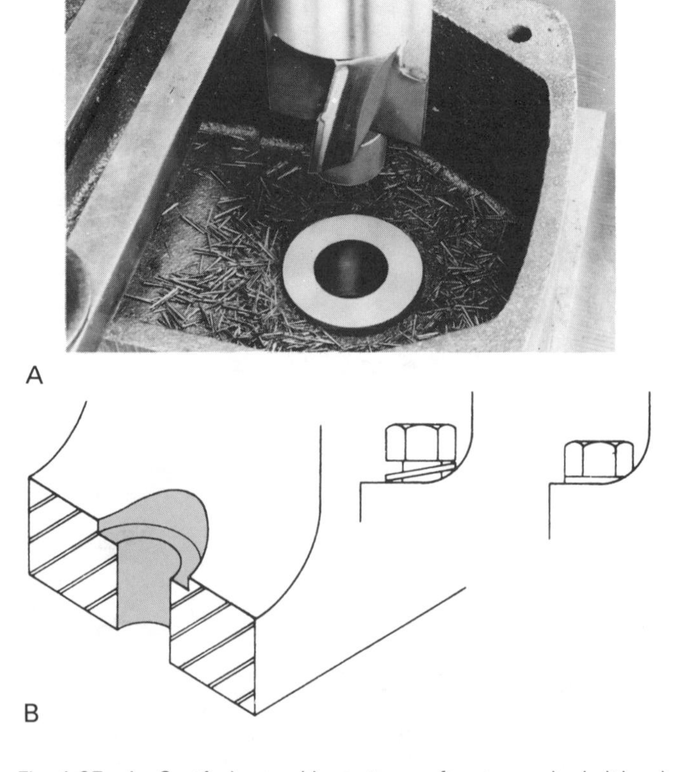

Fig. 4-27. A—Spotfacing machines a true surface to permit a bolt head or nut to bear uniformly over its entire contact area. B—Sectional view of a casting with a mounting hole that has been spotfaced. Smaller drawings show a side view of casting before and after spotfacing. Note that bolt cannot be drawn down tightly until mounting hole has been spotfaced.

A B

Fig. 4-28. A—Drilling is not the same as boring. Here a drill is being used to cut a hole in a casting. B—Boring is a machining operation that is used to enlarge a drilled hole to exact size and to a much closer tolerance than a drill. (Clausing)

Fig. 4-29. This is a shaper. (Rockford Machine Tool Co.)

Planing machines

Planing machines are used primarily to machine flat surfaces. The shaper, planer, slotter, and broach are classified this way.

Shaper

While primarily used to machine flat surfaces, the *shaper* and a skillful machinist can manipulate it to cut curved and irregular shapes, slots, grooves, and keyways. Refer to Fig. 4-29.

Because of the way a shaper operates (cutting tool travels back and forth over the work), the cutting stroke is normally limited to a maximum length of 36 in. (approx. 0.9 m). Fig. 4-30 shows the action of a shaper.

Planer

A *planer* can handle work that is too large to be machined on milling machines. Look at Fig. 4-31. Planers are large machine tools and many are capable of machining surfaces up to 20 feet (approx. 6 m) wide and twice as long.

The cutting tool on a planer remains stationary. The work travels back and forth under the cutter, Fig. 4-32.

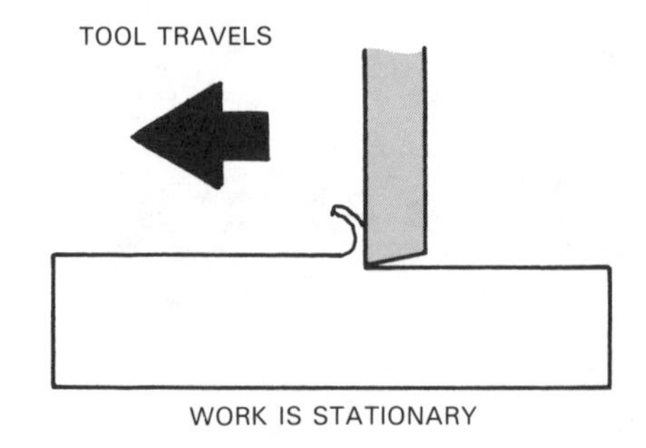

Fig. 4-30. Note how a shaper works. Work is stationary and cutting tool moves across it to remove metal.

Slotter

The *slotter* is a vertical shaper. One is pictured in Fig. 4-33.

The slotter operates similar to the shaper except that the cutting tool moves VERTICALLY rather than horizontally. See Fig. 4-34. The work is held stationary.

The slotter is used to cut slots, keyways (both internal and external), and for such jobs as machining internal and external gears.

Fig. 4-31. People look small standing on large (144 in. x 126 in. x 40 ft.) double-housing planer. (G.A. Gray Co.)

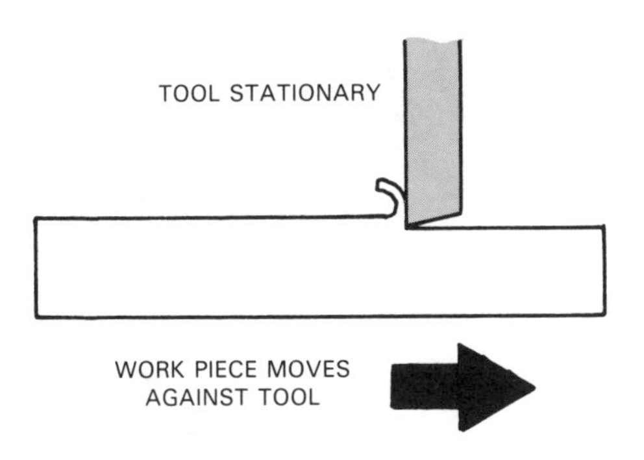

Fig. 4-32. Study how planer works. Tool remains stationary while work moves against it.

Fig. 4-33. This job is being done on a vertical shaper or slotter.

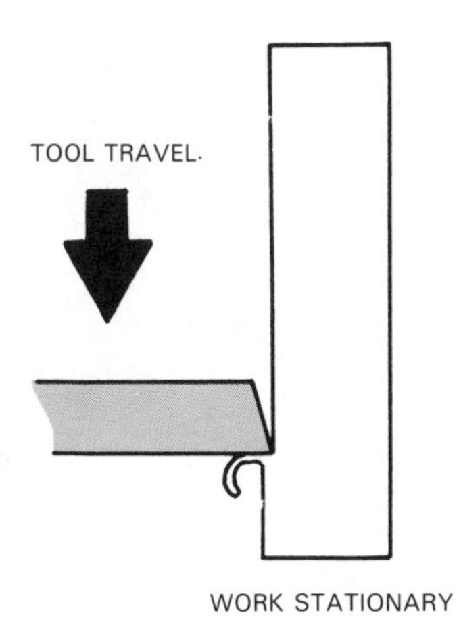

Fig. 4-34. Note how vertical shaper works. Machine operates in a manner similar to shaper, however, tool moves vertically. Work is held stationary.

Broach

Broaching is similar to shaping, but instead of a single cutting tool advancing slightly after each stroke, the broach is a long tool with many cutting teeth. Refer to Fig. 4-35.

Each broach tooth has a cutting edge that is a few thousandths of an inch (hundredths of a millimeter) higher than the one before it. Each tooth also increases in size to the exact finished dimension required.

The broach is pushed or pulled over the surface being machined. Fig. 4-36 shows a modern broaching machine.

Milling machines

The *milling machine* removes metal by moving a rotating multi-tooth cutter into the work, Fig. 4-37. It can be used to machine flat and irregularly shaped surfaces, drill, bore, or cut gears and splines.

With a *vertical milling machine*, the cutter is mounted VERTICAL to the work table, Fig. 4-38.

The *horizontal milling machine* is designed to do peripherial milling and the cutter is mounted PARALLEL to the work table. See Fig. 4-39.

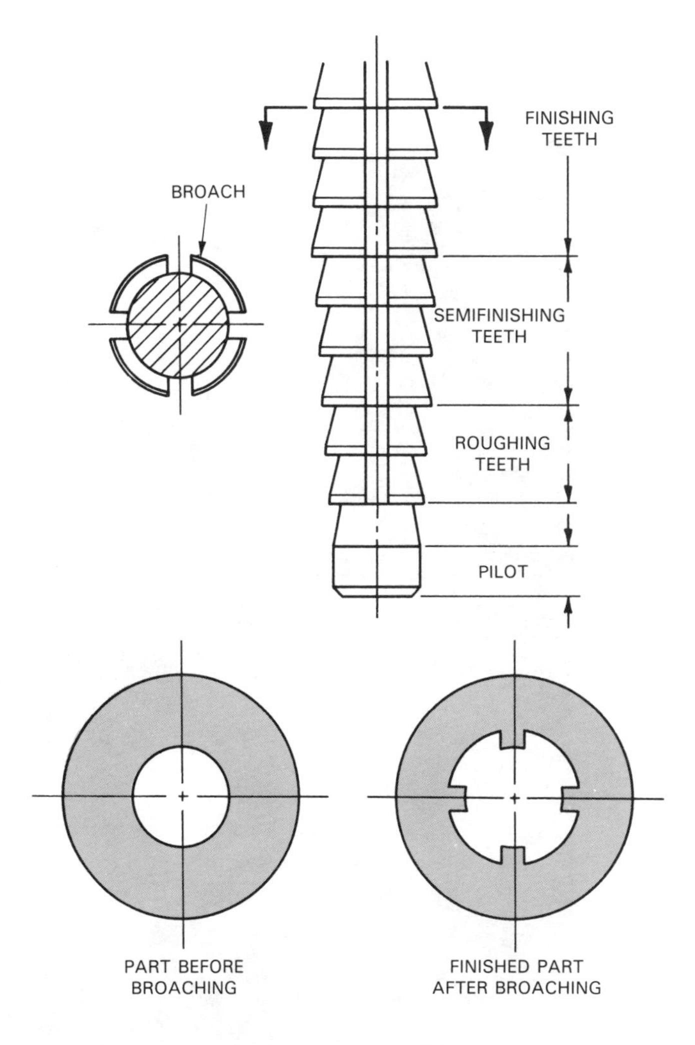

Fig. 4-35. Note short section of a broaching tool and a cross section of spline it cuts. Pilot guides cutter in work. Each cutting tooth increases slightly in size until specified size is attained.

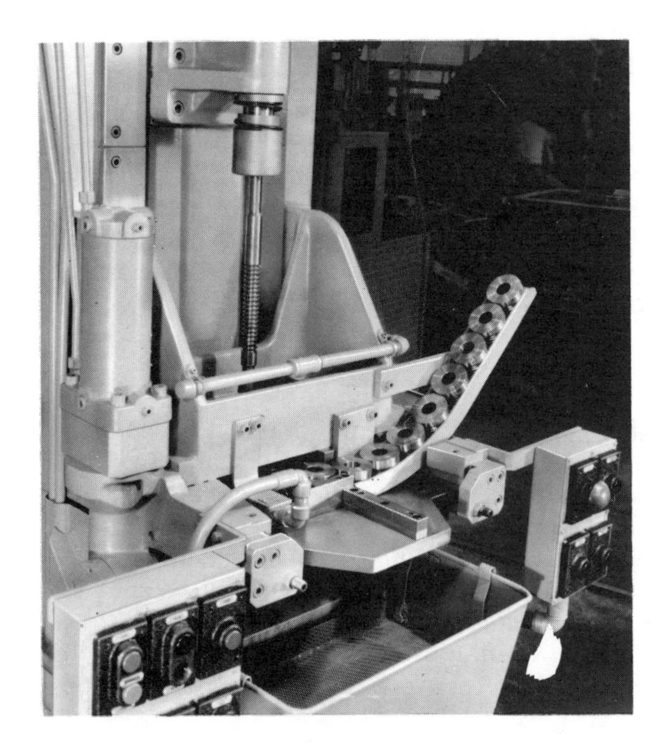

Fig. 4-36. Note how work is moving into position on this broaching machine. (Sundstrand Corp.)

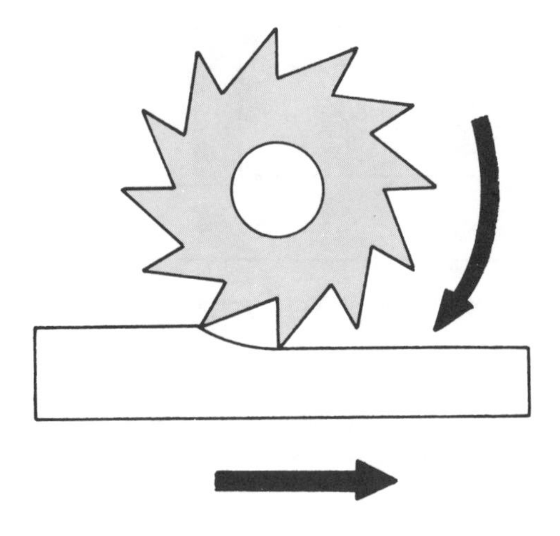

Fig. 4-37. Study action of a milling machine.

There are many different variations of the two types of basic milling machines. They vary in size and by the type of controls. Controls can be manual, semi-automatic, and fully automatic (tape and/or computer controlled).

Fig. 4-38. Cutter is mounted vertical to work table on a vertical milling machine.

Fig. 4-39. Horizontal milling machine has cutter mounted parallel to work table.

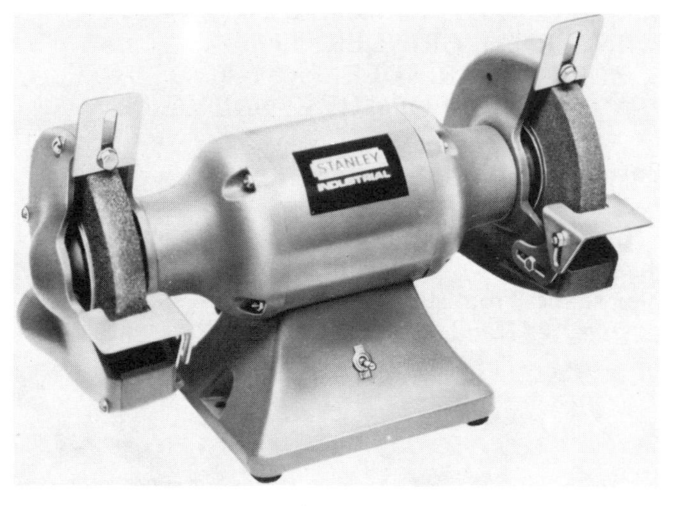

Fig. 4-41. This is a bench grinder. (Stanley Tools)

Grinding

Grinding is an operation that removes material by rotating an abrasive wheel against the work, Fig. 4-40. It is often used to sharpen tools, remove material that is too hard to be machined by other methods, or when fine surface finishes and close tolerances are required.

There are many types of grinding machines:

1. BENCH and PEDESTAL GRINDER, Fig. 4-41.
2. PORTABLE GRINDER, Fig. 4-42.
3. CYLINDRICAL GRINDER, Fig. 4-43.

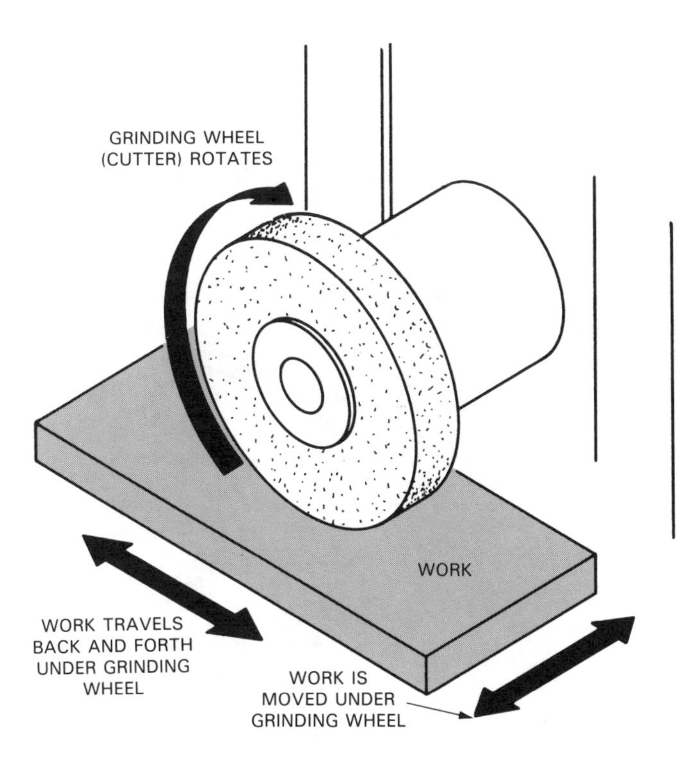

Fig. 4-40. Grinding is operation that removes material by rotating an abrasive wheel against work.

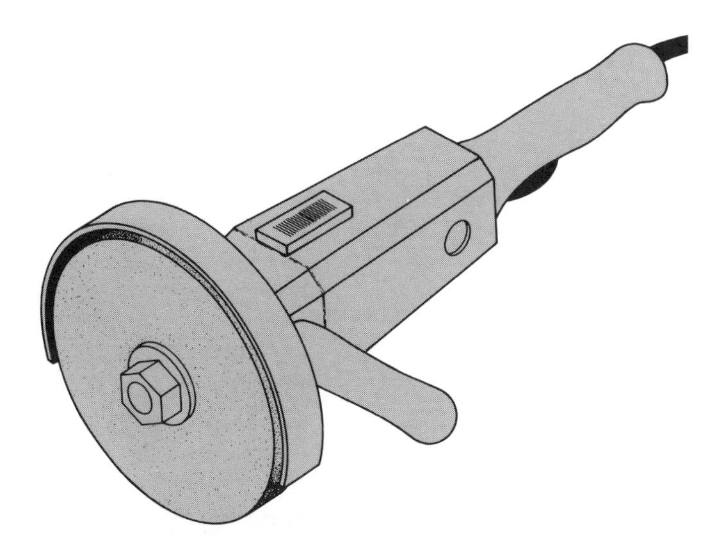

Fig. 4-42. Portable hand held grinder can be taken to job.

Fig. 4-43. Close-up of a cylindrical grinding operation. (Cincinnati Milling Machine Co.)

4. CENTERLESS GRINDER, Fig. 4-44.
5. INTERNAL GRINDER, Fig. 4-45.
6. SURFACE GRINDER, Fig. 4-46.
7. VARIATIONS OF THESE GRINDERS.

Sawing

In addition to cutting stock shapes to length, *band machining* is a sawing technique used to machine complex shapes from standard stock metal shapes. Refer to Figs. 4-47 and 4-48.

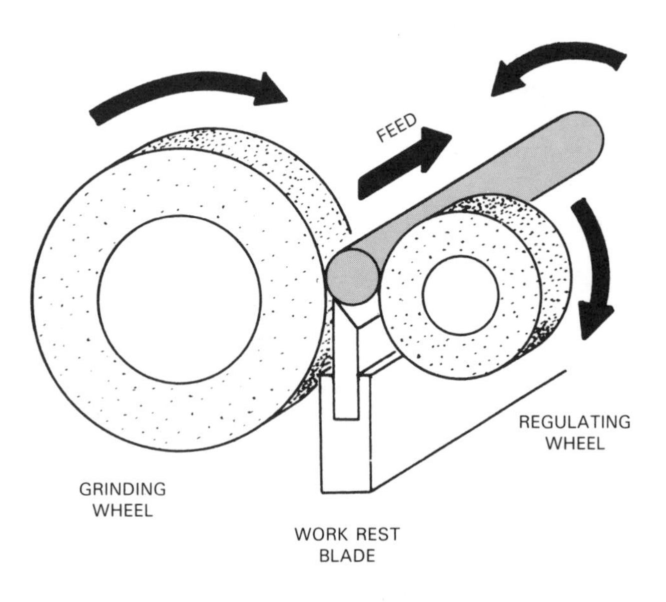

Fig. 4-44. Note how centerless grinding works.

Fig. 4-45. Internal grinding operation is being used to surface inside diameter of part. (Norton Co.)

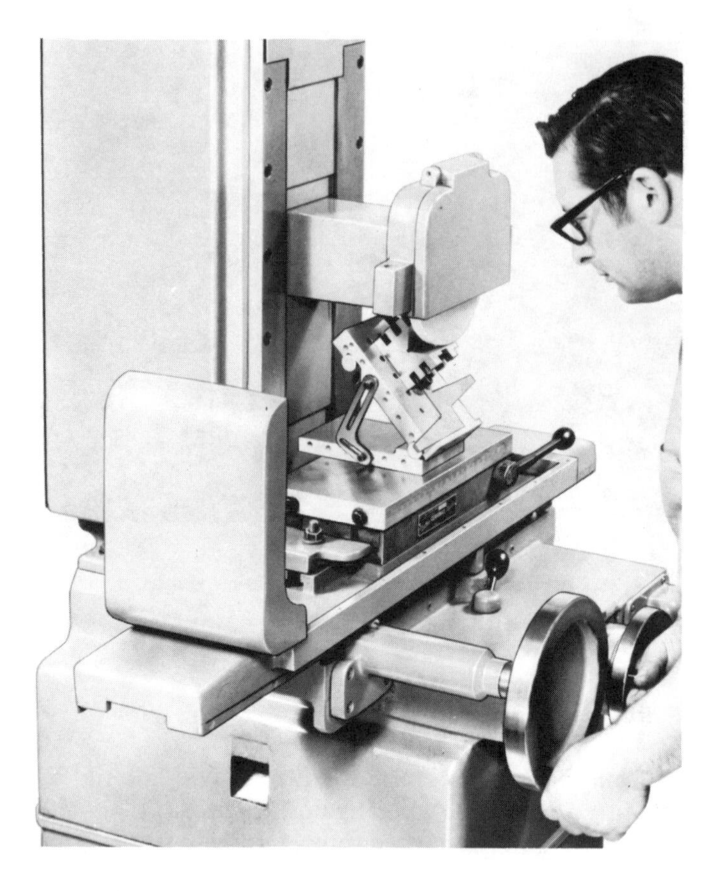

Fig. 4-46. Surface grinder will make a precise finish on workpiece.

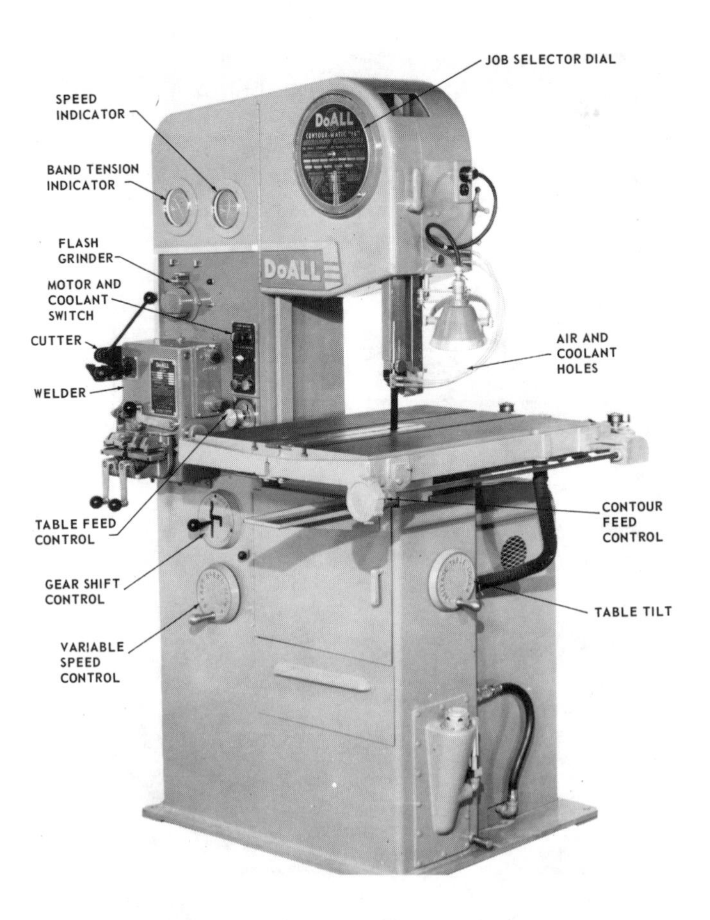

Fig. 4-47. Study parts of this vertical band saw. (DoALL Co.)

Fig. 4-48. Unwanted material is removed by sawing in solid sections, rather than being reduced to chips. (DoALL Co.)

There are three principle types of saws used to cut material to required lengths.

One has a RECIPROCATING (back and forth) cutting action, Fig. 4-49. It uses a blade similar to the one found in a hand hacksaw.

Another saw type uses a CONTINUOUS or BAND BLADE. One is shown in Fig. 4-50.

The third type uses a CIRCULAR BLADE with either a toothed blade or abrasive wheel. One is pictured in Fig. 4-51.

COLD FORMING METAL

Many products are fabricated from metal forms produced by one or more of the following cold forming techniques. Welding plays a very important part in the joining together of these components.

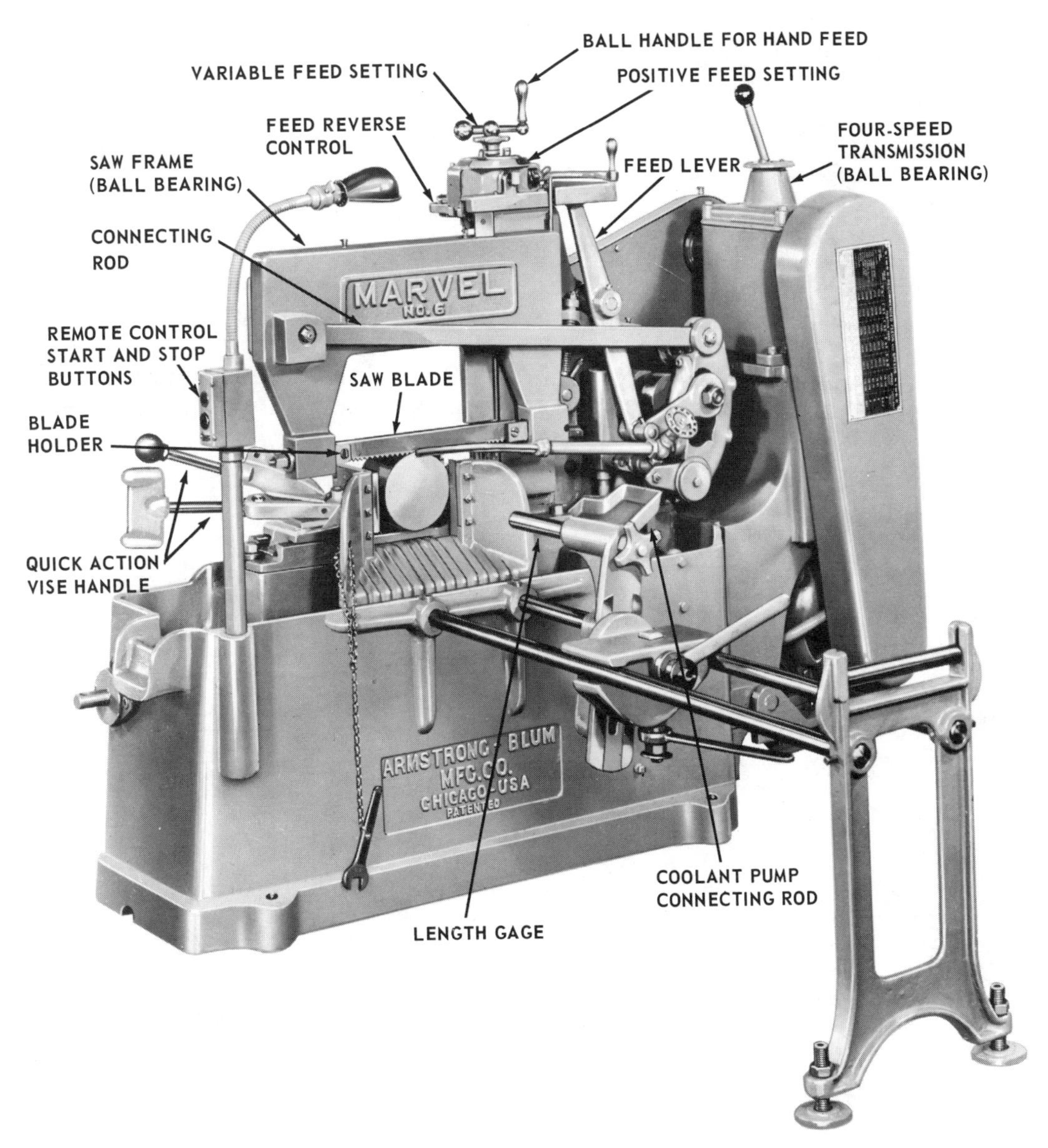

Fig. 4-49. Reciprocating power hacksaw can be found in many welding shops. Note part names. (Armstrong-Blum Mfg. Co.)

Fig. 4-50. Continous or band blade power hacksaw is being used to cut channel for weldment.

Fig. 4-52. A 42 in. by 50 in. shear spinning machine is producing head of a rocket motor case. When fully formed, it will be 13 1/2 in. deep and 40 in. in diameter.
(Meta-Dynamics, Cincinnati Milling Machine Co.)

Fig. 4-51. This is a dry type abrasive cutoff saw.
(Allison-Campbell Div., American Chain and Cable Co.)

Spinning

Spinning is a method of working metal sheet into three-dimensional shapes. An example is in Fig. 4-52.

Spinning involves, in its simplest form, rotating a disc of metal with a forming block (chuck). The forming block is made to the dimensions specified for the inside of the finished object. As the lathe rotates the metal disc and forming block, pressure is applied. The metal is gradually worked around the form until the disc assumes its shape and size.

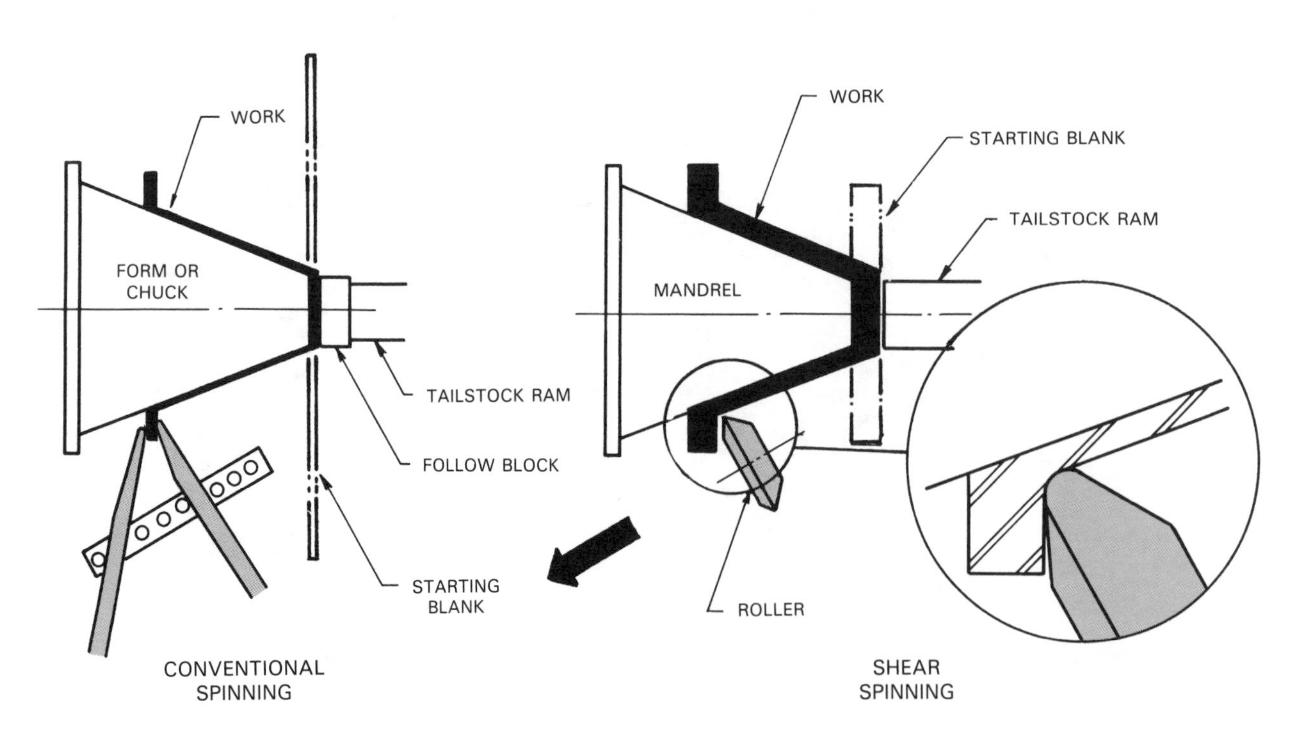

Fig. 4-53. Note difference between conventional spinning and shear spinning.

Shear spinning

Shear spinning is a metalworking technique that is similar to conventional spinning in outward appearance only. Refer to Fig. 4-53.

Shear spinning is a cold extrusion process where the parts are shaped by rollers that exert tremendous pressures on a starting blank or preform. This displaces the metal parallel to the centerline of the workpiece.

The metal for shear spinning is taken from the thickness of the blank. Whereas in spinning, the metal is taken from the diameter of the blank.

Explosive forming

Explosive forming uses a high energy pressure pulse of very short duration to shape metal sheet and plate. It is also known as *high energy rate forming* (HERF). Look at Fig. 4-54.

The pressure pulse for explosive forming can be generated by a chemical explosive or an electrical discharge. These two methods are shown in Fig. 4-55.

Stamping

Stamping is the term used for many press forming operations. Included is a cutting operation called blanking, Fig. 4-56.

Fig. 4-54. A few of the many parts shaped by explosive forming. (Ryan Aeronautical Co.)

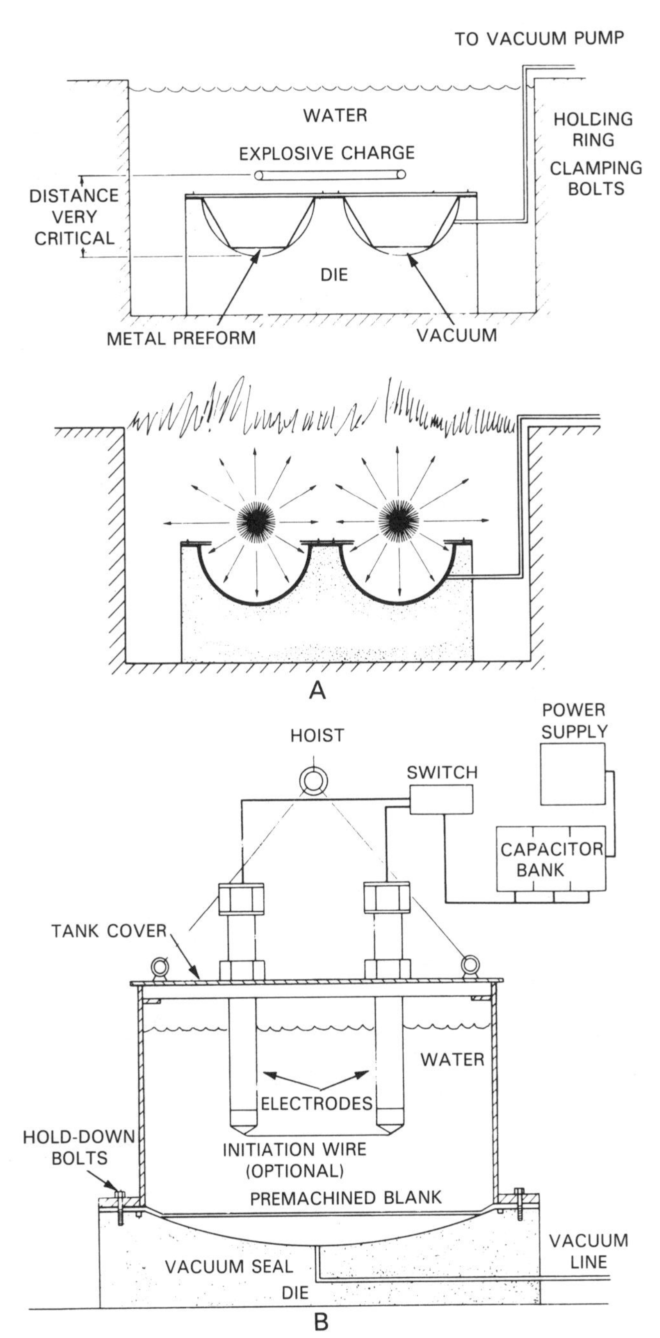

Fig. 4-55. A—Diagram showing principle of explosive forming process. A small charge of explosive material is used to form metal sheet. B—Diagram showing setup when using electrical energy as a source of power. (NASA)

Blanking involves cutting flat sheets to the shape of the finished part. Cutting is done with a punch and die.

Drawing, Fig. 4-57, takes the flat metal blanks and forms or draws them into three-dimensional shapes. The technique makes use of a draw press and a matched punch and die. There are many variations of the drawing technique.

Stretch forming

Stretch forming is a process where a metal blank is gripped by inserting opposite edges in clamps and subjecting it to a light pull. This causes the blank to hug or wrap around a form block of the required shape. This is illustrated in Fig. 4-58.

The piece is trimmed to final shape after the forming operation.

Bending

Bending is an operation in which the surface area of the work is not appreciably changed. A *forming brake*, Fig. 4-59, or *forming rolls* are used.

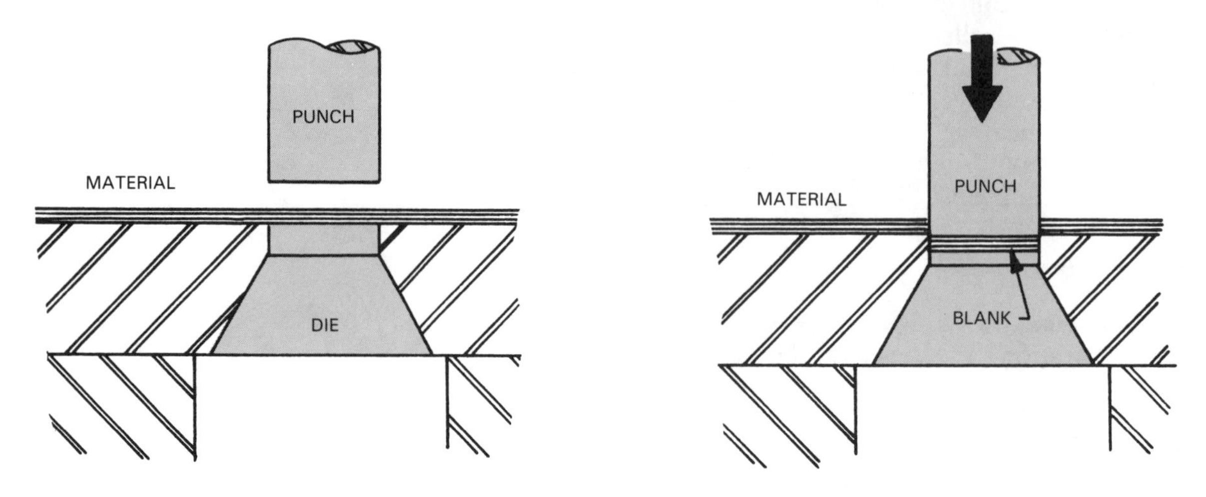

Fig. 4-56. Blanking is punch and die cutting of flat sheet metal to shape of finished part.

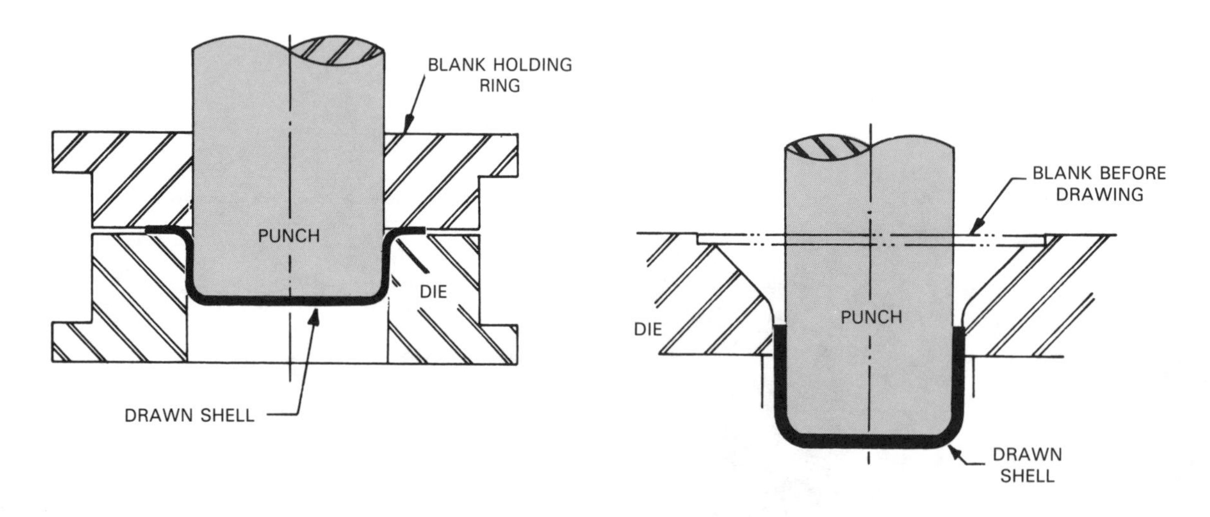

Fig. 4-57. Study basic drawing process.

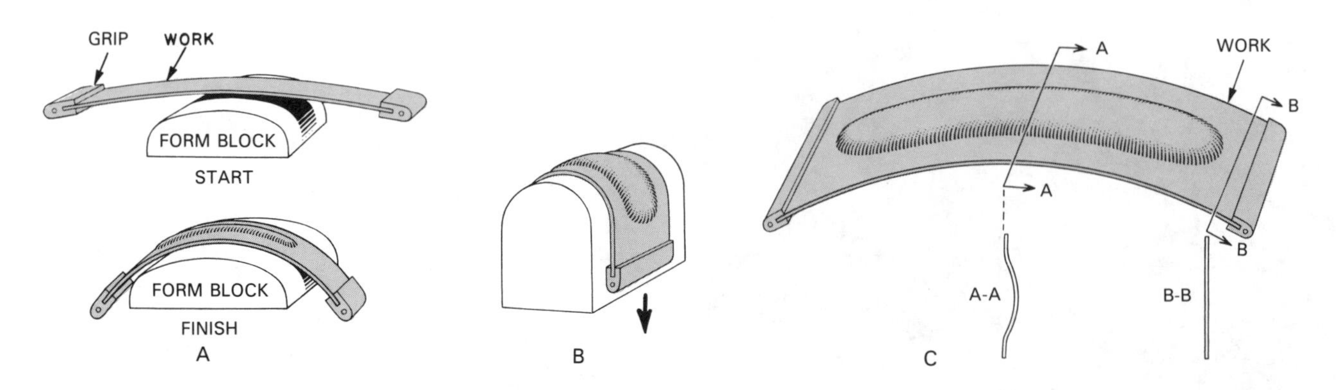

Fig 4-58. Stretch forming process. A—Relative positions of work. Form block and clamps at start and finish of a typical stretch forming operation producing a raised rib section. B—A typical shape produced by stretch forming over a form block. C—A stretch formed shape similar to that of bottom of a canoe. (Allegheny Ludlum Steel Corp.)

Fig. 4-59. This is a forming brake. Note how it has bent large piece of plate. (Niagara Machine and Tool Works)

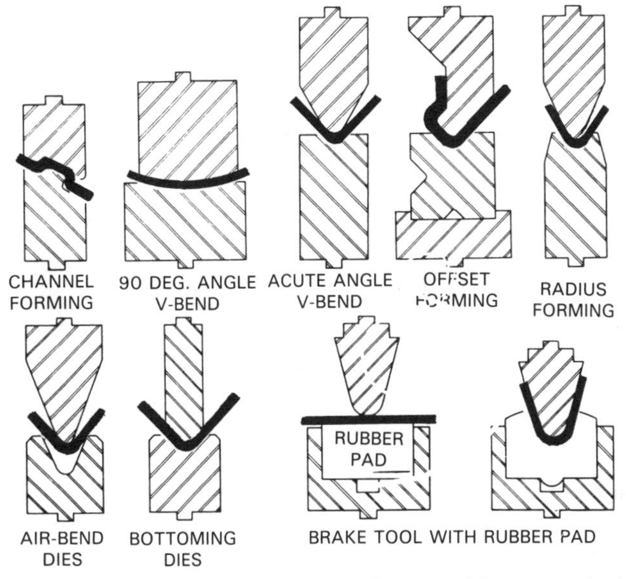

Fig. 4-60. Study typical bending operations possible on press brake equipment.

A few typical bending operations that are possible on a forming brake are shown in Fig. 4-60.

Roll forming

Roll forming takes a flat metal strip and passes it through a series of rolls that progressively form it into the required shape. See Fig. 4-61. The process is extremely rapid and almost any desired configuration is possible.

Tube bending

Tube bending is used to form radii in tubing without causing the tubing to collapse. Very complex bends on several axes are possible. Fig. 4-62 shows this process.

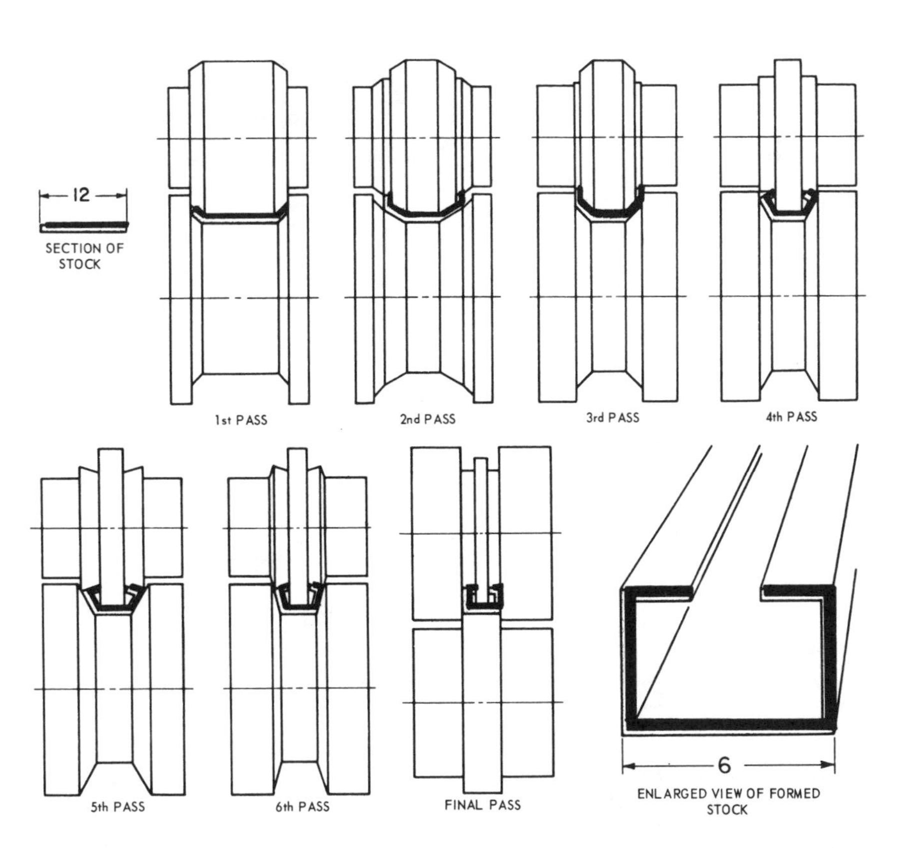

Fig. 4-61. Sequence followed for roll forming channel. Note how rolls progressively form sheet in one continuous operation.

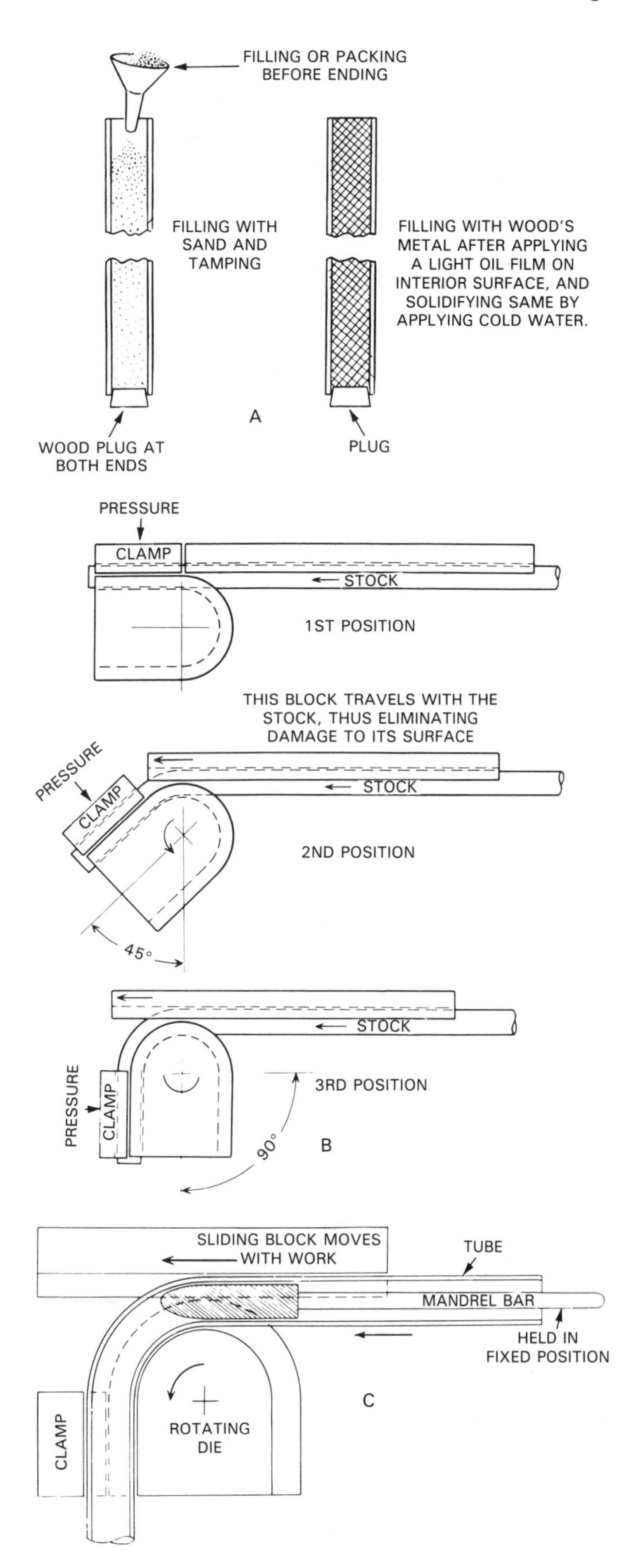

Fig. 4-62. Diagrammatic layout of rotating die tube bending technique. A—Preparing tube for bending. Filler material prevents tube from collapsing during forming operation. B—Steps in bending a tube. C—It is not necessary to fill tube before bending if reinforcing mandrel is used. (Allegheny Ludlum Steel Corp.)

UNIT 4—TEST YOUR KNOWLEDGE

1. Casting is a metalworking process for making metal objects by ____________________
____
____
____

2. Briefly describe the following casting techniques:
   a. Permanent mold casting ____________
   ____
   ____
   ____
   ____
   b. Centrifigual casting ____________
   ____
   ____
   ____
   ____
   c. Die casting ____________
   ____
   ____
   ____
   ____
   d. Investment casting ____________
   ____
   ____
   ____
   ____
   e. Sand casting ____________
   ____
   ____
   ____
   ____

The following questions are of the matching type. In the blank on the left, place the letter of the sentence that best describes that metalworking process.

3. ____ Forging. a. Uses a high energy pressure pulse of very short duration to shape metal sheet and plate. Pressure can be generated chemically or electrically.

4. ____ Extrusion. b. Takes flat sheet metal blanks and forms them into three-dimensional shapes.
5. ____ Spinning. c. Process where a metal blank is gripped by inserting opposite edges into clamps and subjecting it to a light pull. This causes blank to hug or wrap around a form block of required shape.
6. ____ Shear spinning. d. Used to form radii in hollow metal sections without causing section to collapse.
7. ____ Explosive forming. e. Process that uses pressure to shape metal. Metal may or may not be heated, but NOT to melting point. Pressure is usually applied in a hammering action.
8. ____ Stamping. f. Metal in this process is heated to a plastic state and pressure is applied to squeeze it through die of desired shape.
9. ____ Drawing. g. Method of working sheet metal disc into a three-dimensional shape by forcing it against a rotating forming block (chuck). Pressure is applied and metal is gradually worked around form until it assumes correct shape and size.
10. ____ Stretch forming. h. Part is shaped by rollers that exert tremendous pressures on a blank or preform. Rollers displace metal parallel to centerline of workpiece. Metal is taken from thickness of blank; whereas in another similar process, metal is taken from diameter of blank.
11. ____ Roll forming. i. An operation in which surface area of metal sheet or plate being worked is NOT appreciably changed.
12. ____ Bending. j. A term used for many press forming operations.
13. ____ Tube bending. k. A process that takes flat metal strip and passes it through a series of rolls that progressively form it into required shape.

14. A machine tool is a: (check correct answer or answers.)
    a. ____ A power driven, hand portable cutting tool.
    b. ____ A power driven machine used to shape metal by a cutting process.
    c. ____ Manufactured in a large range of shapes and sizes.
    d. ____ All of the above.
    e. ____ None of the above.

15. Prepare a sketch showing how a lathe operates.

16. The drill press uses a cutting tool called a ______.
    It operates by ______________________________
    ____

17. How does boring differ from drilling? ________
    ____
    ____

18. Planing machines are used primarily to machine __________ surfaces.

19. How does a milling machine operate? ________
    ____
    ____

20. There are two basic types of milling machines. List them. How do they differ? (Use sketches if necessary.)
    a. ______________________________
    ____
    ____
    b. ______________________________
    ____
    ____

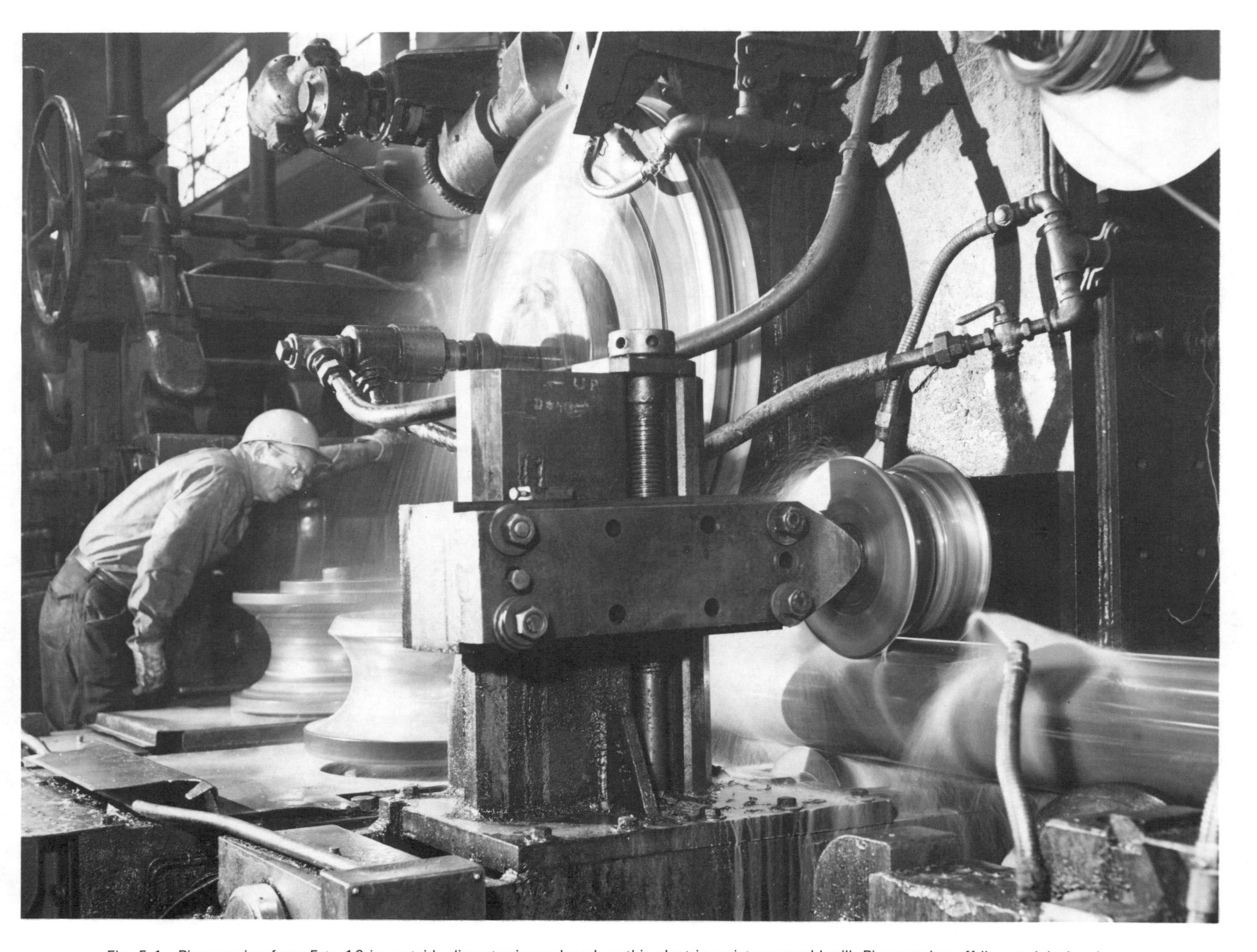

Fig. 5-1. Pipe ranging from 5 to 16 in. outside diameter is produced on this electric resistance weld mill. Pipe running off line at right has been rolled into circular form as it comes down line from left. It is welded together by large rotating circular electrodes in center of picture. (Bethlehem Steel Corp.)

Unit 5

WELDING PROCESSES

There is sometimes confusion about what the terms soldering, brazing, and welding mean. They all use heat to join metals. The differences, however, must be fully understood.

SOLDERING

Soldering is a method of joining metals with a nonferrous metal filler WITHOUT having to heat them to a point where the base metals melt. Soldering is carried out at temperatures lower than 800°F (427°C). The process is sometimes called *soft soldering.*

The strength of solder is relatively low. It is used for low stress, low pressure applications.

BRAZING

Brazing is a group of joining processes that use nonferrous alloys that have melting temperatures above 800°F (427°C). However, the filler metal's melting point is lower than the melting point of the metals being joined.

The strength of brazed joints depend on the alloying (resulting metal content in weld) of the filler metal with the base metal.

WELDING

Welding is a method of joining metals by heating to a suitable temperature to cause them to melt and fuse together. This may be with or without the application of pressure, and with or without the use of filler material having similar composition and melting point of the base metal.

MODERN WELDING TECHNIQUES

Many different welding processes have been developed since the days of simple welding done by the blacksmith at a forge (forge welding). Today, there is hardly an industry or business which does not depend in some way upon welding. Refer to Fig. 5-1.

Each welding process described has its disadvantages and disadvantages and may be more adaptable to

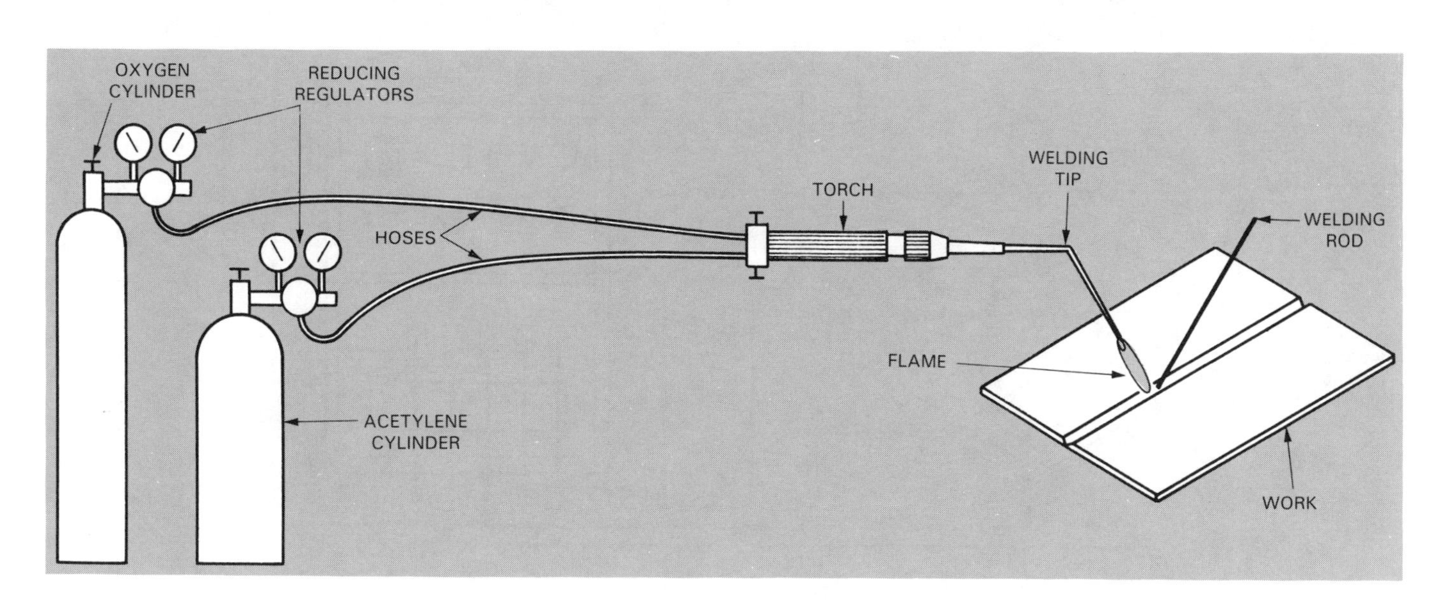

Fig. 5-2. Gas welding makes use of burning gases to produce heat to melt and fuse base metal. Study components of gas welding outfit.

tain applications than other welding processes. Many factors affect the selection of one welding process over another. A few of these factors include:
1. Materials to be joined.
2. Size and/or shape of the part.
3. Material thickness.
4. Type of material.
5. Production requirements.

Gas welding

Gas welding includes a group of welding processes that use burning gases, such as acetylene or hydrogen mixed with oxygen, to produce heat to cause the base metal to melt and fuse. Filler material of a similar composition and melting temperature as the base metal may or may not be used. Fig. 5-2 shows a gas welding setup.

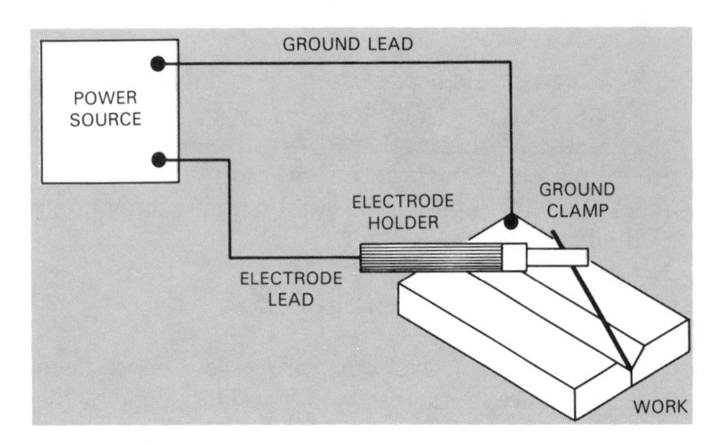

Fig. 5-3. Arc welding is a joining technique where electric current makes heat to melt and fuse base metal.

Arc welding

Arc welding is a joining technique that uses an electric arc to produce the heat necessary to cause the base metals to melt and fuse together. Filler metal in the form of an electrode may be added to the joint, Fig. 5-3.

GTAW and GMAW

Gas Tungsten Arc Welding (GTAW) and Gas Metal Arc Welding (GMAW) are used in this text as the professional terms for these processes. On the job, the informal terms such as TIG (Tungsten Inert Gas) and MIG (Metal Inert Gas) will be encountered. TIG refers to GTAW and MIG to GMAW. These convenient conversational terms have the same meaning to the welder as the professional terms in this text.

GTAW uses an electrode that is NOT consumed in the welding process. Fig. 5-4 illustrates the equipment for GTAW.

GMAW uses a consumable electrode that melts and contributes filler metal to the joint. Refer to Fig. 5-5.

The electrode, arc, and molten pool of the weld are protected from atmospheric contamination by a soft stream of inert gas (helium or argon). The gas is directed to the weld area by a tube that surrounds the electrode. When properly done, a solid joint, that requires little or no additional finishing, is produced.

Submerged arc welding

Submered arc welding is a technique where coalescence (fusion) is produced by heating with an electric arc between a bare metal electrode and the work. The welding arc is shielded by a blanket of flux

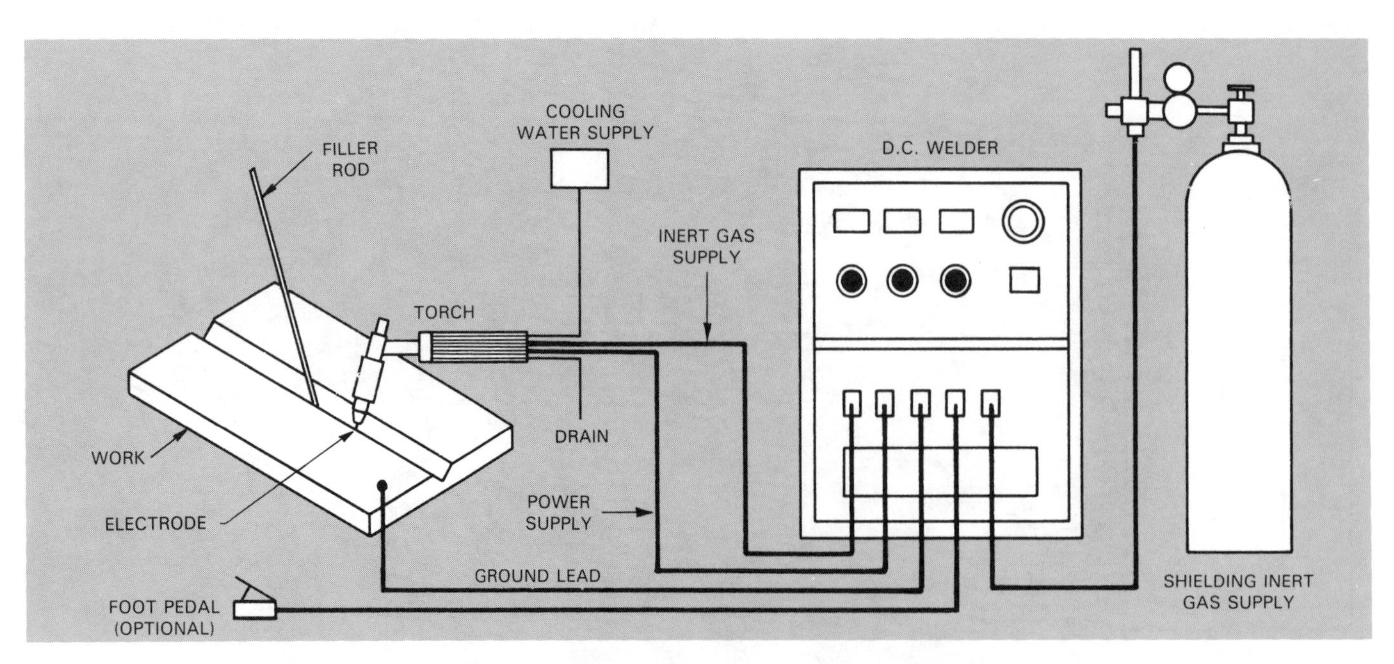

Fig. 5-4. TIG or GTAW welding is a gas shielded arc welding technique accomplshed with a permanent electrode that is not consumed. Study this welding process.

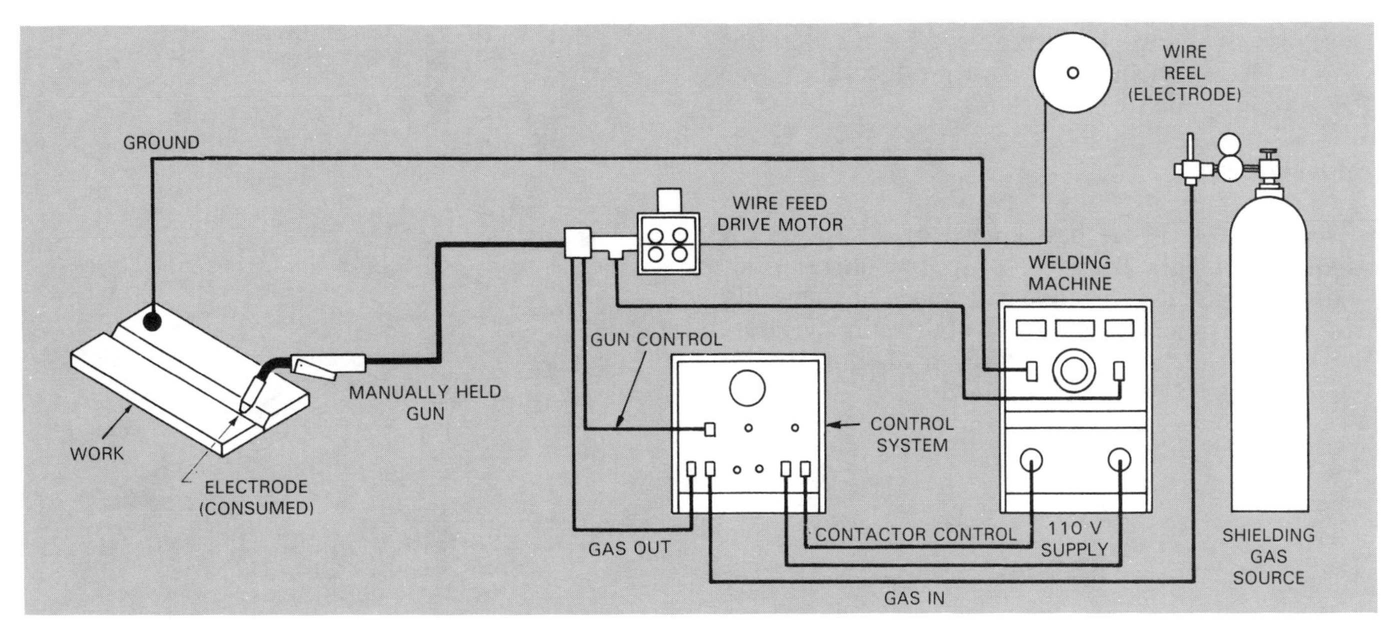

Fig. 5-5. MIG or GMAW welding is a gas shielded arc welding technique that uses an electrode that is consumed in welding process. Electrode contributes filler metal to joint.

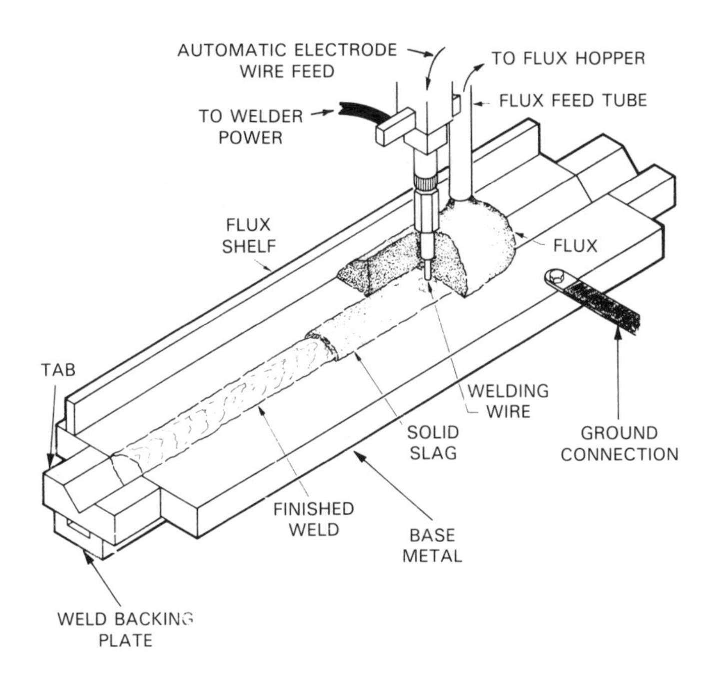

Fig. 5-6. Study submerged arc welding process.

on the work. Pressure is NOT used and filler metal is obtained from the electrode and sometimes from a supplementary welding rod.

The process produces no smoke, arc rays, radiant heat, or spatter. To weld, the operator fills the flux cone, points the gun into the joint, allows a pile of flux to accumulate and then strikes an arc under the flux with the electrode.

Once the arc is struck, the electrode automatically feeds into the arc as the gun is moved over the work. This is illustrated in Fig. 5-6.

Stud welding

Stud welding, Fig. 5-7, is also an arc welding process. Fusion is produced by an electric arc between a metal stud, or similar part, and the other work part. The surfaces to be joined are heated; then they are brought together under pressure. No shielding is used.

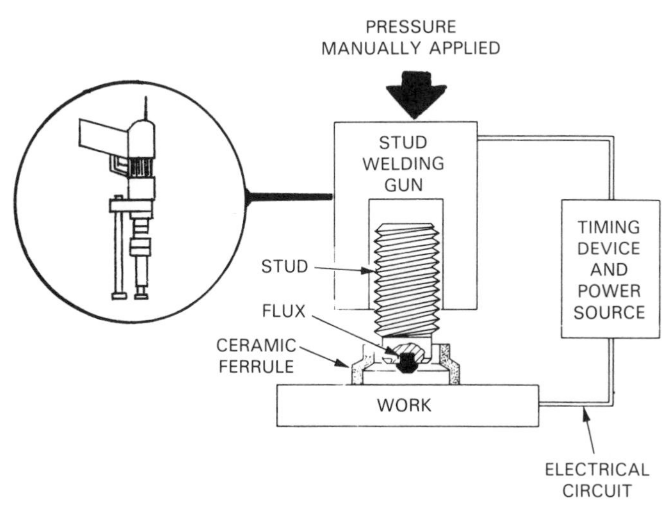

Fig. 5-7. Stud welding is an arc welding technique that welds a metal stud or similar part to another metal surface.

Resistance welding

Resistance welding is a group of welding processes where fusion heat is obtained from the electrical resistance of the work. Electric current through the welding electrodes and work parts produces the weld. Pressure is applied when welding.

Spot welding

Spot welding is the best known of the resistance welding techniques. See Fig. 5-8. Spot welding is widely used because it saves time and weight as the welds can be made quickly and they do not require the addition of filler metal. The welds are made directly between the metal parts being joined.

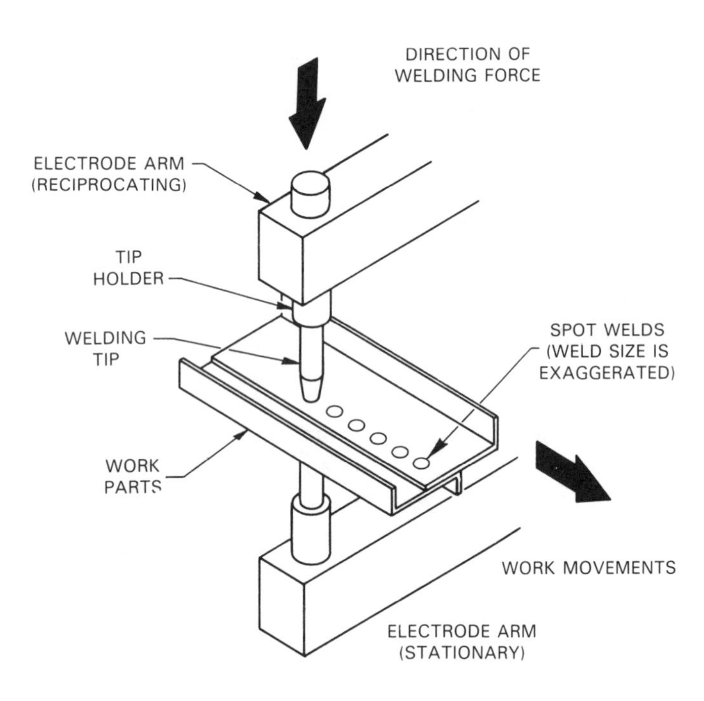

Fig. 5-8. Spot welding is a resistance welding technique.

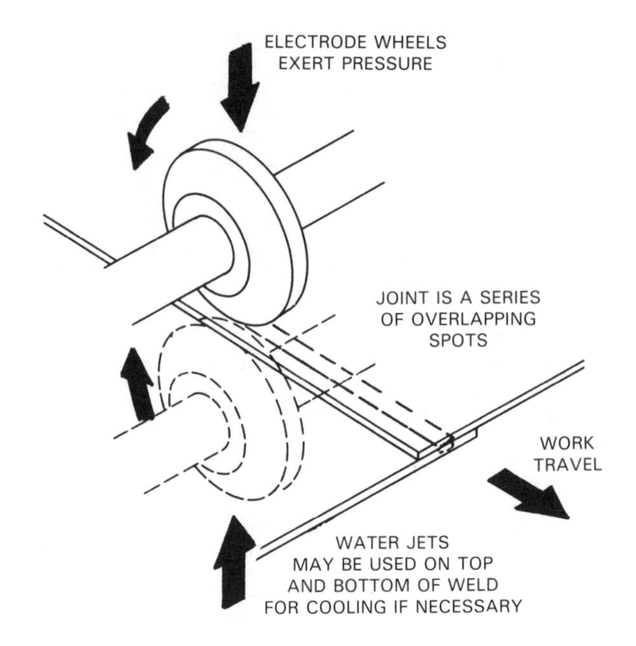

Fig. 5-9. Study seam welding process. It is a technique that produces a series of overlapping spot welds along joint.

Seam welding

Seam welding is a resistance welding process where fusion is produced by the heat obtained from resistance to the flow of an electric current through the work parts. The work parts are held together under pressure by circular electrodes. The resulting weld is a series of overlapping spot welds made progressively along the joint by rotating the electrodes. Fig. 5-9 pictures this welding process.

Projection welding

Projection welding is also a resistance welding technique, as shown in Fig. 5-10. Heat is produced by the flow of an electric current through the work parts that are held under pressure by the electrodes. The welds are localized at predetermined points by the design of the parts to be welded.

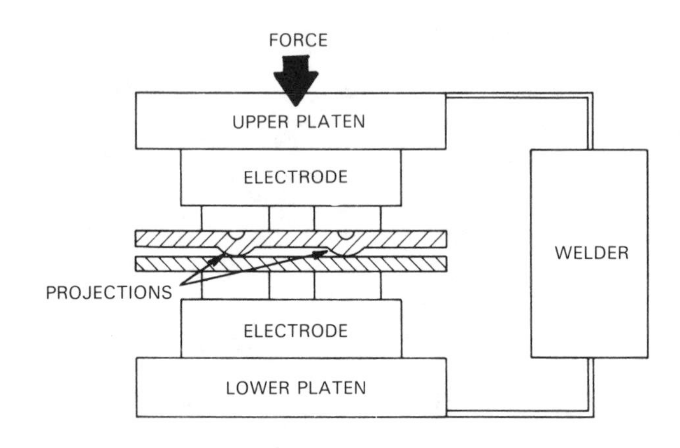

Fig. 5-10. Projection welding also uses heat produced by current through the work parts held under pressure by electrodes. Note projections on parts.

Flash welding

In *flash welding,* fusion is produced simultaneously over the entire area of abutting surfaces, Fig. 5-11. Welding heat is obtained from electric current between the two surfaces, and by the application of pressure after heating is substantially completed. Flashing is accompanied by the ejection of molten metal from the joint.

Upset welding

Upset welding, Fig. 5-12, is also a resistance welding process very similar to flash welding. However, the pieces to be joined are first butted together under pressure and then current is allowed to flow through the pieces until the joint is heated to the fusion point. Pressure is continued after the power is turned off. This causes an UPSETTING (bulging) action at the point where fusion occurs.

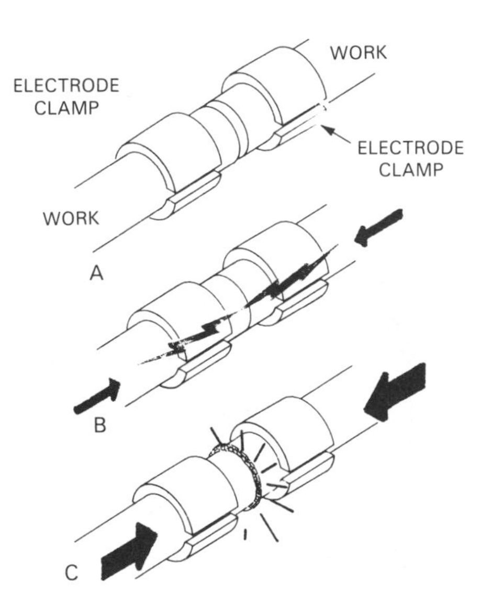

Fig. 5-11. In flash welding, two parts are moved into each other. Fusion is produced over entire area of abutting surfaces by heat obtained from resistance to current between the two surfaces, and by pressure after heating is almost completed.

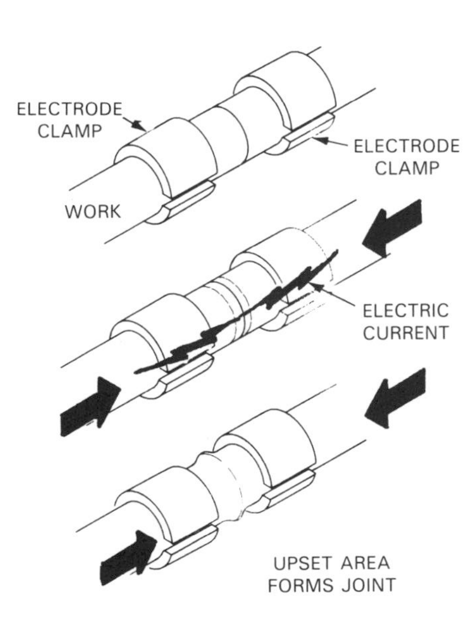

Fig. 5-12. During upset welding, pieces to be joined are butted together under pressure. Then current is allowed to flow through pieces until joint is heated to fusion point.

NEW WELDING PROCESSES

All areas of welding have made significant advances in recent years. The use of "exotic" metals for aerospace and nuclear applications has resulted in the development of new welding techniques.

Electron beam welding

The *electron beam welding process* makes use of a beam of fast moving electrons to supply the energy to melt and fuse the base metals. This is illustrated in Fig. 5-13. Welds must be made in a vacuum of 10^{-3} to 10^{-5} mm Hg, which practically eliminates the contamination of weld metal by atmospheric gases. The electron beam is capable of melting any known metal.

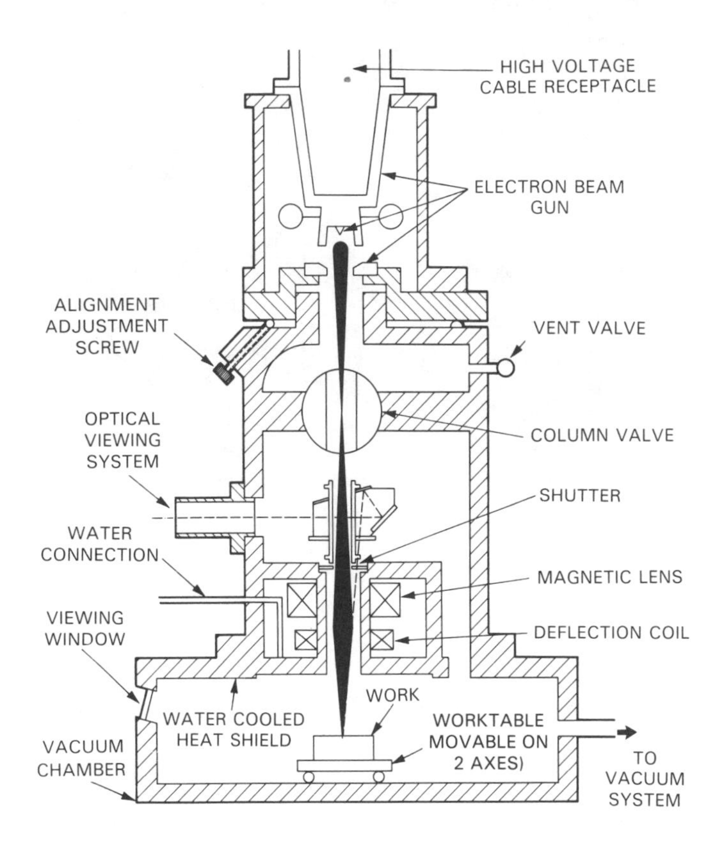

Fig. 5-13. With electron beam welding, beam of fast moving electrons supply energy to melt and fuse base metals. Welds must be made in a vacuum.

Inertia welding

Inertia welding is one of the simplest and most unique of the new welding processes. The technique utilizes frictional heat and pressure to produce full strength welds in a matter of seconds, Fig. 5-14.

The parts to be joined may be bar or tubular in shape; however, flat plates or formed shapes can be joined if the INTERFACE (portions forming joint) is generally circular. See Fig. 5-15.

Laser beam welding

Laser is the abbreviation for Light Amplification by Stimulated Emission of Radiation. The laser produces a narrow and intense beam of coherent monochromatic light that can be focused onto an area only a few microns (million of an inch) in diameter. Fig. 5-16 shows laser welding.

With *laser beam welding,* the light beam is used to vaporize the work at its point of focus. Molten metal surrounds the point of vaporization when the beam is moved along the path to be welded. A vacuum is not required as with electron beam welding.

A—In inertia welding, the flywheel, chuck, and one of the parts to be welded are accelerated to a preset speed.

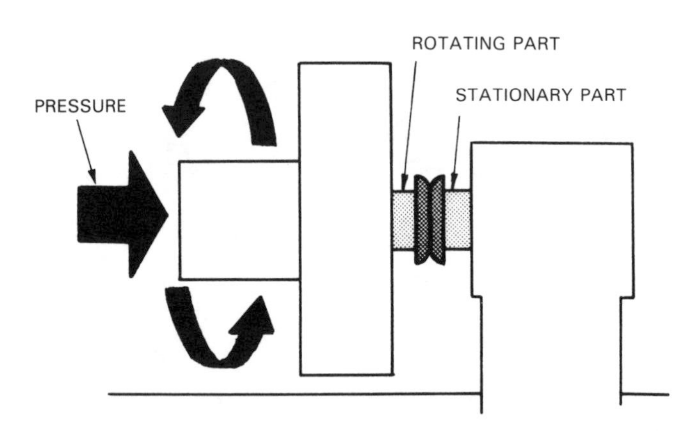

B—Upon reaching required speed, drive is disengaged and rotating part is thrust against stationary part. Energy in flywheel is discharged into interface (joint) and makes weld.

Fig. 5-14. Study inertia welding process.

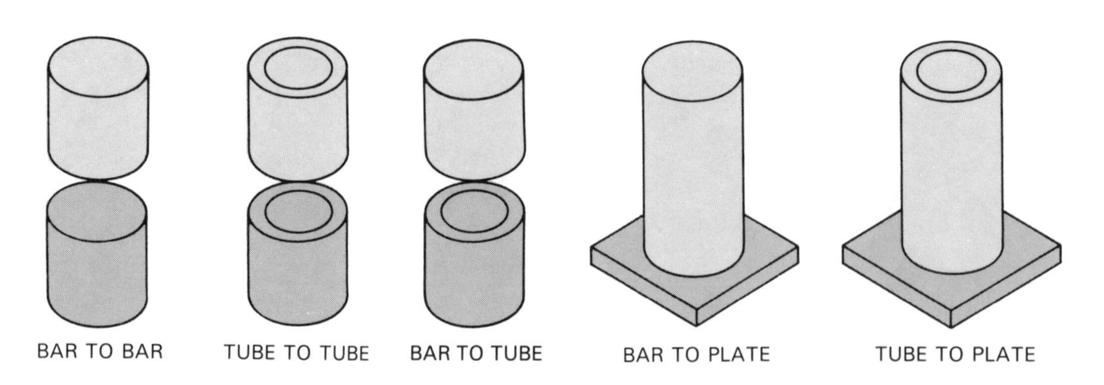

Fig. 5-15. Inertia welding requires that joint face of at least one part be essentially round.

Ultrasonic welding

Ultrasonic welding is a process for joining metals without the use of solders, fluxes, or filler metals and usually without the application of external heat. As in Fig. 5-17, the metals to be joined are clamped lightly between SONOTRODES (welding tips), and ultrasonic energy is introduced for a brief time (usually 1-3 seconds). A strong metallurgical bond is produced. There is little or no external deformation which is characteristic of pressure welding and without the heat affected zones found in resistance welding.

Cold pressure welding

Pressure alone is used to join the two metal surfaces in *cold welding,* Fig. 5-18. The process, however, involves more than pressure. Special tools were developed to produce the deformation required to direct the flow of metal into a true weld.

Cold welding is especially adaptable to aluminum. However, dissimilar metals like aluminum to copper, silver, lead, or nickel can be joined with this process. Thin sheet can be readily bonded to thick sections.

Micro welding

Technology has made it possible to fit 10,000 transistors on a CHIP (microscopic sized electronic circuit).

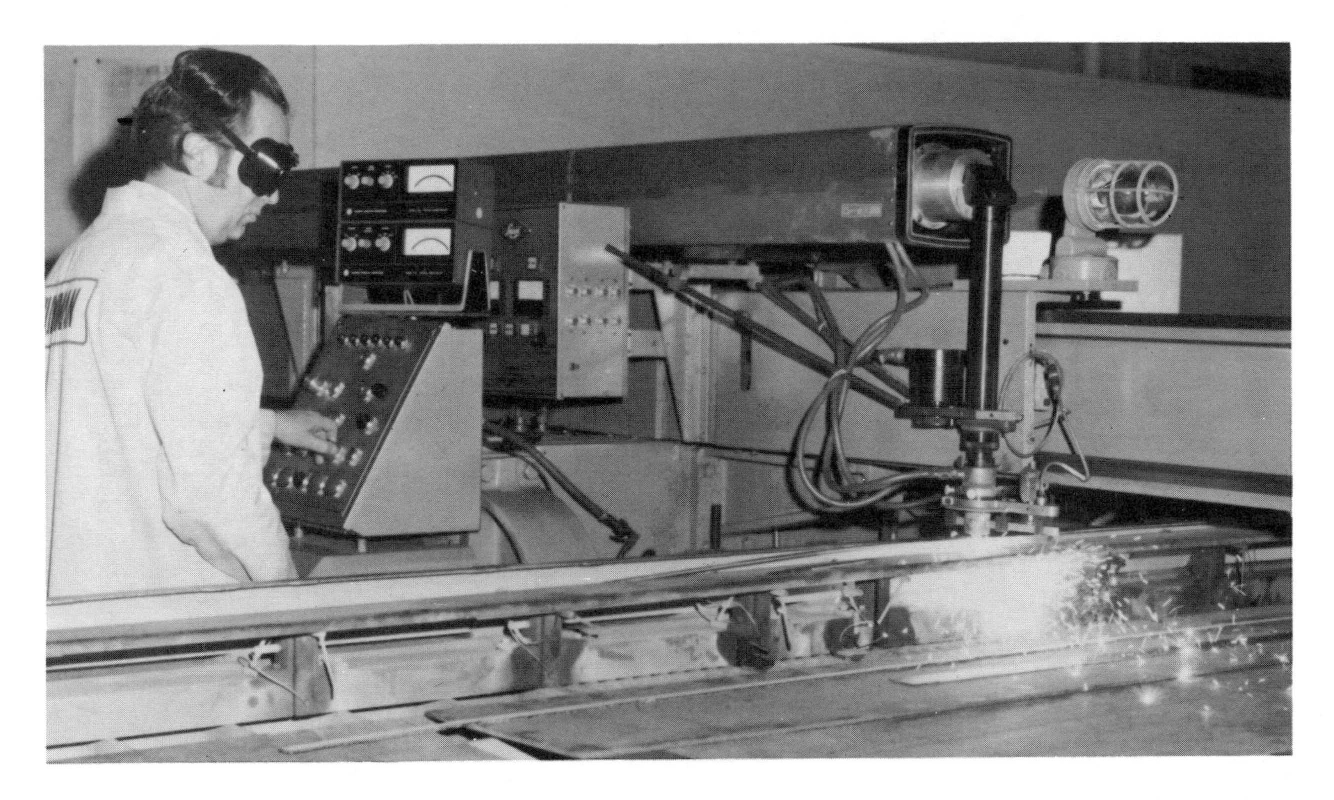

Fig. 5-16. In laser welding, a narrow and intense beam of light is used to vaporize work. Joint forms as molten metal surrounds point of vaporization when beam is moved along weld joint. (Grumman Aerospace Corp.)

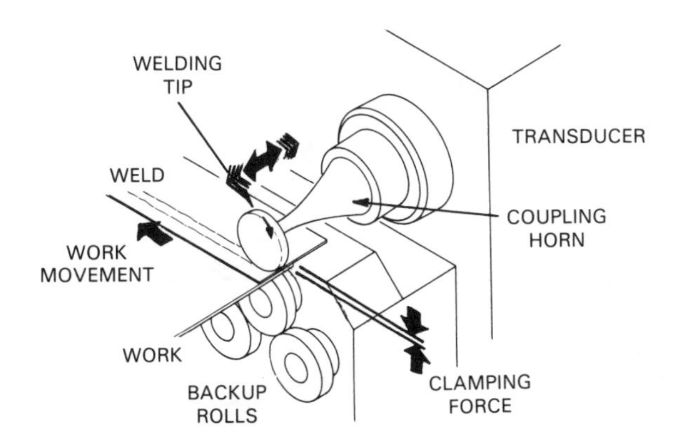

Fig. 5-17. Ultrasonic welding joins metals without solders, fluxes, or filler metals. Ultrasonic or high frequency wave energy is used to make strong metallurgical bond.

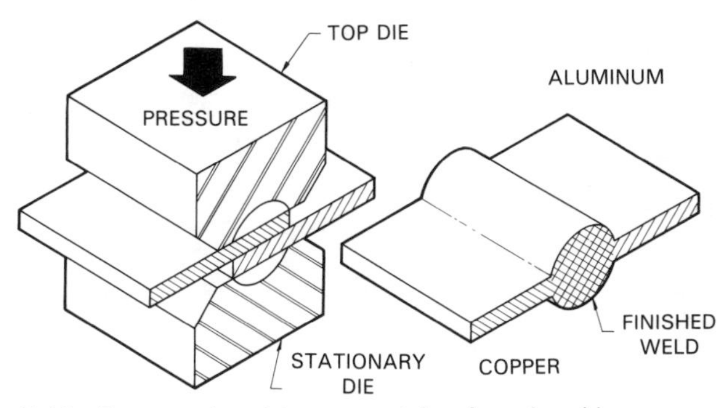

Fig. 5-18. Pressure alone joins two metal surfaces in cold or pressure welding. Similar or dissimilar metals can be joined.

The chip may only by 0.025 in. thick and 0.20 in. square (0.64 mm thick and 5.00 mm square). An example is given in Fig. 5-19.

Micro welding was developed to attach leads to these microcircuits or chips.

The first welding stage attaches 0.0015 in. (0.04 mm) gold wire to the chip and to heavier leads. This is done in an untraclean room, Fig. 5-20.

A microwelding machine can attach 40 wires in 15 sec. Very little heat is produced in making the welds. This prevents chip overheating and damage. After checking each weld, chips are enclosed in protective covers.

Flame spraying

Flame spraying is the term used to describe the process where a metal is brought to its melting temperature and sprayed onto a surface to produce a coating. The AWS (American Welding Society) recognizes this technique as a welding process although it is never used to join two metal sections. These processes include:

1. METALLIZING.
2. THERMO SPRAY.
3. PLASMA FLAME.

The term *metallizing* describes the flame-spraying process that involves the use of metal in wire form. Specially made wire is drawn through a unique spray gun by a pair of powered rolls. The wire is melted in the gas flame and atomized by compressed air. The

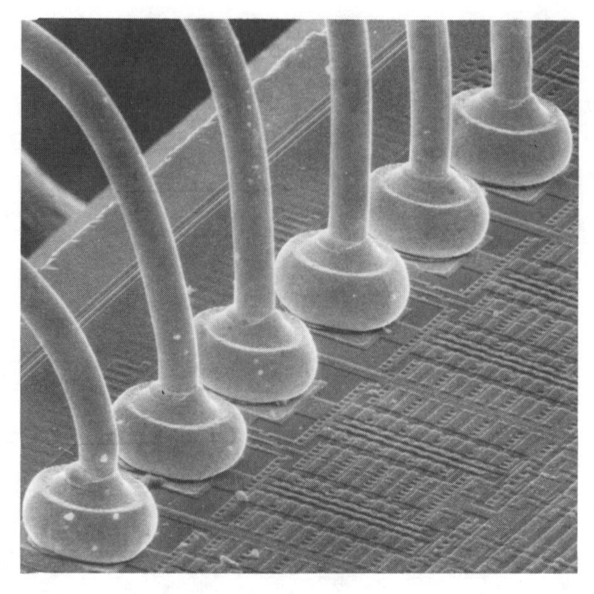

Fig. 5-19. Microwelding process. A—This machine spends less than 30 seconds to ''stitch'' forty gold threads between microcircuit and larger external leads. B—Close up view of tiny gold wires (99.99 percent pure) used to connect integrated circuit to lead frame. (Delco Electronics Div., General Motors Corp.)

Fig. 5-20. Microwelding must be done in an ultraclean room; otherwise, dust might ruin microcuits. (Delco Electronics Div., General Motors Corp.)

compressed air forces the molten metal onto the previously prepared work surface, Fig. 5-21.

Upon striking the surface, the particles interlock or mesh to produce a coating of the desired metal. Receiving surfaces must be cleaned and roughened before spraying, or the metal will not bond to it. Theoretically, there is no limit to the thickness that can be built up by flame spraying.

Any metal that can be drawn into wire form can be sprayed by this process. Fig. 5-22 shows a modern metallizing gun.

Metallizing has been employed to apply protective corrision resistant coatings of zinc and aluminum to steel surfaces. It is also used to build up worn bearings and shafts that might otherwise have to be discarded.

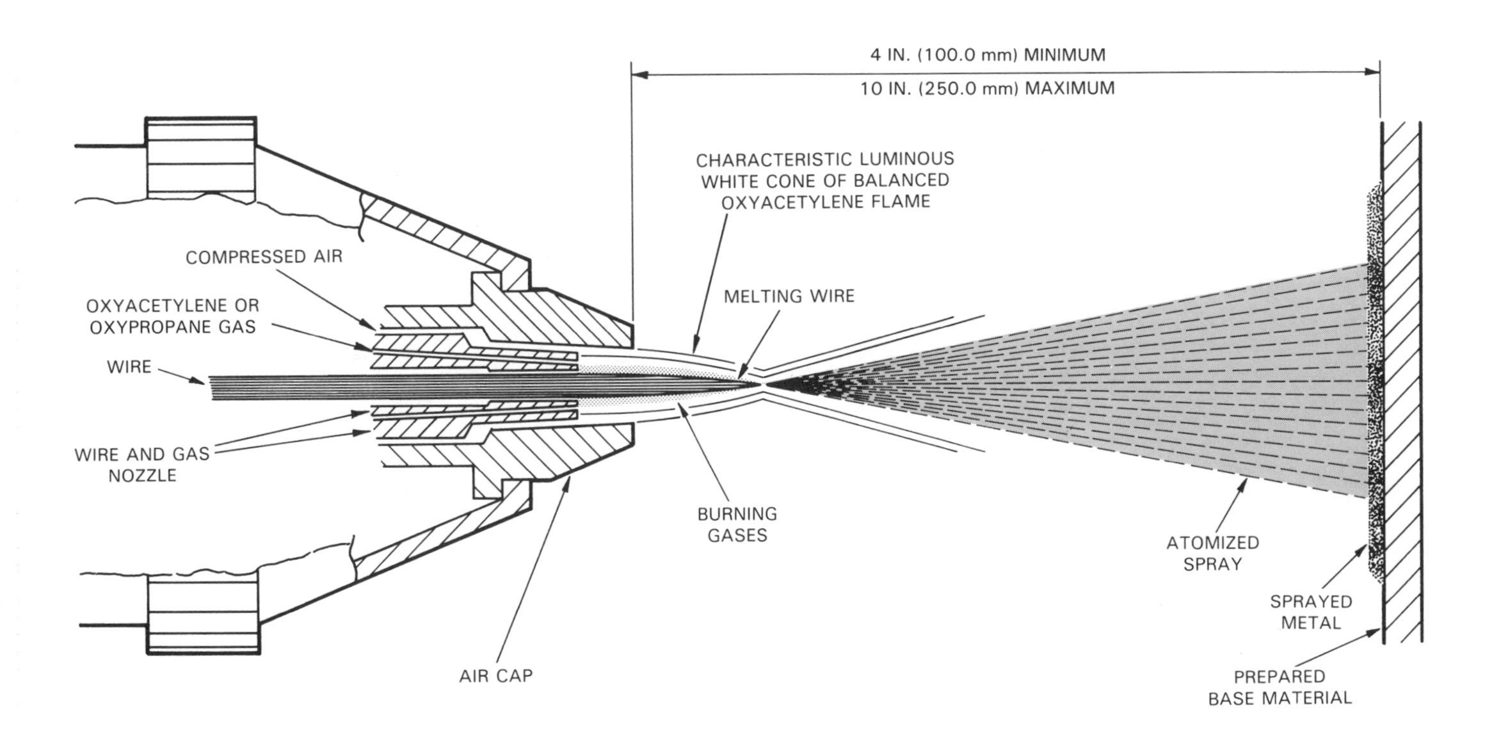

Fig. 5-21. During metallizing process, wire is fed through a special spray gun, melted by a gas flame, atomized, and sprayed onto work by compressed air.

Fig. 5-22. This metallizing gun is applying stainless steel onto a roll surface. (Metco Inc.)

The term *thermo spray* is used to describe the flame spraying equipment that involves the application of metals, and other materials, that cannot be drawn into wire. They are used in POWDER FORM. See Figs. 5-23 and 5-24.

These special alloys and materials are ideal for hard surfacing critical areas that must operate under severe conditions. High temperature, refractory materials which are also chemically inert can be sprayed.

In the *plasma flame spray process,* the spray gun

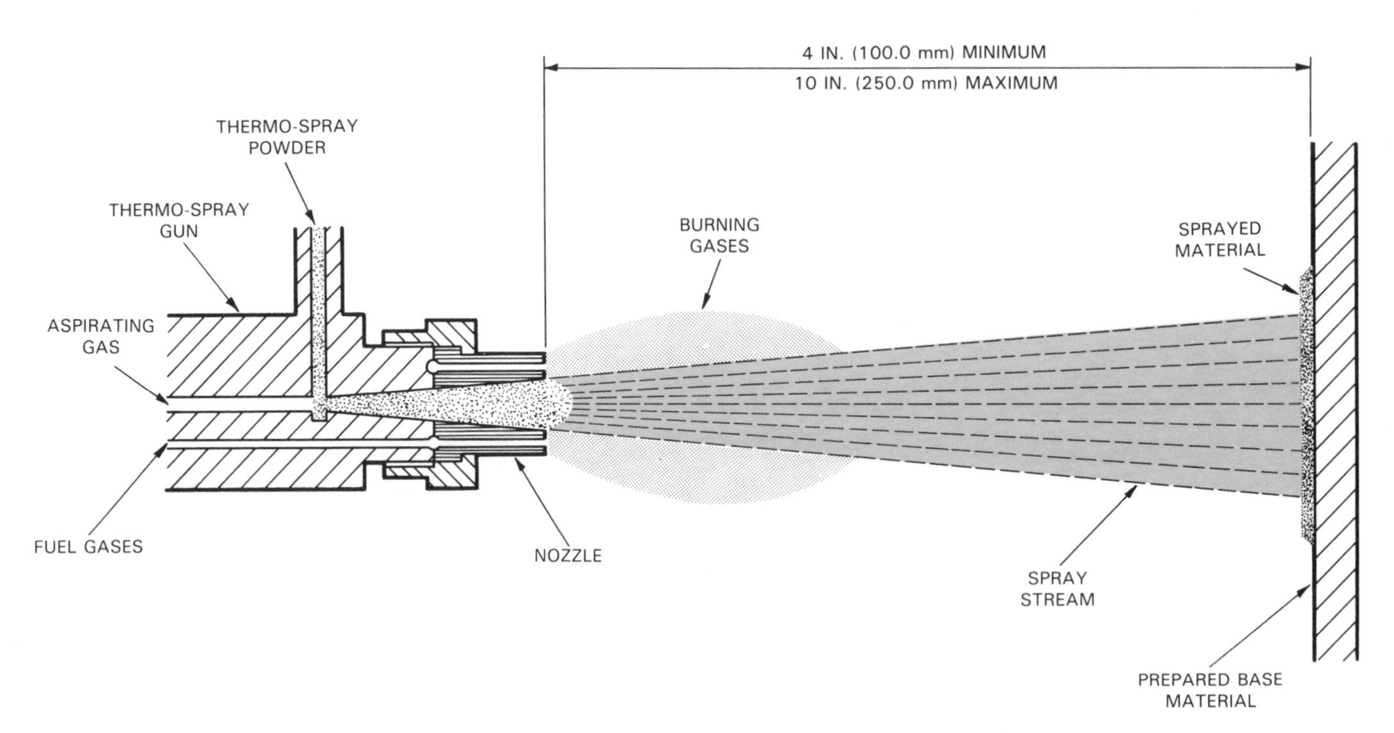

Fig. 5-23. Study diagram of thermo-spray process and how it operates.

Fig. 5-24. Thermo-spray gun uses metal and ceramic materials in powder form and an oxy-fuel gas flame to apply coatings. (Metco Inc.)

utilizes an electric arc that is contained within a water-cooled jacket. Refer to Fig. 5-25.

An inert gas, passed through the arc, is "excited" to temperatures up to 30,000°F (16 662°C). In general, most inorganic materials that can be melted without decomposition can be applied, Fig. 5-26.

Application of this process includes the spraying of rocket nozzles and nose cones with high melting point materials. Jet engine turbine blades are often given a protective coating by the plasma flame process.

AUTOMATIC (ROBOTIC) WELDING SYSTEMS

Automatic or *robotic welding systems* use computer controlled machines to ensure consistent welds and fits between parts. Fig. 5-27 shows an example of robotic welding.

The use or robots in welding is limited to mass production of products. The high cost of the robots makes an automatic system impractical for simple, low volume weldments.

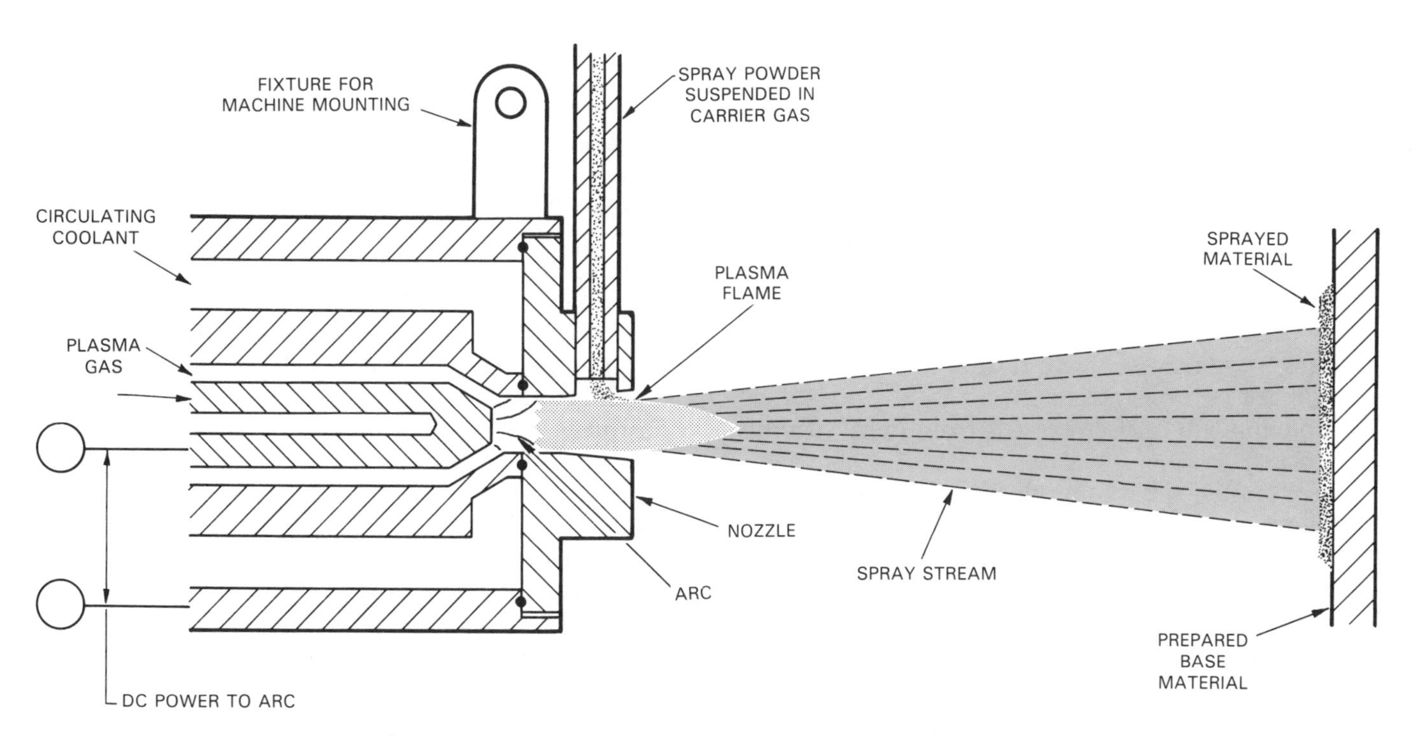

Fig. 5-25. Plasma flame process, capable of "super-high" temperatures, can spray any material that will melt without decomposing.

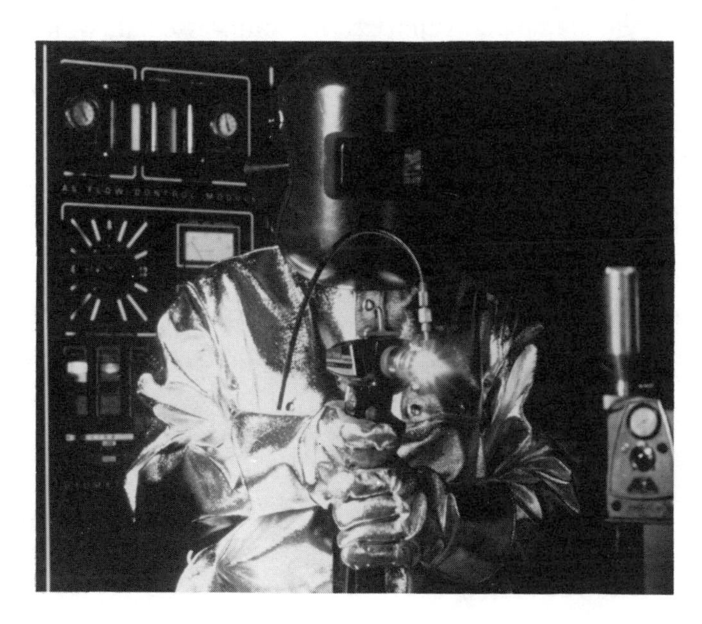

Fig. 5-26. Preparing to use plasma flame process. Note how operator is dressed for protection against heat. (Metco Inc.)

Fig. 5-27. Robotic welding, although fast and accurate, is limited to high volume production, like welding or these modern automobile unitized bodies. (Ford Motor Co.)

UNIT 5—TEST YOUR KNOWLEDGE

The following questions are of the matching type. Place the letter of the correct answer in the appropriate blank space next to the number.

1. ____ Micro welding.
2. ____ Soldering.
3. ____ Metalizing.
4. ____ Brazing.
5. ____ Welding.
6. ____ Gas welding.
7. ____ Arc welding.
8. ____ GTAW welding.
9. ____ GMAW welding.
10. ____ Thermo spray.
11. ____ Submerged arc welding.
12. ____ Stud welding.
13. ____ Plasma spray.
14. ____ Resistance welding.
15. ____ Spot welding.
16. ____ Seam welding.
17. ____ Projection welding.
18. ____ Flash welding.
19. ____ Electron beam welding.
20. ____ Inertia welding.
21. ____ Laser beam welding.
22. ____ Ultrasonic welding.
23. ____ Robotic welding.
24. ____ Cold pressure welding.

a. Joining technique that uses an electric arc to produce heat to melt and fuse base metals. Filler metal from an electrode may be added to joint.
b. A welding technique that uses a bare metal electrode. Welding arc is shielded by a blanket of flux on work.
c. Pressure alone is used to make joint in this welding process.
d. Uses frictional heat and pressure to produce a full strength weld in seconds.
e. Molten metal is often ejected from joint with this process.
f. Uses burning gases to produce heat needed to make weld.
g. Uses a gas shielded permanent electrode that is not consumed in welding process.
h. Uses a gas shielded consumable electrode that melts and contributes filler metal to joint.
i. A welding technique used to attach metal studs or similar objects.
j. Heating joining metals to a suitable temperature to cause them to melt and fuse together. Pressure may or may not be applied, and with or without use of a filler metal.
k. Uses a nonferrous alloy with melting temperatures above 800°F (427°C) but lower than the melting point of metals being joined.
l. Widely used because it saves time and weight as welds can be made quickly and do not require addition of filler metal.
m. Must be used in a vacuum to prevent contamination of weld metal by atmospheric gases.
n. Carried out at a temperature lower than 800°F (427°C) and has relatively low joint strength.
o. Group of welding processes where fusion heat is obtained from resistance to flow of electric current through work and by application of pressure.
p. Form of resistance welding where welds are localized at predetermined points by design of parts to be welded.
q. Process that uses a beam of intense light only a few microns in diameter.
r. Process for joining metals without use of solders, fluxes, or filler metals and usually without application of external heat.
s. Circular electrodes apply pressure to hold metal pieces being welded together. Resulting weld is a series of overlapping spot welds made by rotating electrodes.
t. Technique employed to weld very fine wires to microcircuits (chips).
u. Computer controlled automatic welding systems.
v. Flame spraying process where wire is melted in a gas flame and vaporized by compressed air which carries it to a previously prepared work surface.
w. Flame spray metals in powder form that cannot be drawn into wire.
x. Capable of producing temperatures up to 30,000°F (16 662°C); this flame spraying technique can use any material that will melt without decomposing.

Unit 6

STRUCTURAL METALS

Metal is available in a large range of shapes and sizes. Study Fig. 6-1. It shows the customary methods for designating or billing individual sections on prints.

While these shapes are generally accepted, there may be differences in the way they are used by some firms. Stock structural shapes when cut, machined, bent,

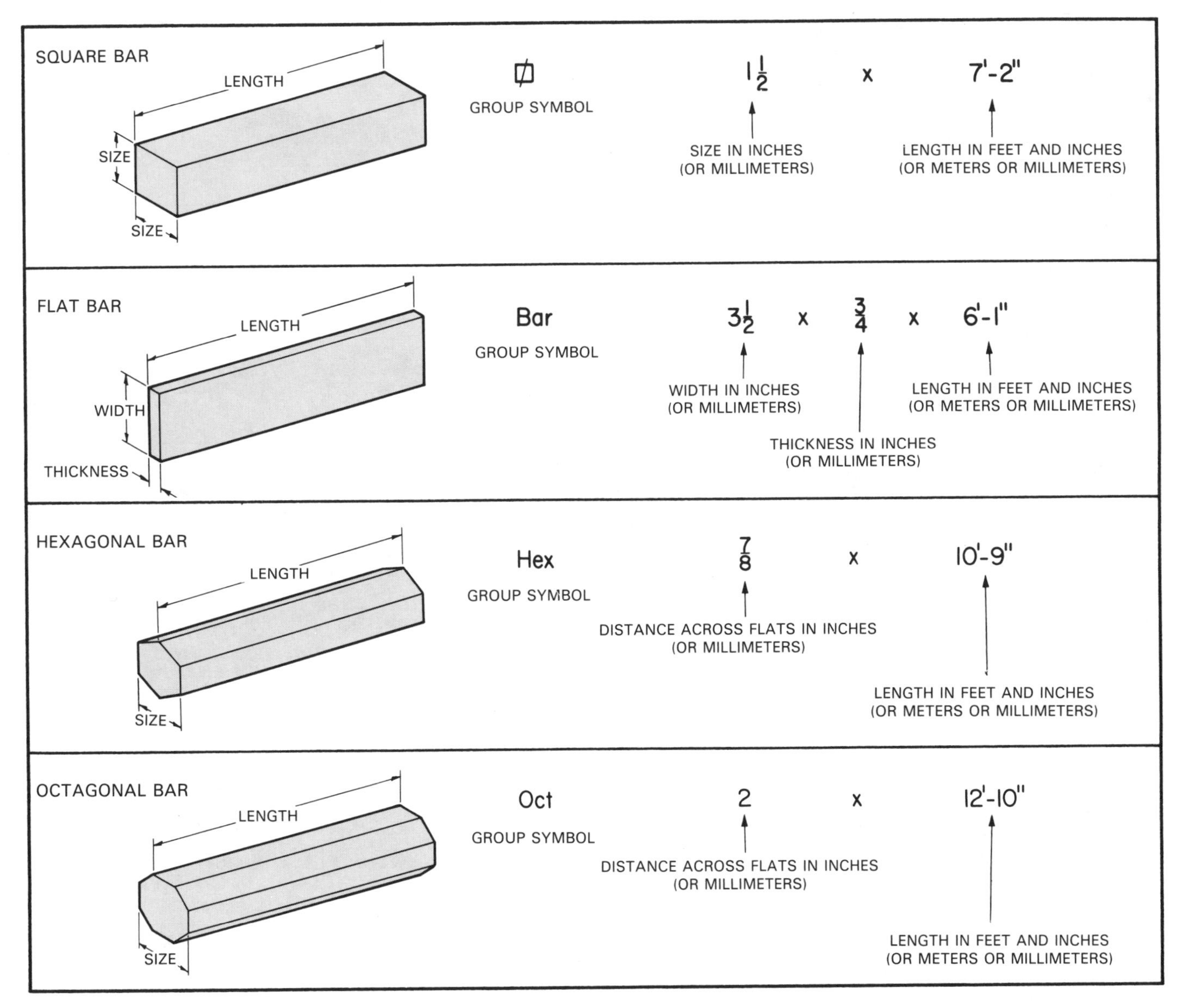

Fig. 6-1. Study usual method of designing structural steel shapes on shop drawings. Dimensions will be given in either English or metric units.

(Fig. 6-1 Continued on Page 64)

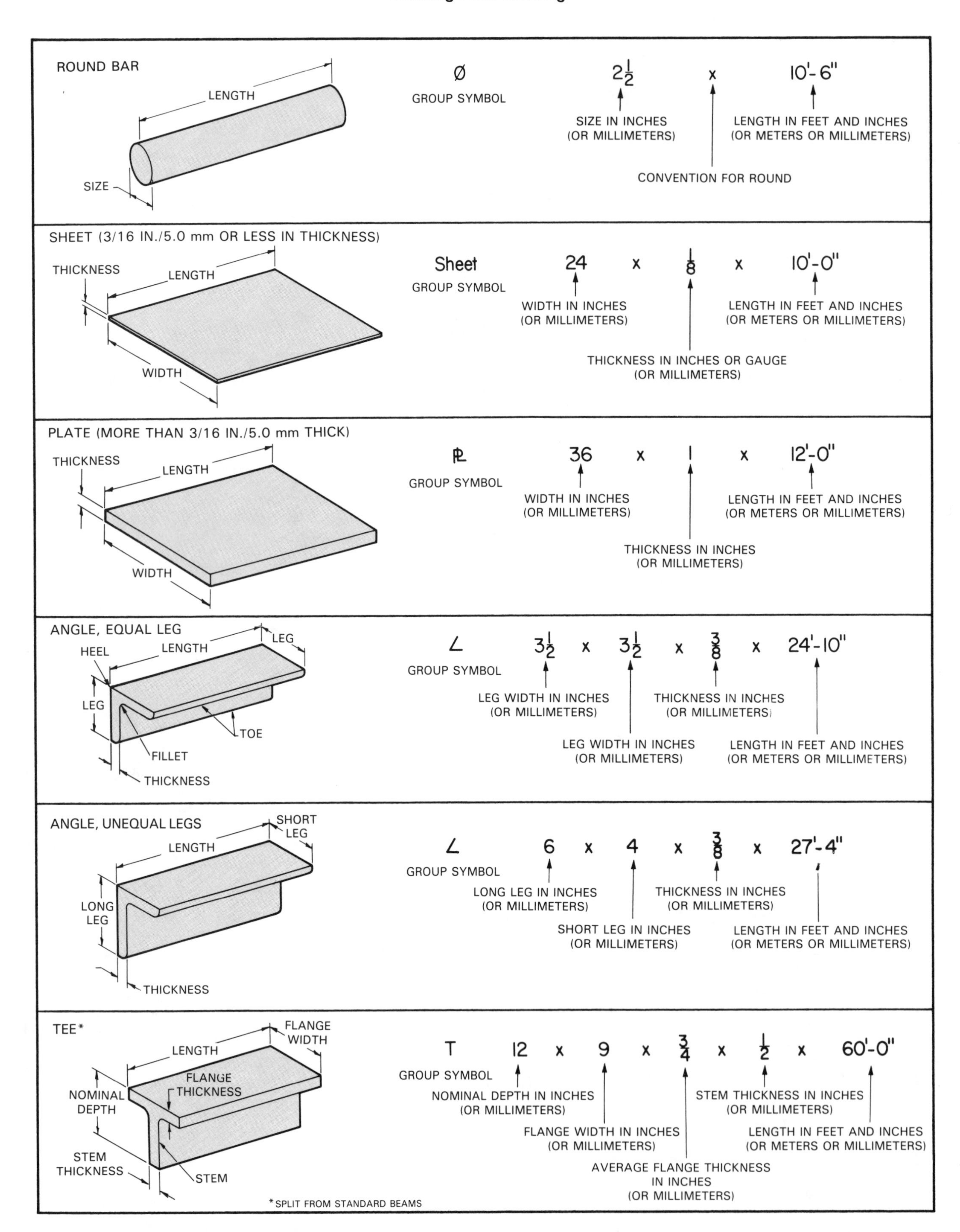

(Fig. 6-1 Continued from Page 63)

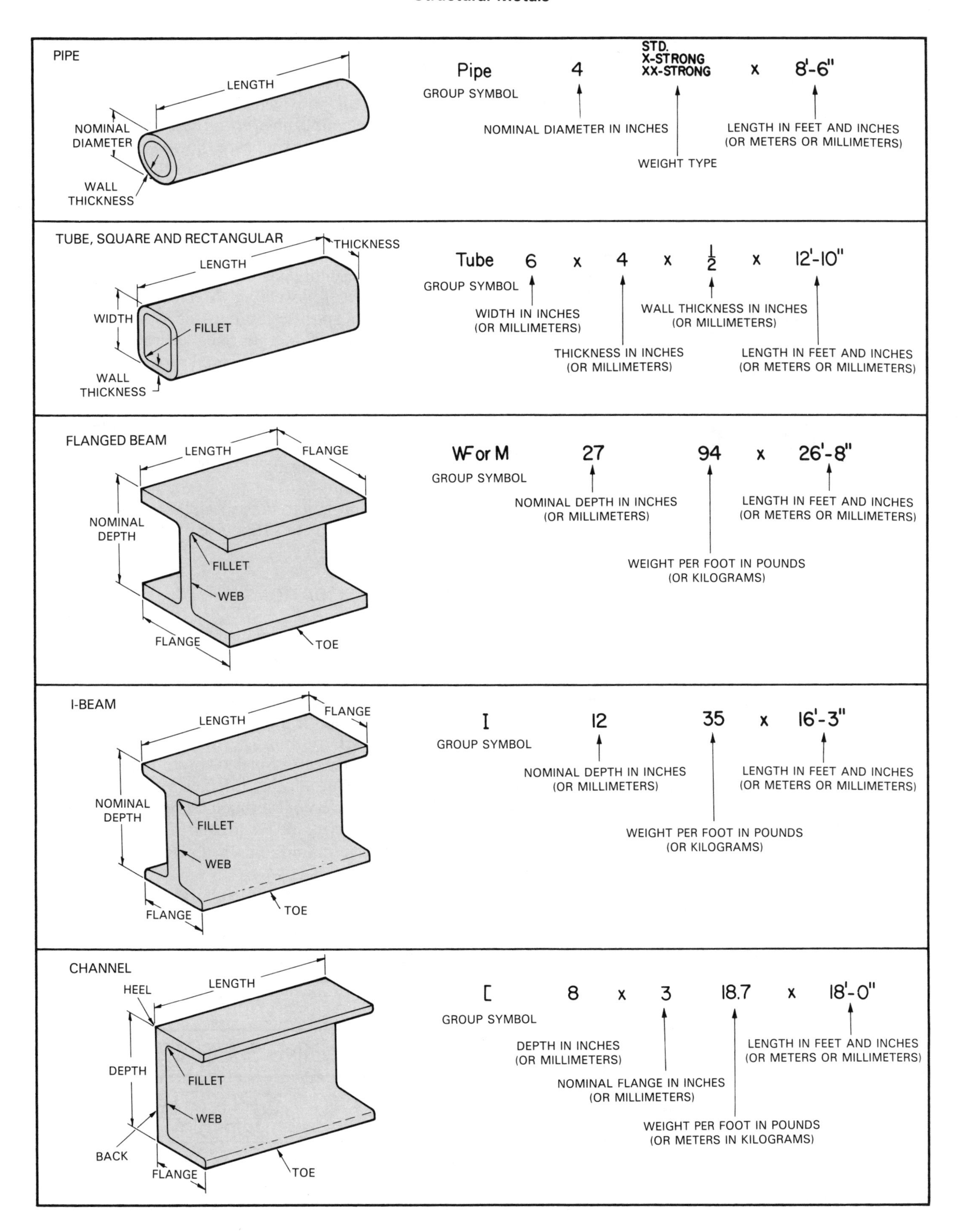

(Fig. 6-1 Continued from Page 63)

rolled, spun, stamped, or drawn and combined with castings, forgings, and extrusions, can be found in all products fabricated by welding. Refer to Fig. 6-2.

Stock shapes are used whenever possible to keep material and machining costs at a minimum.

Fig. 6-2. Many products, like this bridge, are designed to be fabricated from standard stock steel shapes. (Bethlehem Steel Corp.)

WHAT IS METAL?

Since metal plays such an important role in welding, do you know as much about metal as you should? How would you define the term "metal"? If you described it as a material that is tough, malleable, with high tensile strength and able to withstand high temperatures without melting or burning, you would be only partly correct. Only some metals have these characteristics.

There are metals, however, that do NOT have these features. For example: MERCURY is fluid at normal room temperature; body heat will melt GALLIUM in the palm of your hand; LITHIUM is so soft it can be scratched by your fingernail.

Can you tell by looking at a piece of metal whether it is a *ferrous metal* (metal that contains iron) or a *nonferrous metal* (metal that contains no iron)? Could it be an *alloy* (mixture of two or more metals)? Perhaps it is a *base metal* (pure metal like copper, tin, zinc, etc.).

There is no simple answer to these questions. With additional study and experience, however, a welder can develop a working knowledge of metals that will make the job safer and easier.

METALS SUPPLIED FOR WELDING

The welder usually has little or no control over the metal furnished for welding. Metal provided by the employer for a specific job must be assumed to meet print specifications. Metallurgical characteristics of a metal can be certified by the metal supplier or the mill that produced the metal.

As a welder, it is important that you know the characteristics of the metal(s) you will be welding. This is important for two reasons.

1. It will aid in insuring that the welds meet design specifications.
2. It will permit you to take special safety precautions when welding metals that give off toxic fumes and residue.

METAL SPECIFICATIONS

Metal specifications are usually located in a special section of the title block. Fig. 6-3 shows an example.

There are times when the metal specs may be given elsewhere on the drawing. Usually, the specifications are provided in one or more of the following standards:

1. Military (MIL-).
2. American Iron and Steel Institute (AISI-).
3. Society of Automotive Engineers (SAE-).
4. American Society of Mechanical Engineers (ASME-).
5. American Society for Testing and Materials (ASTM-).
6. Some larger manufacturers have their own, like General Motors (GM-).

To secure the complete metallurgical specifications of the metal, including the chemical and mechanical properties, it will be necessary to refer to the appropriate metals handbook.

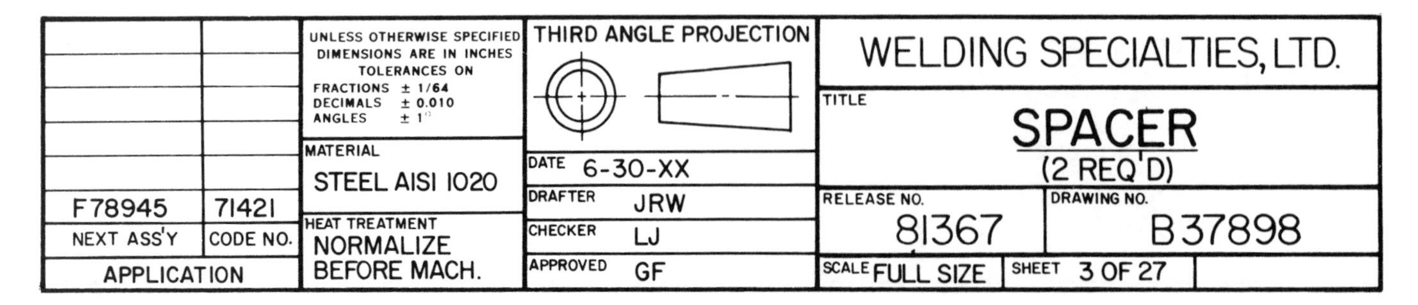

UNLESS OTHERWISE SPECIFIED DIMENSIONS ARE IN INCHES
TOLERANCES ON
FRACTIONS ± 1/64
DECIMALS ± 0.010
ANGLES ± 1°
THIRD ANGLE PROJECTION
WELDING SPECIALTIES, LTD.
MATERIAL: STEEL AISI 1020
DATE 6-30-XX
TITLE: SPACER (2 REQ'D)
F78945 | 71421
NEXT ASS'Y | CODE NO.
APPLICATION
HEAT TREATMENT: NORMALIZE BEFORE MACH.
DRAFTER JRW
CHECKER LJ
APPROVED GF
RELEASE NO. 81367
DRAWING NO. B37898
SCALE FULL SIZE
SHEET 3 OF 27

Fig. 6-3. Metal specifications are usually shown in a special section of drawing's title block.

UNIT 6—TEST YOUR KNOWLEDGE

1. You have seen and handled metal. How would YOU describe what a metal is? ____________

____

____

____

2. Different metals are often used in a product because of their special characteristics. For example, copper is used because it is a good conductor of heat and electricity and can be easily soldered. How many metals can you list and what are their special characteristics?

	METAL	UNUSUAL CHARACTERISTICS
a.	________	______________________
b.	________	______________________
c.	________	______________________
d.	________	______________________
e.	________	______________________

3. How does a ferrous metal differ from a nonferrous metal?

   a. FERROUS METAL ____________

   ____

   b. NONFERROUS METAL ____________

   ____

4. How does a base metal differ from an alloy?

   a. BASE METAL____________

   ____

   b. ALLOY____________

   ____

5. Why is it important for the welder to know the characteristics of the metal to be welded?

____

____

____

Shown below are several structural shapes. What is the name of the shape and the group symbol used to identify it on a print? Print your answers in the blanks.

6. a. ____________ (name) b. ________ (symbol)
7. a. ____________ (name) b. ________ (symbol)
8. a. ____________ (name) b. ________ (symbol)
9. a. ____________ (name) b. ________ (symbol)
10. a. ____________ (name) b. ________ (symbol)
11. a. ____________ (name) b. ________ (symbol)
12. a. ____________ (name) b. ________ (symbol)
13. a. ____________ (name) b. ________ (symbol)
14. a. ____________ (name) b. ________ (symbol)
15. a. ____________ (name) b. ________ (symbol)
16. a. ____________ (name) b. ________ (symbol)
17. a. ____________ (name) b. ________ (symbol)

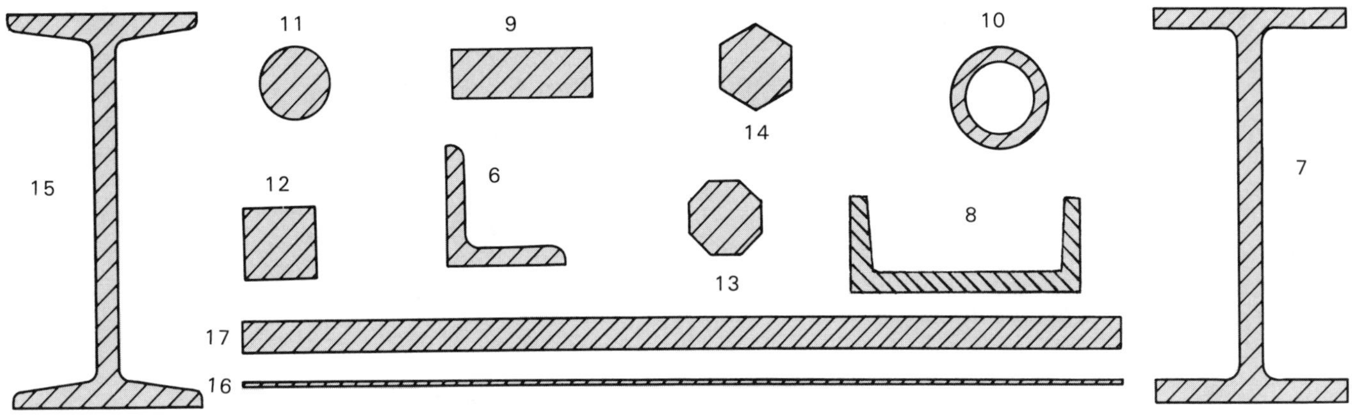

Unit 7

TYPES OF PRINTS

It would not be practical for industry to manufacture a product without using prints or drawings that provide complete manufacturing details. Each part in a product, even the smallest rivet, requires a print.

Skilled welders, whether they work as a production welder or as a job shop welder, must be able to read and interpret drawings.

A *production welder* usually does the same welding procedure over and over. Mass production of a weldment requires this type welder.

The *job shop welder* seldom does the same job twice. He or she commonly makes single, special order weldments for individual jobs.

The drawings and prints welders use vary from a freehand sketch for a simple job, Fig. 7-1, to highly detailed prints for a complex job or product, Fig. 7-2.

WORKING DRAWINGS

Working drawings supply the information needed to make and assemble the many pieces and parts that make up a product. Welders and related workers rely on these prints when doing their assigned tasks.

There are two types of working drawings: detail drawings and assembly drawings.

A *detail drawing* includes a print of the part (one or more views) with dimensions and other information needed to make the part. Fig. 7-3 shows an example of a detail drawing.

An *assembly drawing* shows where and how the parts described in the detail drawings fit into the complete assembly of the product. One is pictured in Fig. 7-4.

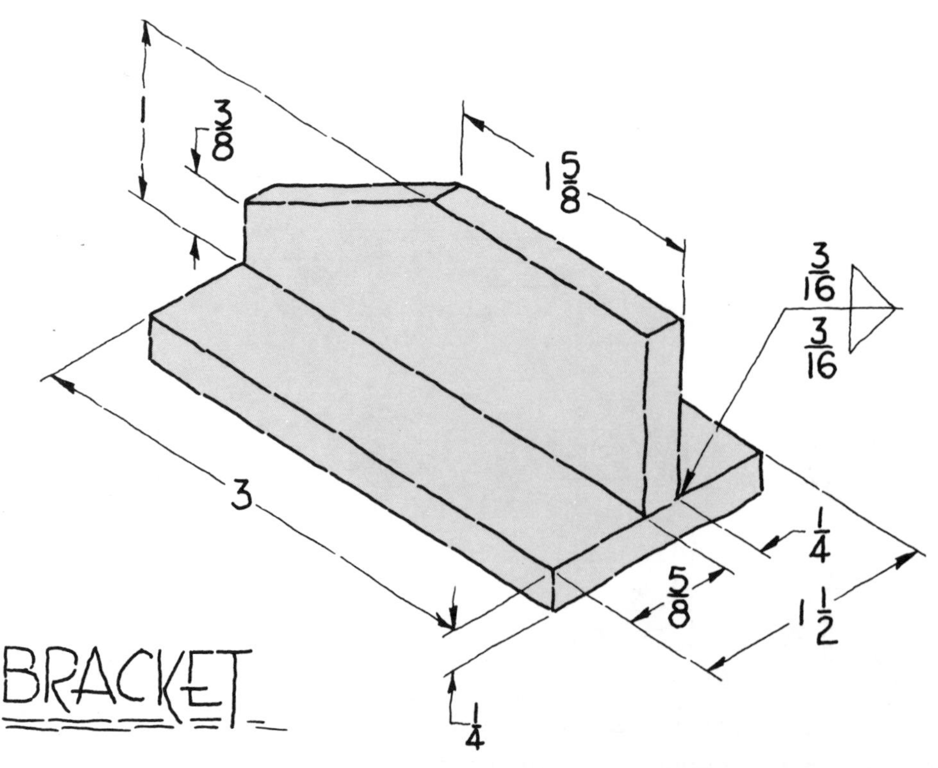

Fig. 7-1. A welder may work from a sketch when weldment is relatively simple.

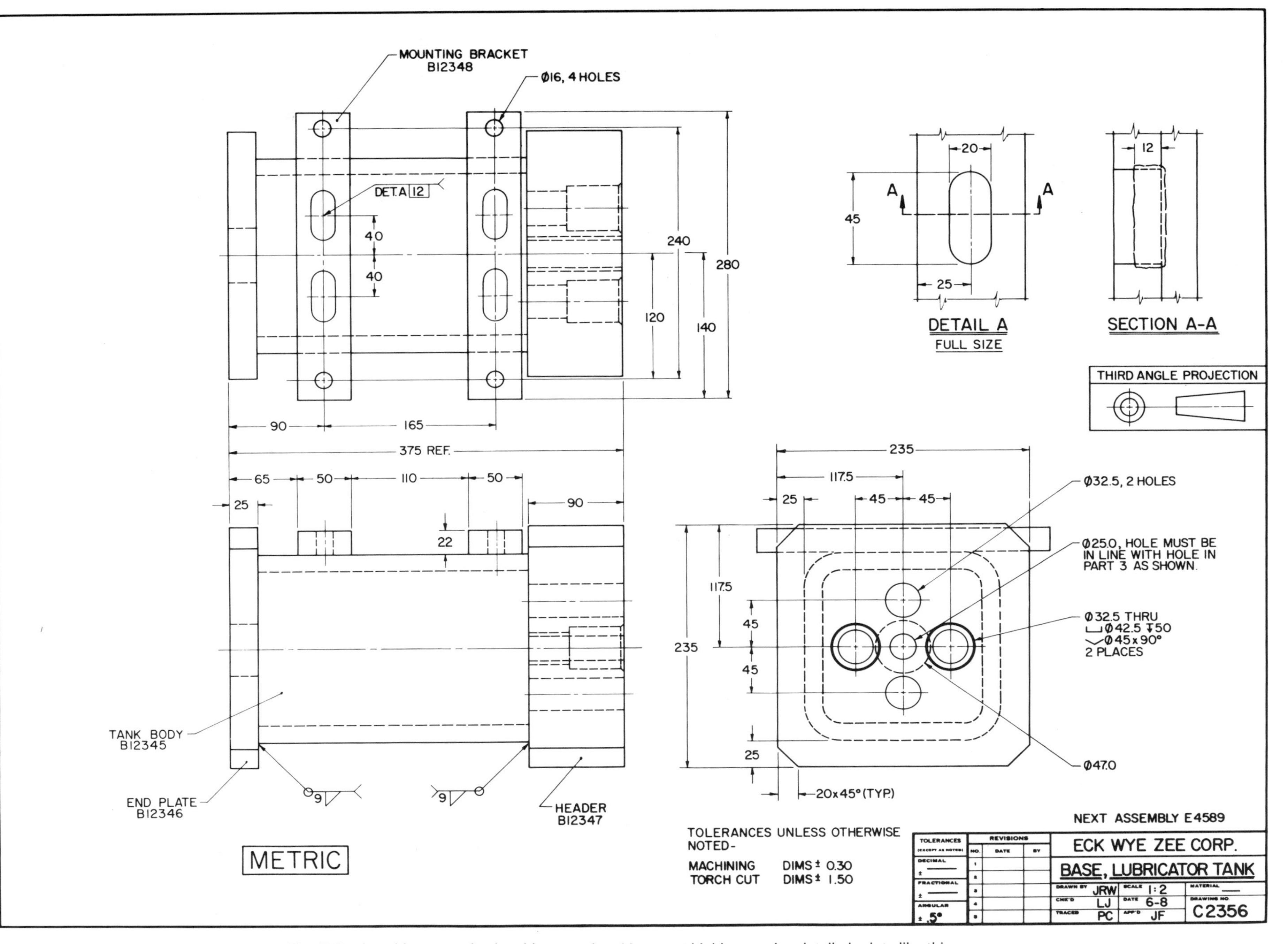

Fig. 7-2. A welder must also be able to read and interpret highly complex detailed prints like this one.

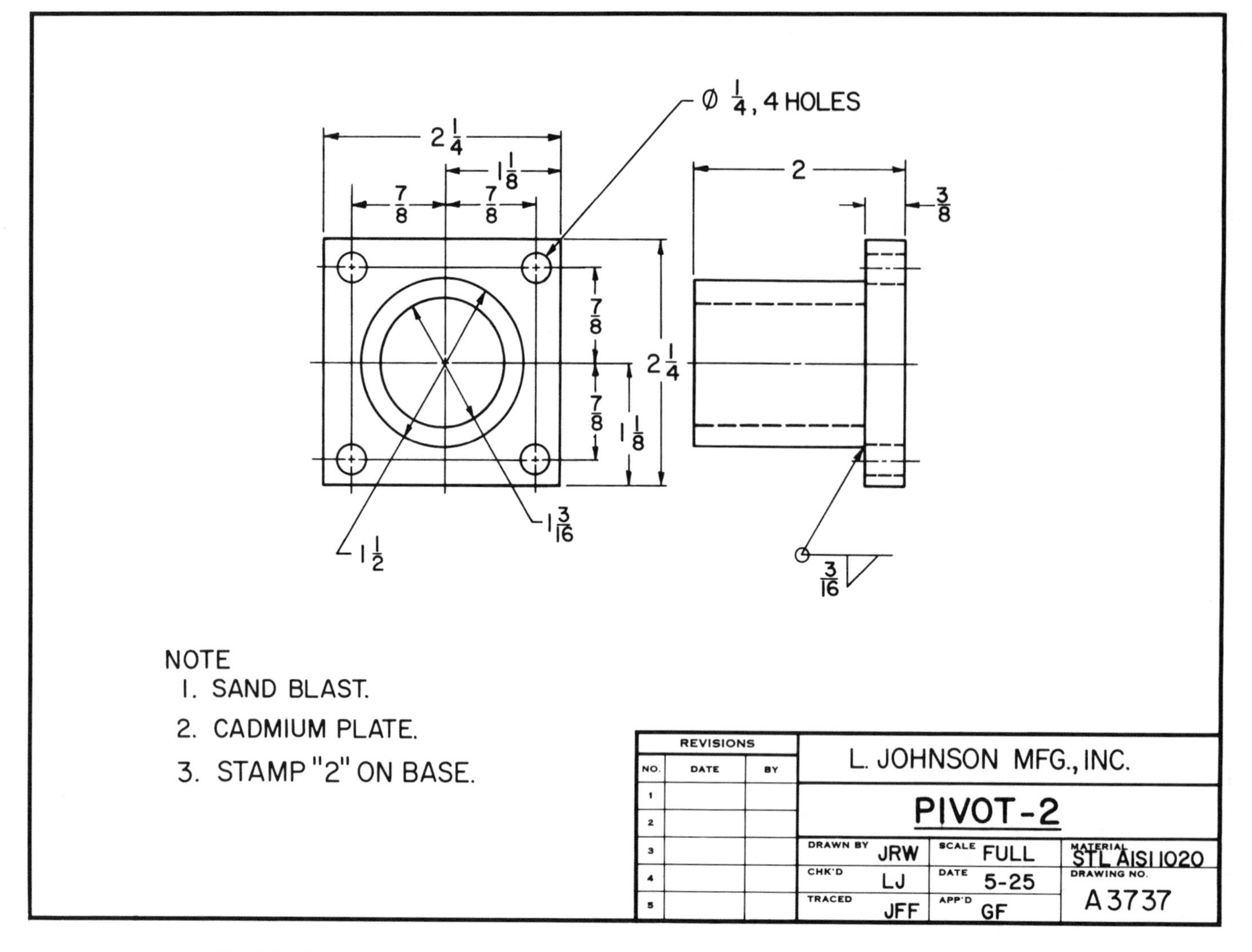

Fig. 7-3. A detail drawing includes all information needed to manufacture or fabricate one part.

Subassembly drawings

A *subassembly drawing* is frequently used on large or complicated products. Refer to Fig. 7-5. Each drawing shows the assembly of a small portion or section of the complete product.

Detail assembly drawing

The detail drawing, in most instances, gives information on only one item. If the mechanism, however, is small in size or if it is composed of only a few parts, the details and assembly may appear on the same print, Fig. 7-6. This type drawing is often referred to as a *detail assembly drawing*.

SPECIALIZED DRAWINGS

When a product is made in quantity, separate detail drawings are usually prepared for each specific manufacturing process. Fig. 7-7, 7-8, and 7-9 show individual detail drawings of the same object for cutting stock, welding, and machining.

Erection drawing

An *erection drawing* is a type of assembly drawing. For the welder, an erection drawing provides the information needed to fabricate (usually in the field) and erect structures such as buildings, bridges, transmission lines towers, etc. Fig. 7-10 is an example of an erection drawing.

Parts list

A *parts list*, also called a *material list, schedule of parts*, or *bill of materials*, includes all of the PARTS required in the manufacture of a product. Also included is a description of each part, the quantity of each part needed per assembly, part number, and the number of the drawing used to manufacture each part.

If the product is fabricated from parts provided by a number of different manufacturers, the name of the manufacturer of each component is also included. See Fig. 7-11 for a typical parts list.

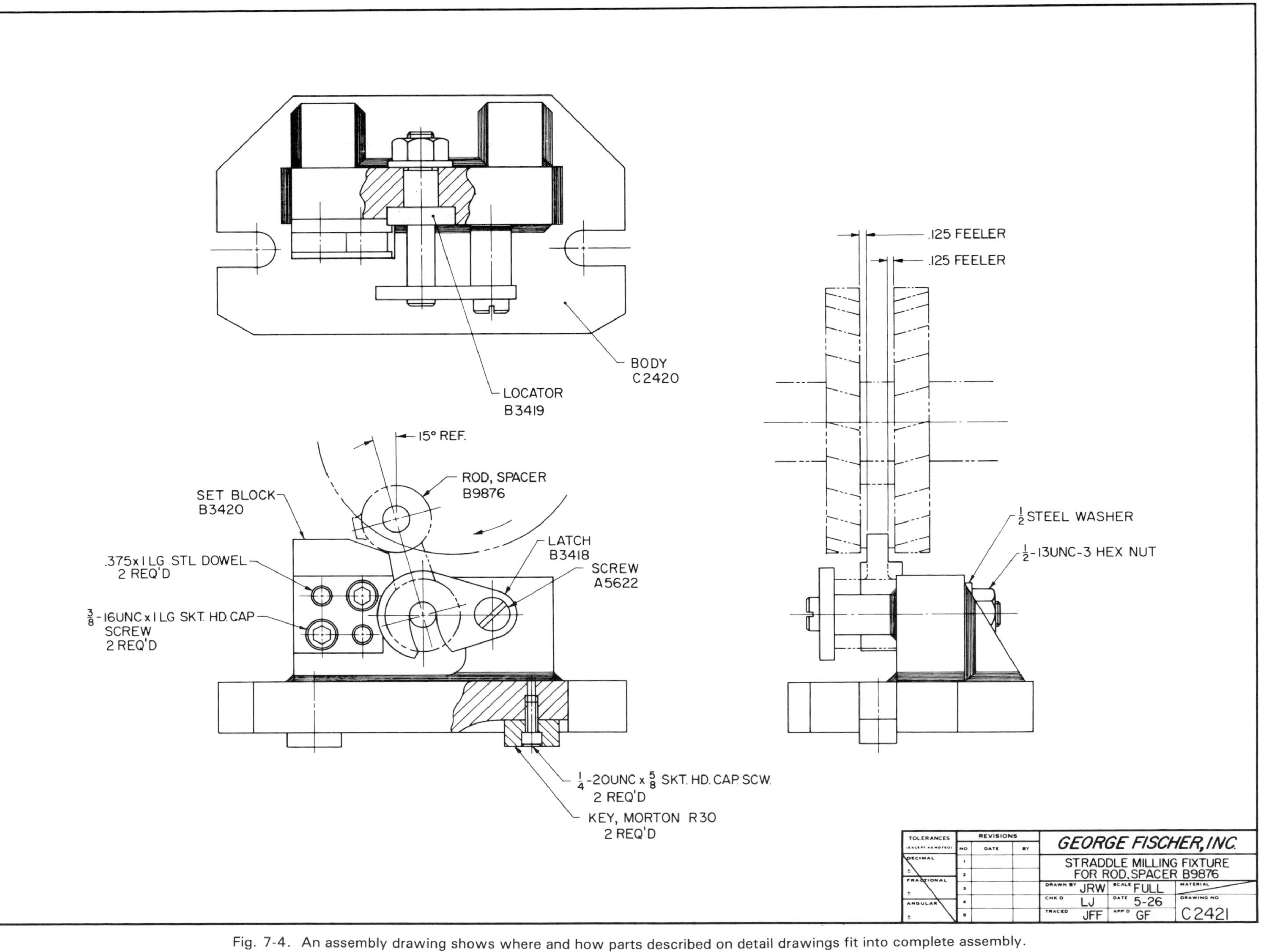

Fig. 7-4. An assembly drawing shows where and how parts described on detail drawings fit into complete assembly.

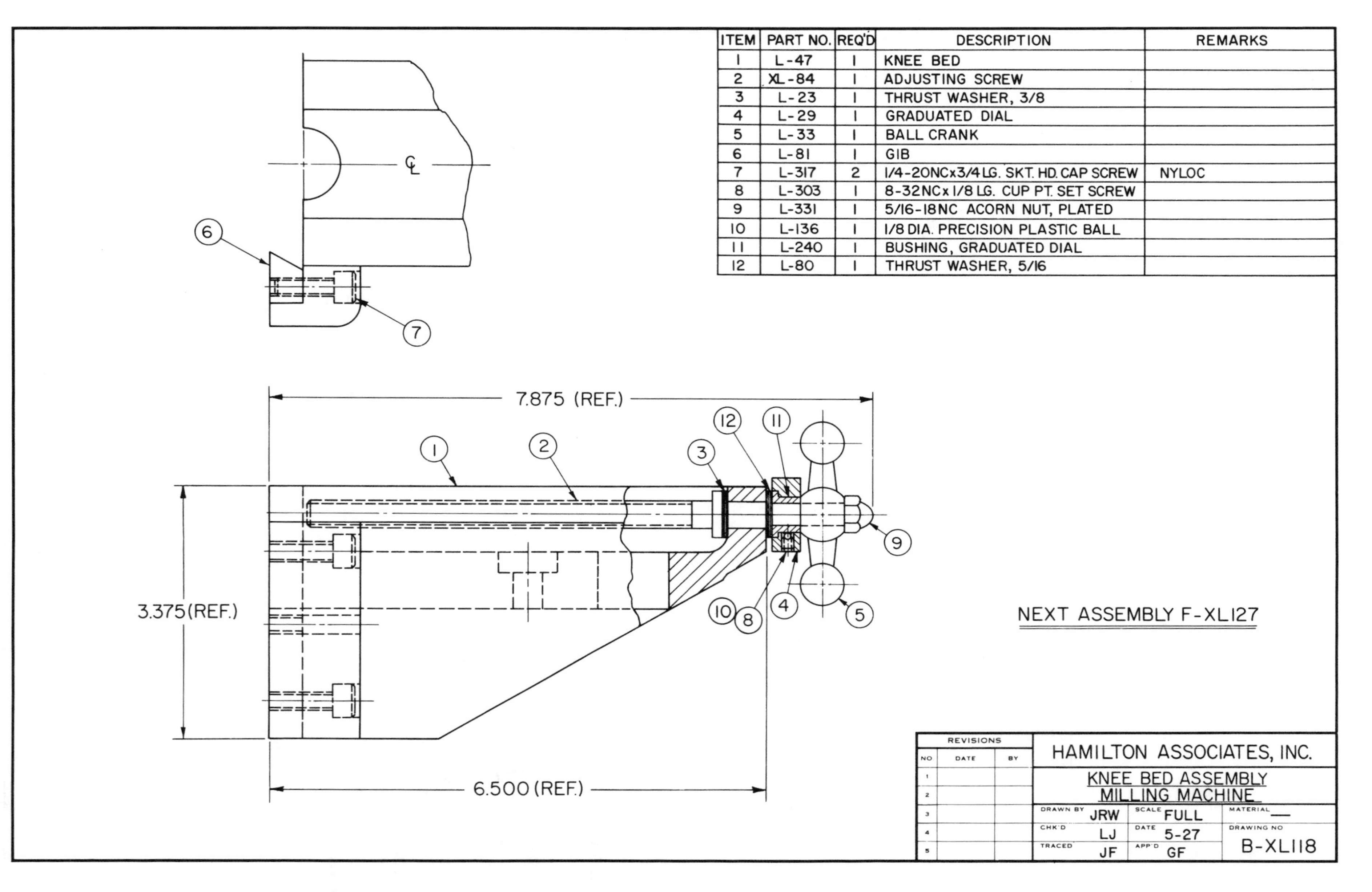

Fig. 7-5. Subassembly drawings are often required for large or complex products. Each subassembly drawing shows small portion or section of completed product. Note how numbers denote names of parts.

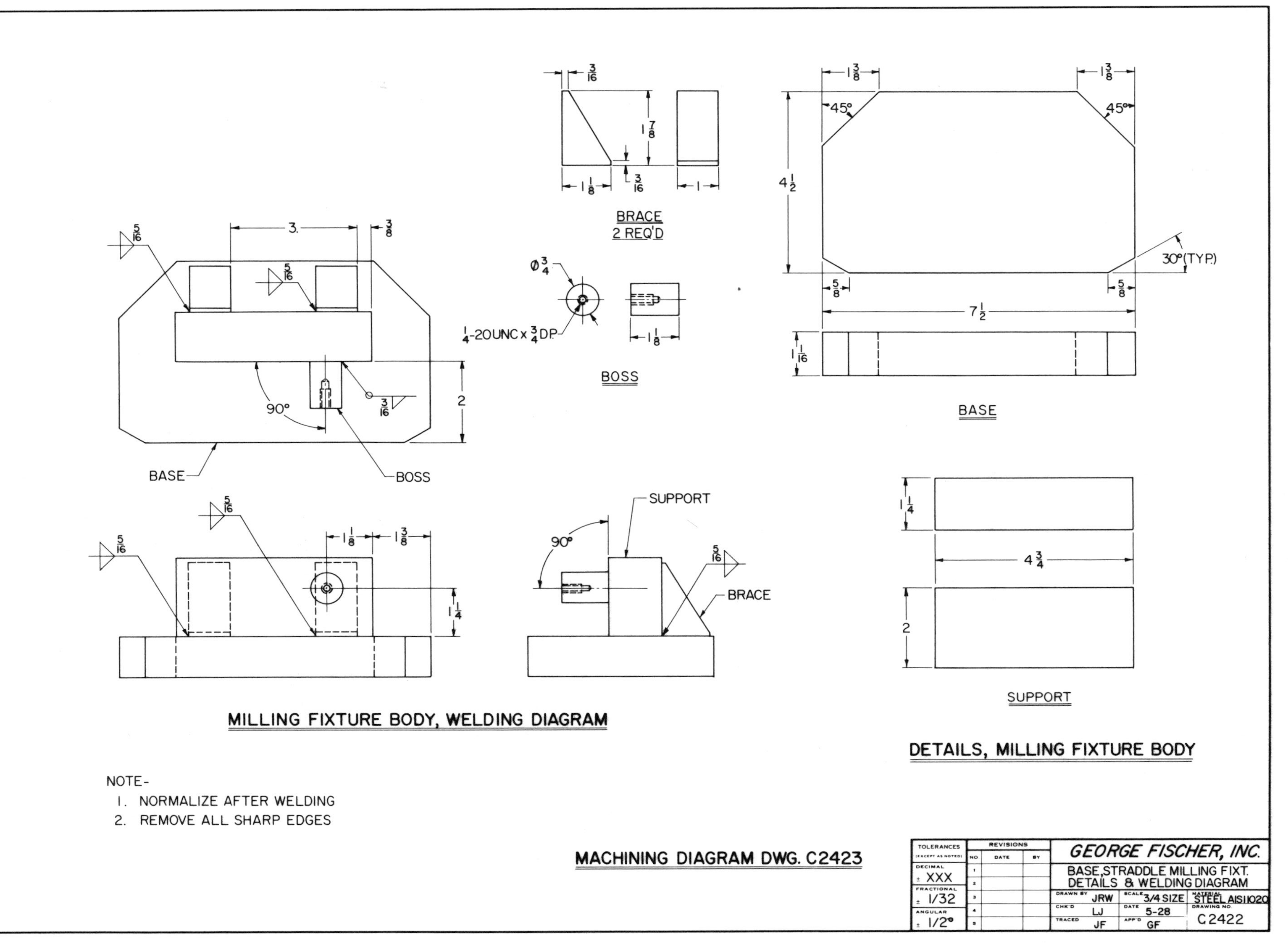

Fig. 7-6. A detail assembly drawing includes details and assembly of a simple part on same print.

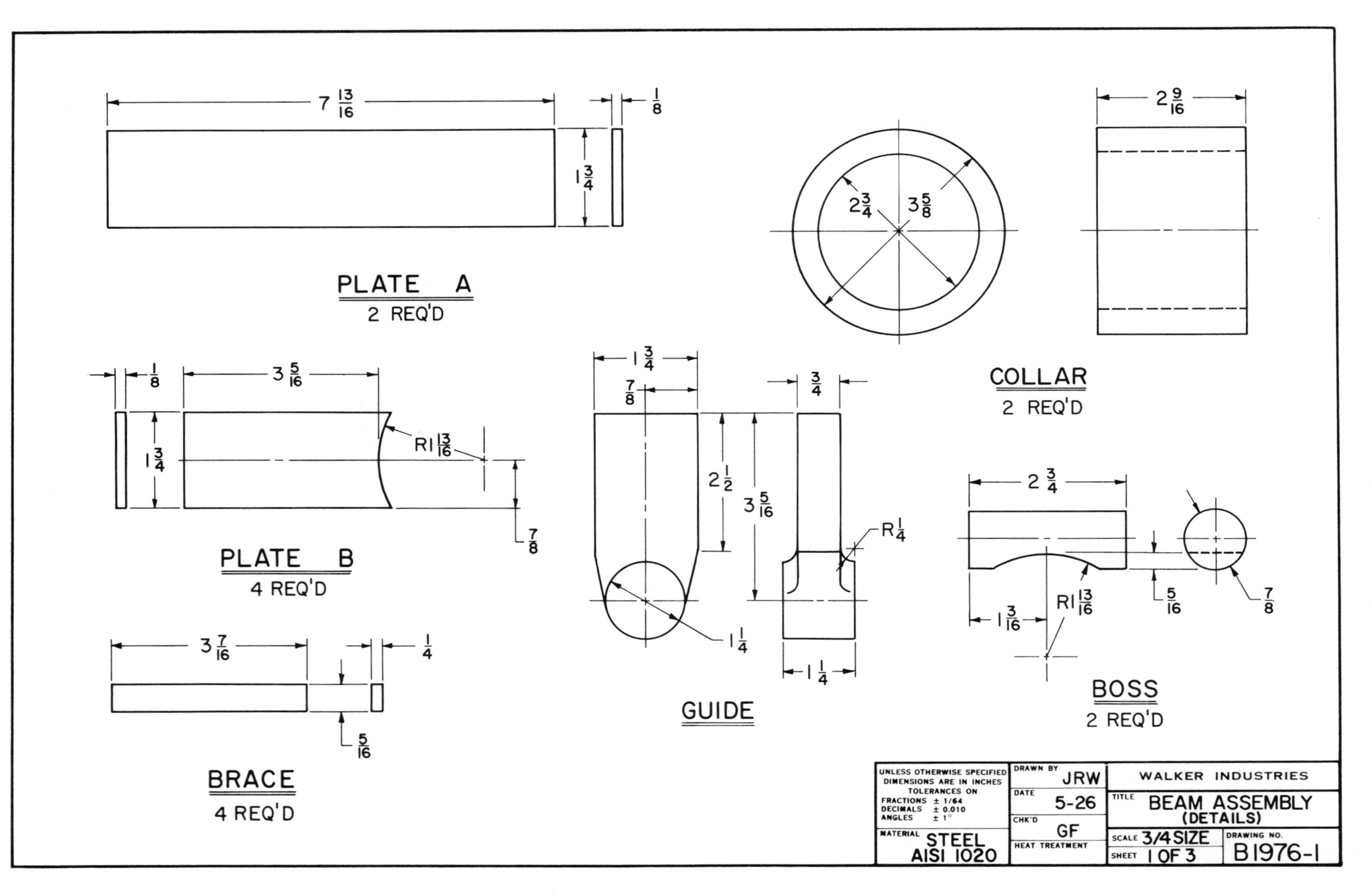

Fig. 7-7. Detail drawing showing how parts should be cut to manufacture a beam assembly. Note how drawing is numbered.

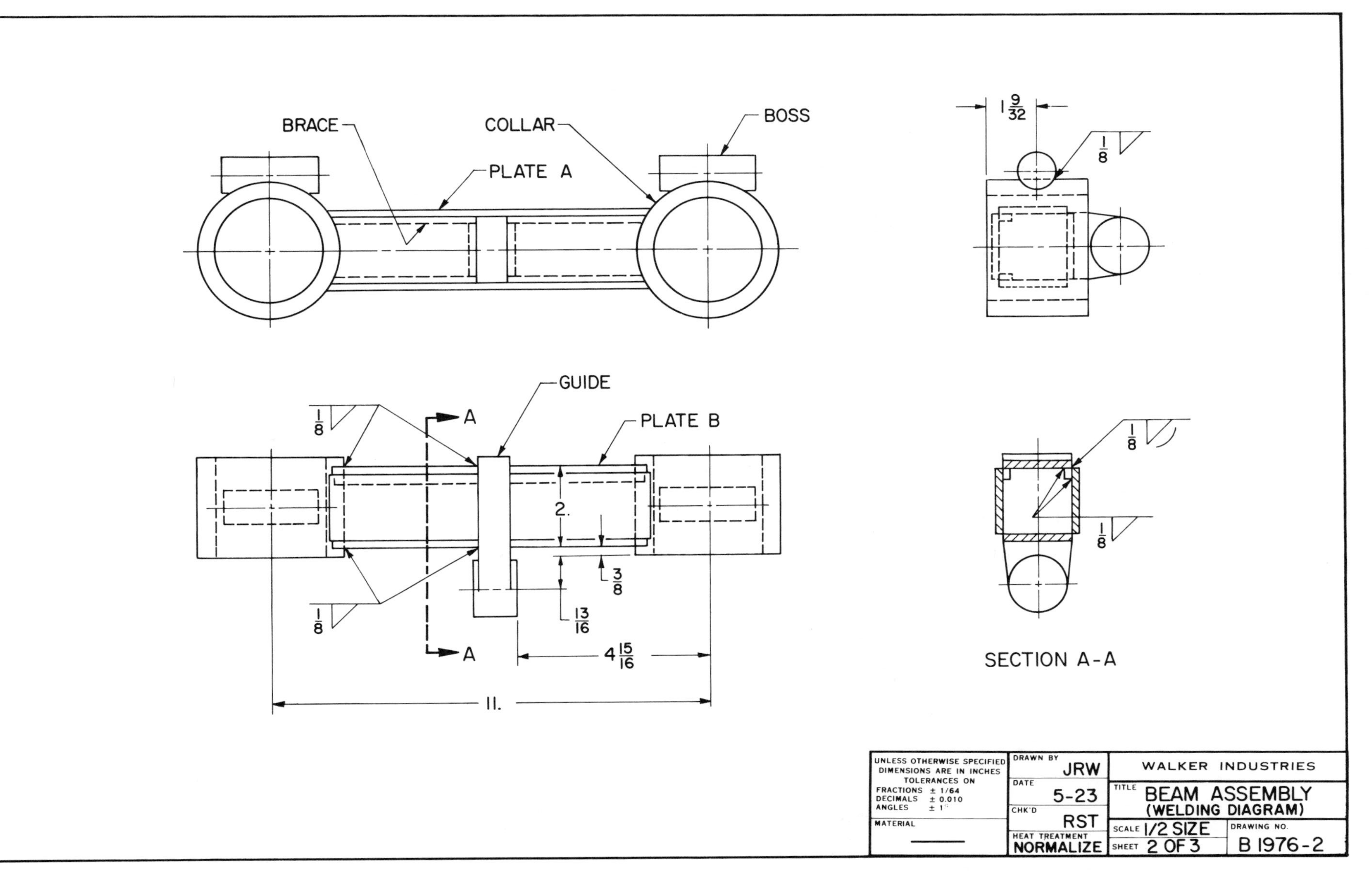

Fig. 7-8. Welding diagram of how parts described in detail drawing of beam assembly are to be fabricated.

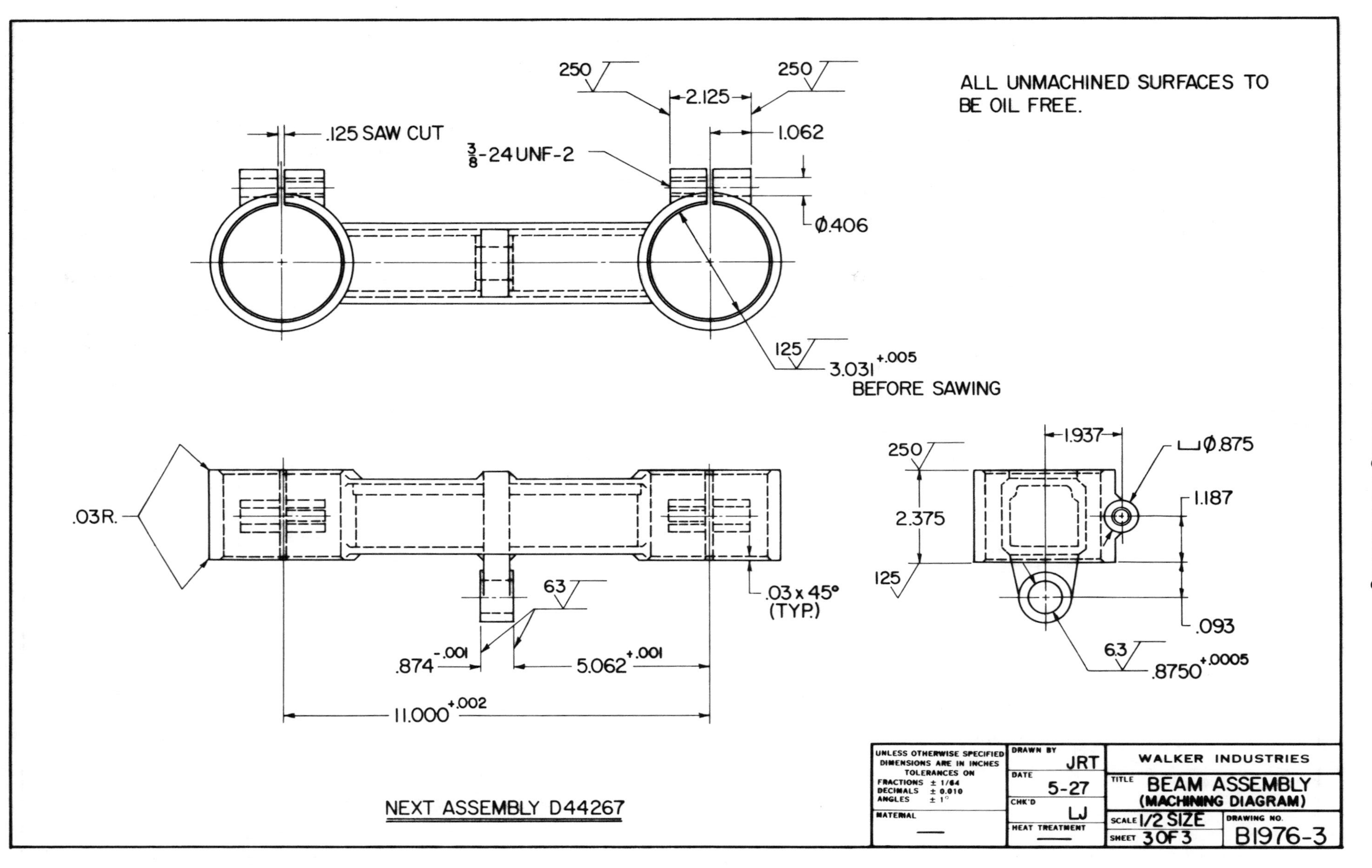

Fig. 7-9. Machining diagram of beam assembly. Welder may not see this drawing as it describes work to be done by a machinist.

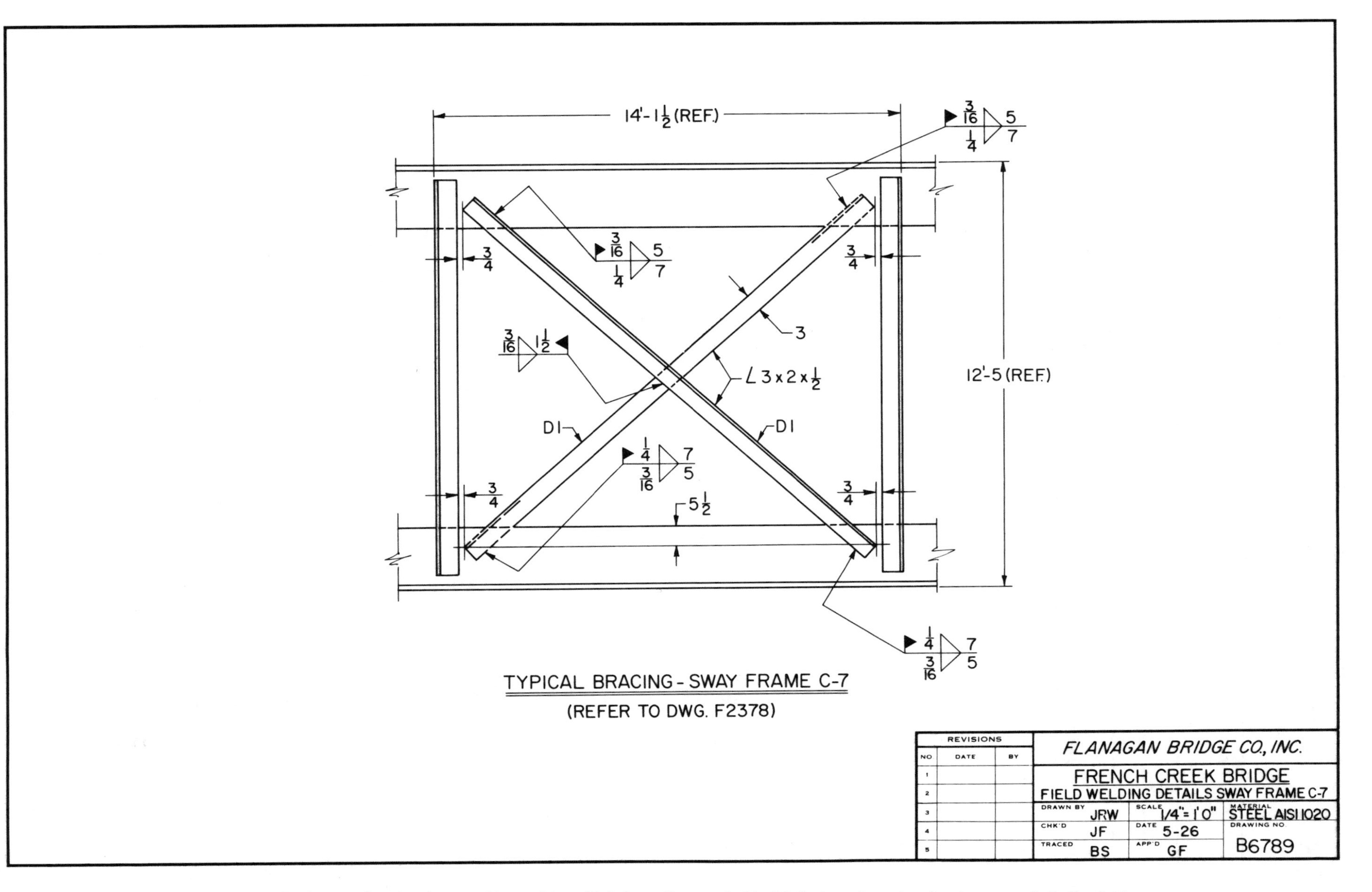

Fig. 7-10. An erection drawing provides welder with information needed to fabricate and erect a structure, usually in the field.

PARTS LIST	AMERICAN UNCIAL	CODE 81361	DRAWING NUMBER	
	DATE		SHEET OF SHEETS	APP.

	SPECIFICATION	DESIGNATION	USED ON	CODE	UNIT DESCRIPTION
GOV'T					
CONTR					
CONTRACT NO.					

DWG. SIZE	DRAWING NUMBER	PART NUMBER	REV.	PART DESCRIPTION	LINE NO.
					1
					2
					3
					4
					5
					6
					7
					8

Fig. 7-11. A parts list indicates, down to the smallest washer, parts required to make a product. Also included is description of parts, quantity of parts, part numbers and number of drawings.

UNIT 7—TEST YOUR KNOWLEDGE

Each word in the left column matches one of the sentences. Place the letter of the sentence in the appropriate blank space.

1. ____ Production welder.
2. ____ Job shop welder.
3. ____ Working drawing.
4. ____ Detail drawing.
5. ____ Assembly drawing.
6. ____ Sub-assembly drawing.
7. ____ Detail assembly drawing.
8. ____ Specialized drawing.
9. ____ Erection drawing.

a. Shows assembly of a small portion of complete product.
b. Drawing of part that includes dimensions and information needed to make it.
c. Seldom does same welding job twice.
d. Shows where and how parts fit into complete unit.
e. Provides information needed to fabricate and set up structures such as bridges, transmission line towers, etc.
f. Drawing used to show specific manufacturing process for a product that will be produced in quantity.
g. Does same welding job over and over.
h. Supply information needed to make and assemble many pieces and parts that make up a product.
i. Contains information to make individual parts and fit them together on a single print.

10. A parts list is also known as a ____________, ____________ or ____________.
11. What is a parts list and what is included on it?

____

____

____

____

Unit 8

ALPHABET OF LINES

As you study a print like the one in Fig. 8-1, you will see that lines of different weights (thicknesses) are used to describe and size the object. Each type of line has a particular meaning. Contrast between the different line weights also helps make a print easier to read or understand.

To be able to interpret a print, it is essential that YOU know the characteristics of the various lines and how they are used on a drawing. These lines, known as the *alphabet of lines,* are universally used throughout industry.

VISIBLE OBJECT LINE

The *visible object line,* also called *visible line,* is used to outline the visible edges of the object. It is a thick, continuous line, as shown in Fig. 8-2.

HIDDEN OBJECT LINE

A *hidden object line,* also termed *hidden line,* Fig. 8-3, shows the hidden features of the object. It is of medium weight and is composed of short dashes that are about 1/8 in. or 3.0 mm long. The dashes are typically separated by 1/16 in. or 1.5 mm spaces. However, hidden object line size may vary according to the size of the drawing.

CENTERLINE

The *centerline* is used to indicate the center of symmetrical objects. Shown in Fig. 8-4, is a fine, dark line composed of alternate long and short dashes with spaces between the dashes.

On a same size drawing, the long section of the centerline is about 3/4 in. or 19.0 mm long. The shorter dashes are about 1/8 in. or 3.0 mm long. The spaces are 1/16 in. or 1.5 mm long.

CUTTING PLANE LINE

The *cutting plane line* indicates the location of an imaginary cut made through the object to reveal its interior characteristics. Refer to Fig. 8-5.

The extended lines are capped with arrowheads that indicate the direction of sight to view the section.

Letters A-A, B-B, etc., identify the section if it is moved to another position on the print, or if several sections are used on a single print.

The cutting plane line is a heavy line. Two forms are recommended for general use.

SECTION LINE

Section lines are used when the interior features of the object are shown. They may also indicate the type of material cut by the cutting plane line. However, a GENERAL PURPOSE section line (symbol for CAST IRON) is often used on a drawing when the material specifications ("specs") are shown elsewhere.

Section lines are fine dark lines, as in Fig. 8-6.

BREAK LINE

Break lines break out or remove sections for clarity. Also, when the object is uniform in cross section for its entire length, they are used to conserve space on the drawing. Fig. 8-7 shows three types of break lines.

A heavy irregular freehand line is used to indicate a SHORT BREAK. A thin light ruled line with freehand zigzags is used to show LONG BREAKS. A thick "S" break is used to indicate the BREAK IN ROUND STOCK.

DIMENSION AND EXTENSION LINES

The dimension line is usually capped at each end with arrowheads and is placed between two extension lines. This is shown in Fig. 8-8. With few exceptions, it is broken with the dimension placed midpoint between the arrowheads.

Dimension and extension lines are fine solid lines. The *dimension line* is used to indicate the extent and direction of dimensions. *Extension lines* indicate the termination of a dimension.

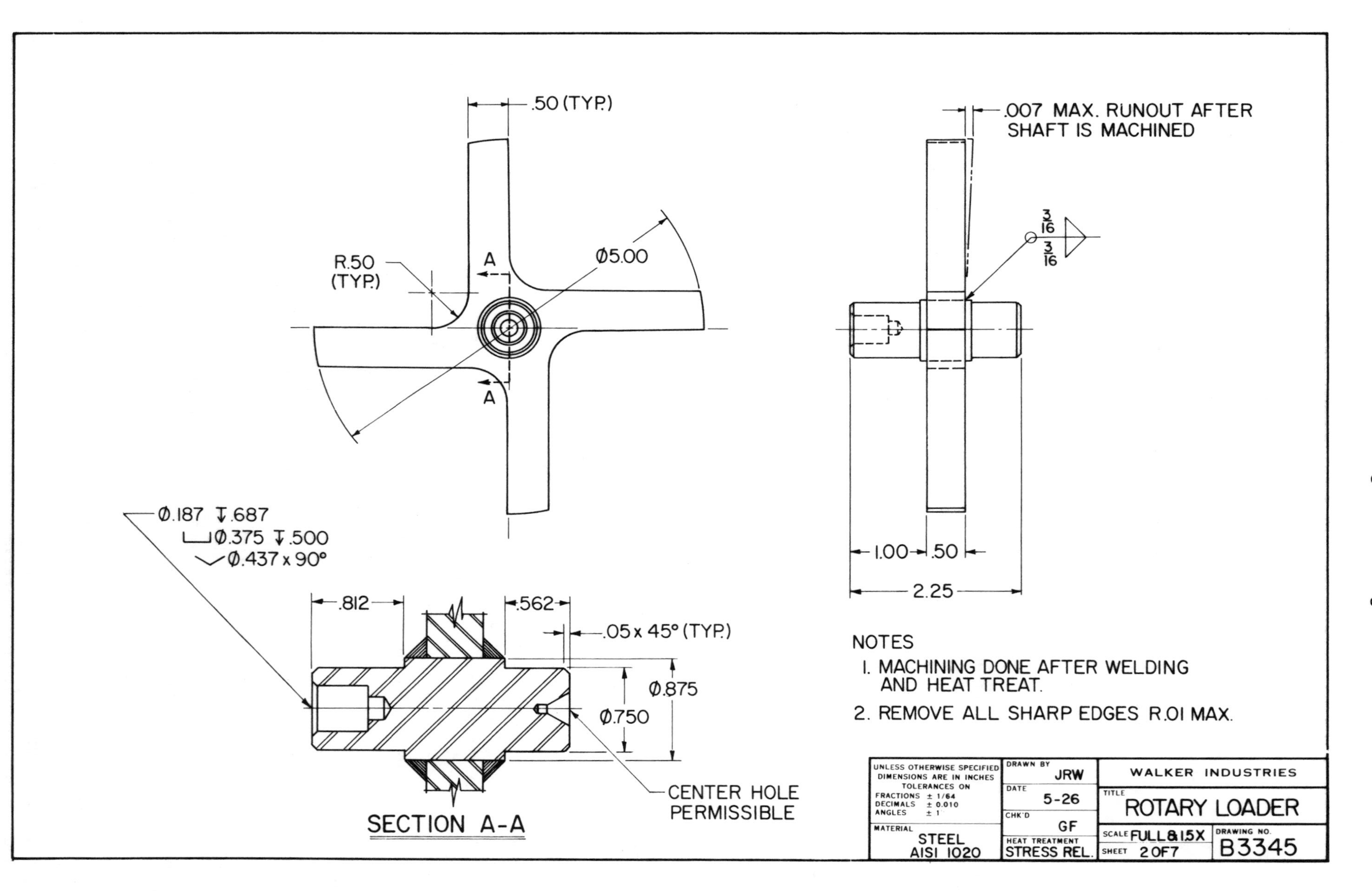

Fig. 8-1. Lines of different weights or thicknesses describe and size object shown on drawing.

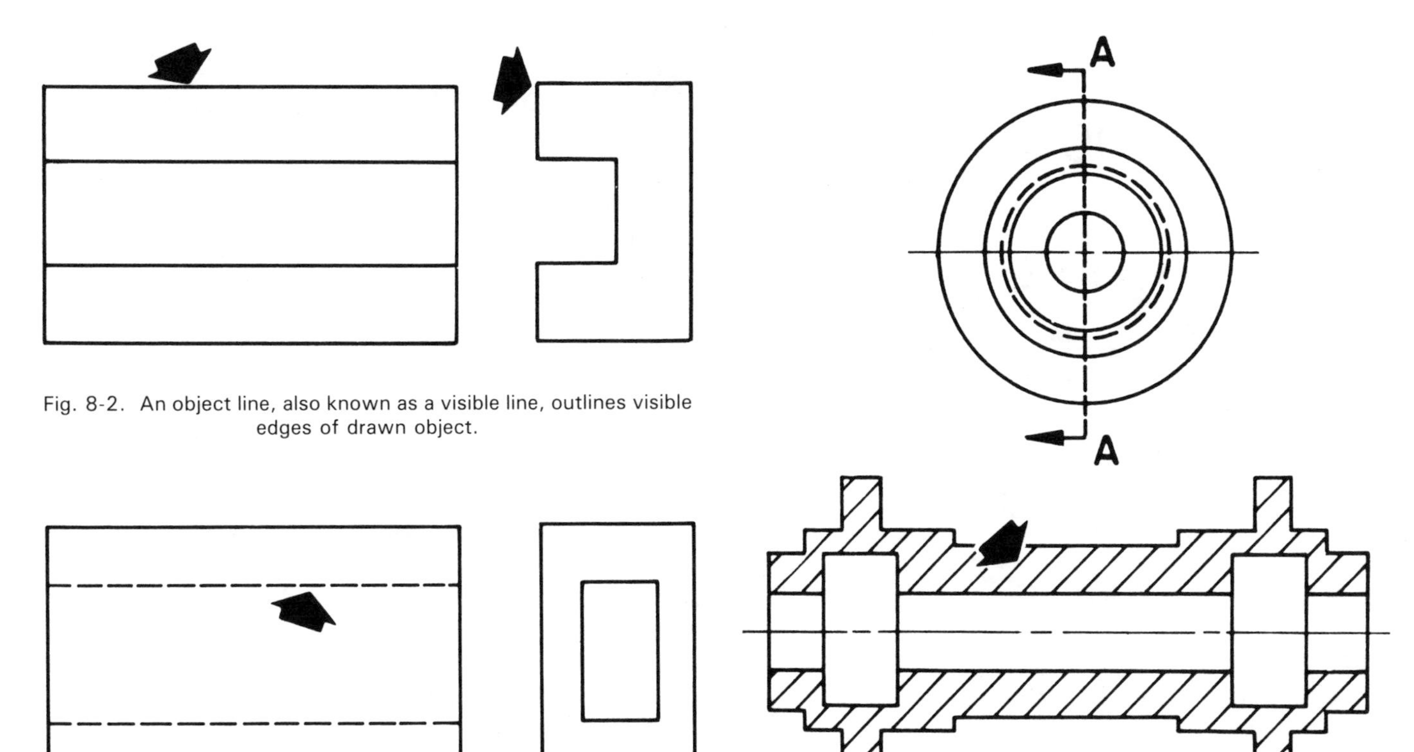

Fig. 8-2. An object line, also known as a visible line, outlines visible edges of drawn object.

Fig. 8-3. A hidden object line or invisible line shows hidden features of object.

Fig. 8-6. Section lines are used when interior features of object are shown. They may also indicate material of object where it has been cut by cutting plane line.

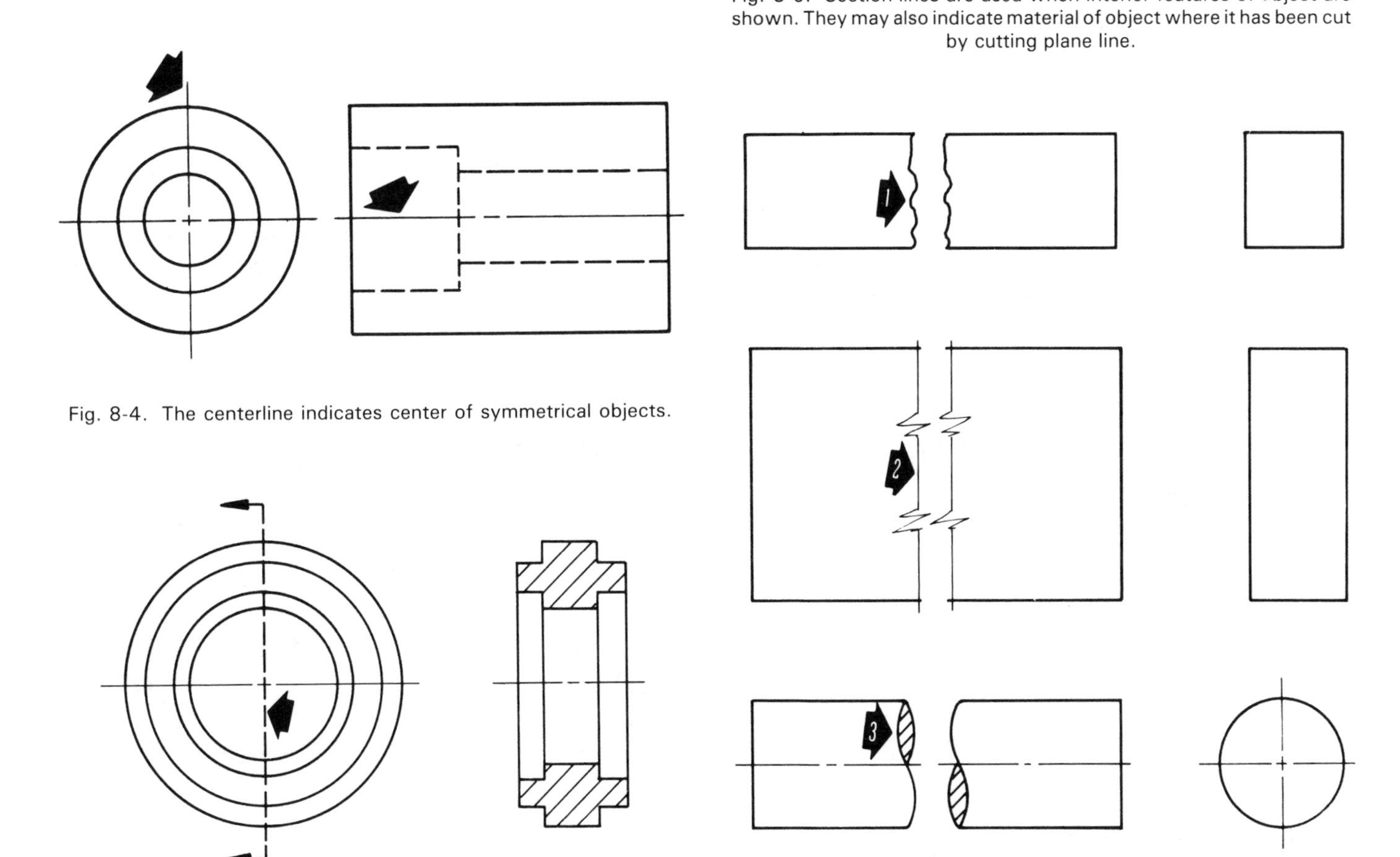

Fig. 8-4. The centerline indicates center of symmetrical objects.

Fig. 8-5. A cutting plane line shows location of imaginary cut made through object to reveal its interior details.

Fig. 8-7. Break lines are employed to limit a partial view of a broken section. 1—Short break. 2—Long break. 3—Round stock break.

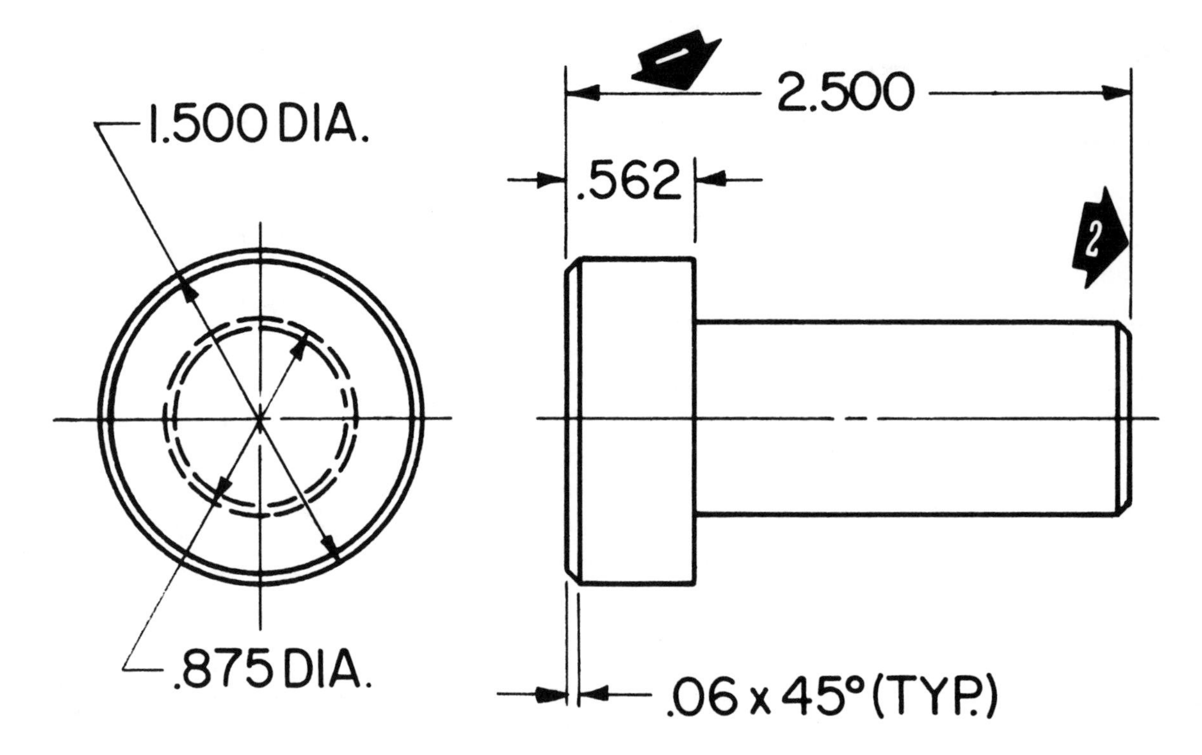

Fig. 8-8. Dimension line usually has arrowheads at each end. It is used to show direction and extent of dimensions. 1 — Dimension line. 2 — Extension line.

A dimension line with an arrowhead on only one end is used as a *leader,* Fig. 8-9. It is used to denote a dimension or for adding a note.

PHANTOM LINE

A *phantom line,* Fig. 8-10, shows adjacent parts and alternate positions of moving parts. It is also used to show repeated details, like threads and springs. It is a thin, dark line made of long dashes alternated with pairs of short dashes. On a same size drawing, the long dashes are 3/4 to 1 1/2 in. (19.0 to 38.0 mm) long. The short dashes are typically 1/8 in. (3.0 mm) long, with 1/16 in. (1.5 mm) spaces.

HOW THE ALPHABET OF LINES IS USED

The sample print, shown in Fig. 8-11, illustrates how the ALPHABET OF LINES is used on a drawing. Study it closely!

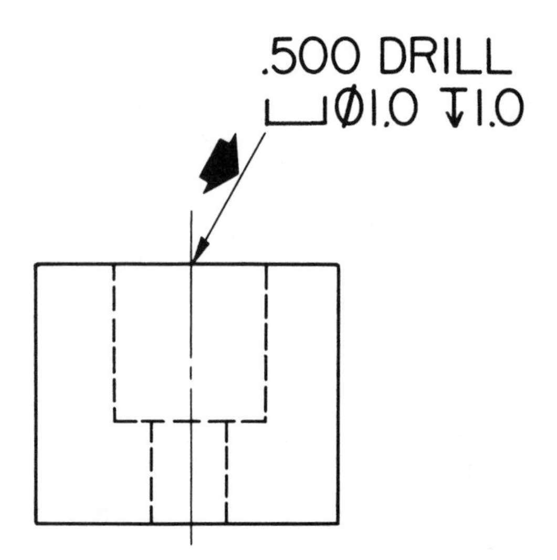

Fig. 8-9. A leader is a dimension line with an arrowhead on only one end. It denotes a dimension or a note.

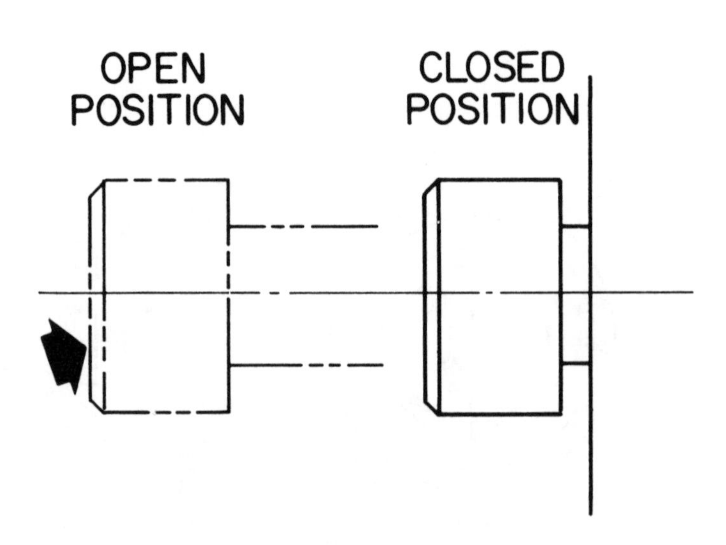

Fig. 8-10. A phantom line shows adjacent parts and alternate positions of a moving part.

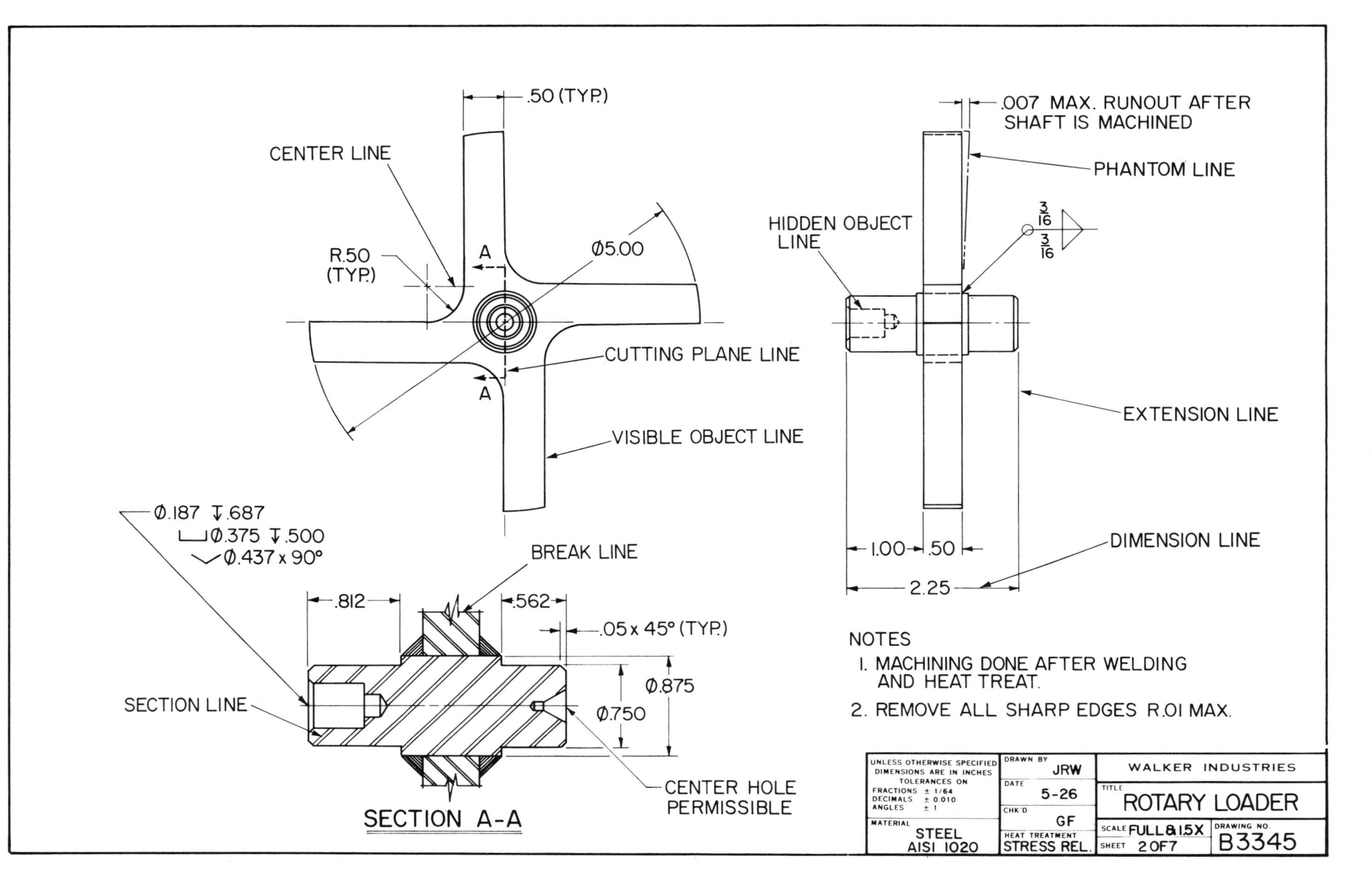

Fig. 8-11. Note how alphabet of lines is used on a drawing.

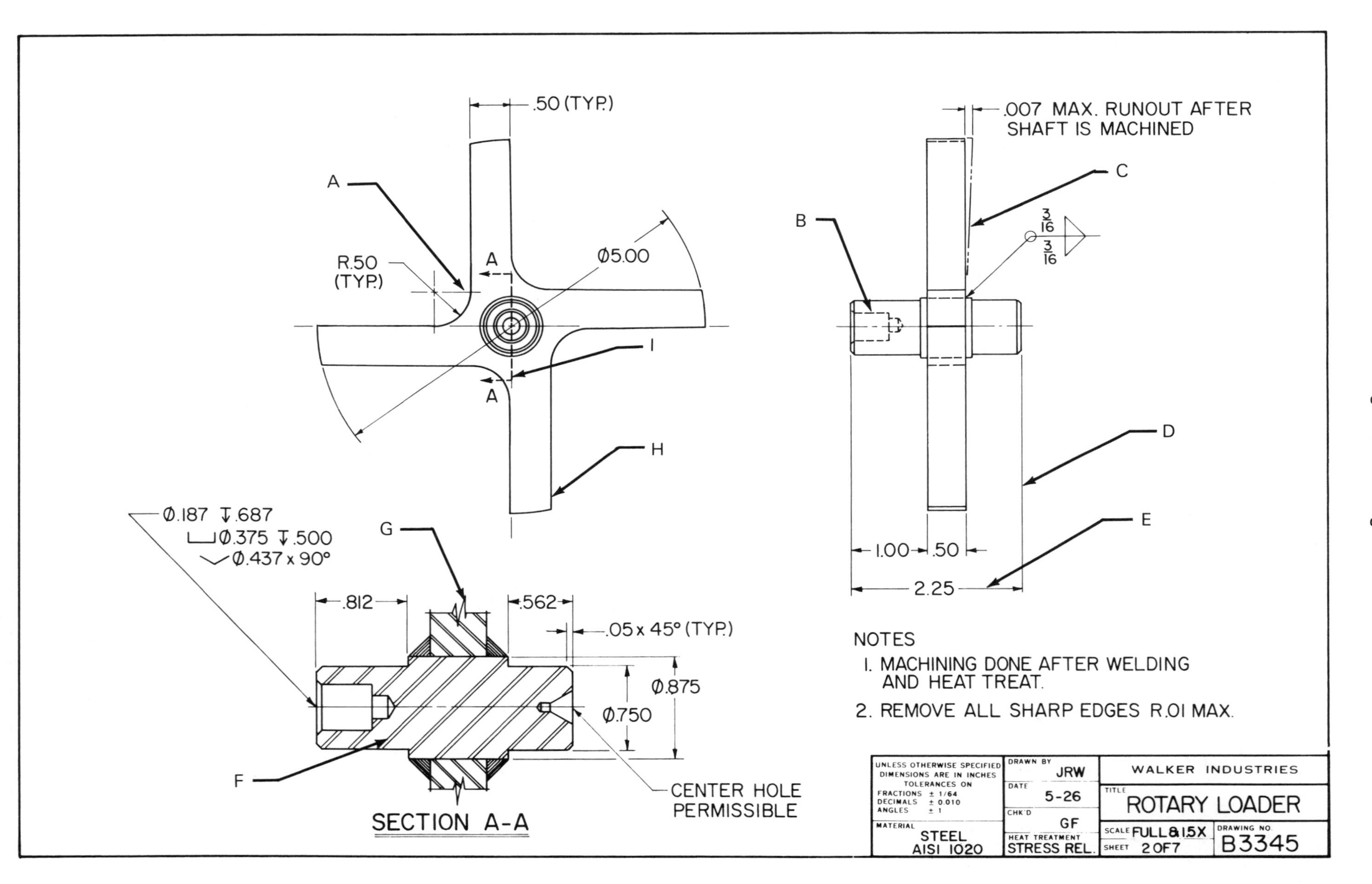

Use this drawing (B3345) to complete Part I of Test Your Knowledge for Unit 8.

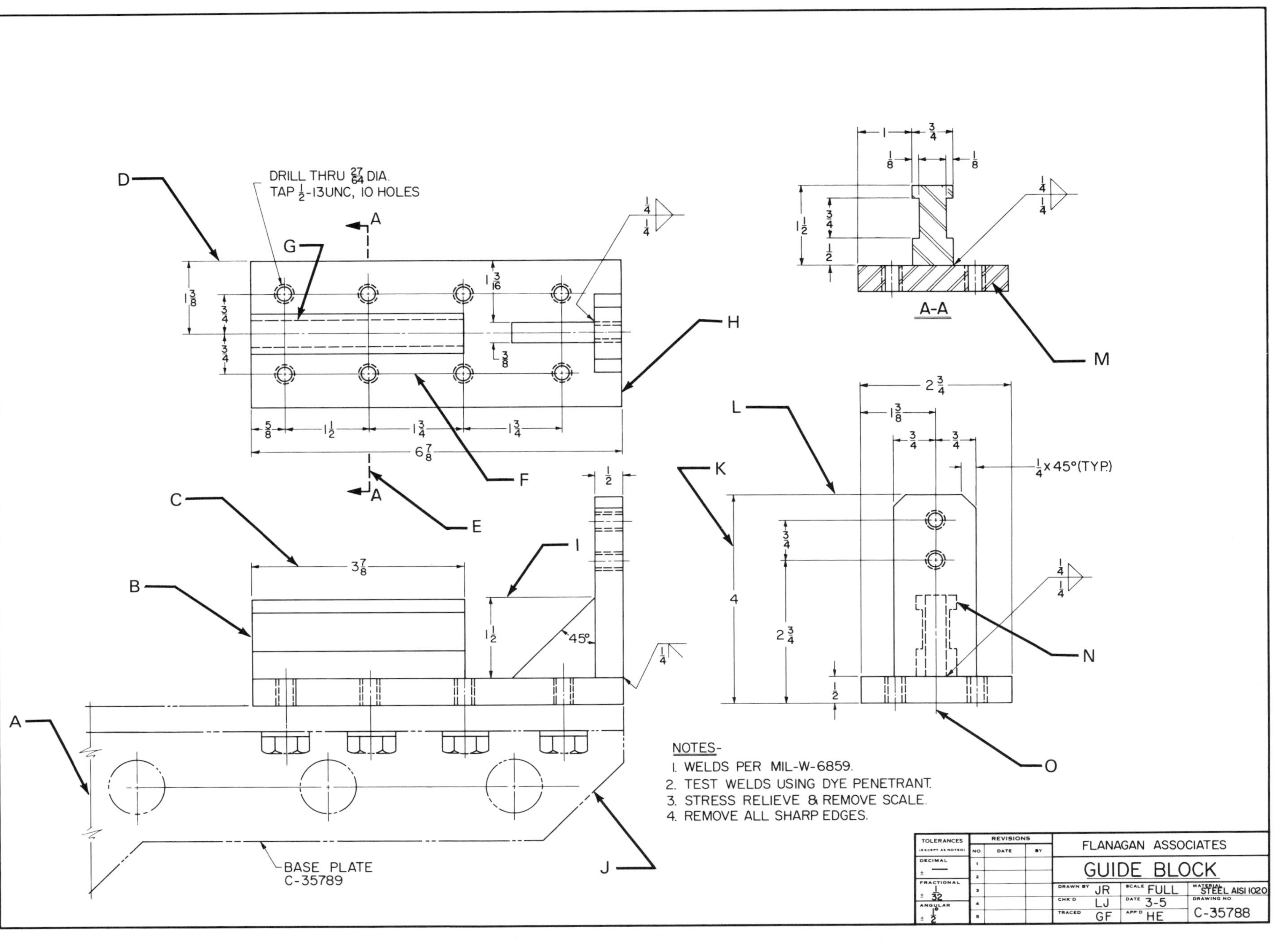

Use this drawing (C-35788) to complete Part II of Test Your Knowledge for Unit 8.

UNIT 8—TEST YOUR KNOWLEDGE

Part I

Refer to the drawing (B 3345) and place the name of the line in the appropriate space.

A. ______________________

B. ______________________

C. ______________________

D. ______________________

E. ______________________

F. ______________________

G. ______________________

H. ______________________

I. ______________________

Part II

Refer to the drawing (C-35788) and place the name of the line in the appropriate space.

A. ______________________

B. ______________________

C. ______________________

D. ______________________

E. ______________________

F. ______________________

G. ______________________

H. ______________________

I. ______________________

J. ______________________

K. ______________________

L. ______________________

M. ______________________

N. ______________________

O. ______________________

Part III

Match the words in the left column with the sentences. Place the letter of the sentence in the appropriate blank.

1. ____ Visible object line.
2. ____ Hidden object line.
3. ____ Centerline.
4. ____ Cutting plane line.
5. ____ Section lines.
6. ____ Break lines.
7. ____ Dimension line.
8. ____ Extension line.
9. ____ Leader.
10. ____ Phantom line.

a. Shows adjacent parts and alternate positions of moving parts.
b. Used when interior features of an object are shown.
c. Indicates location of an imaginary cut made through object to reveal its interior characteristics.
d. Used to outline visible edges of object.
e. Indicates center of symmetrical objects.
f. Shows hidden features of object.
g. Used for purpose of breaking out sections for clarity, or when object is uniform in cross section for its entire length.
h. Usually capped at each end with arrowheads and is placed between two extension lines.
i. Used to denote a dimension or for adding a note. Has an arrowhead on only one end.
j. Indicates termination of a dimension.

Unit 9

PRINT FORMAT

In order to complete the job it is designed to do, a drawing or print must include a "picture" of the product and all of the information needed to make or assemble the product.

DIMENSIONS

A properly dimensioned drawing includes all dimensions, in correct relation to one another, needed to give a complete SIZE description of the product.

The United States is in the process of converting from English measure (inch, pound, quart, etc.) to metric measure (meter, millimeter, kilogram, liter, etc.). Total conversion to the metric system is expected to take many years. During this conversion period, you must be familiar with several print formats. A welder may have to work from prints that have dimensions in:

1. Fractions, Fig. 9-1.
2. Decimal fractions, Fig. 9-2.
3. Metrics, Fig. 9-3.
4. Dual dimensions, Fig. 9-4.

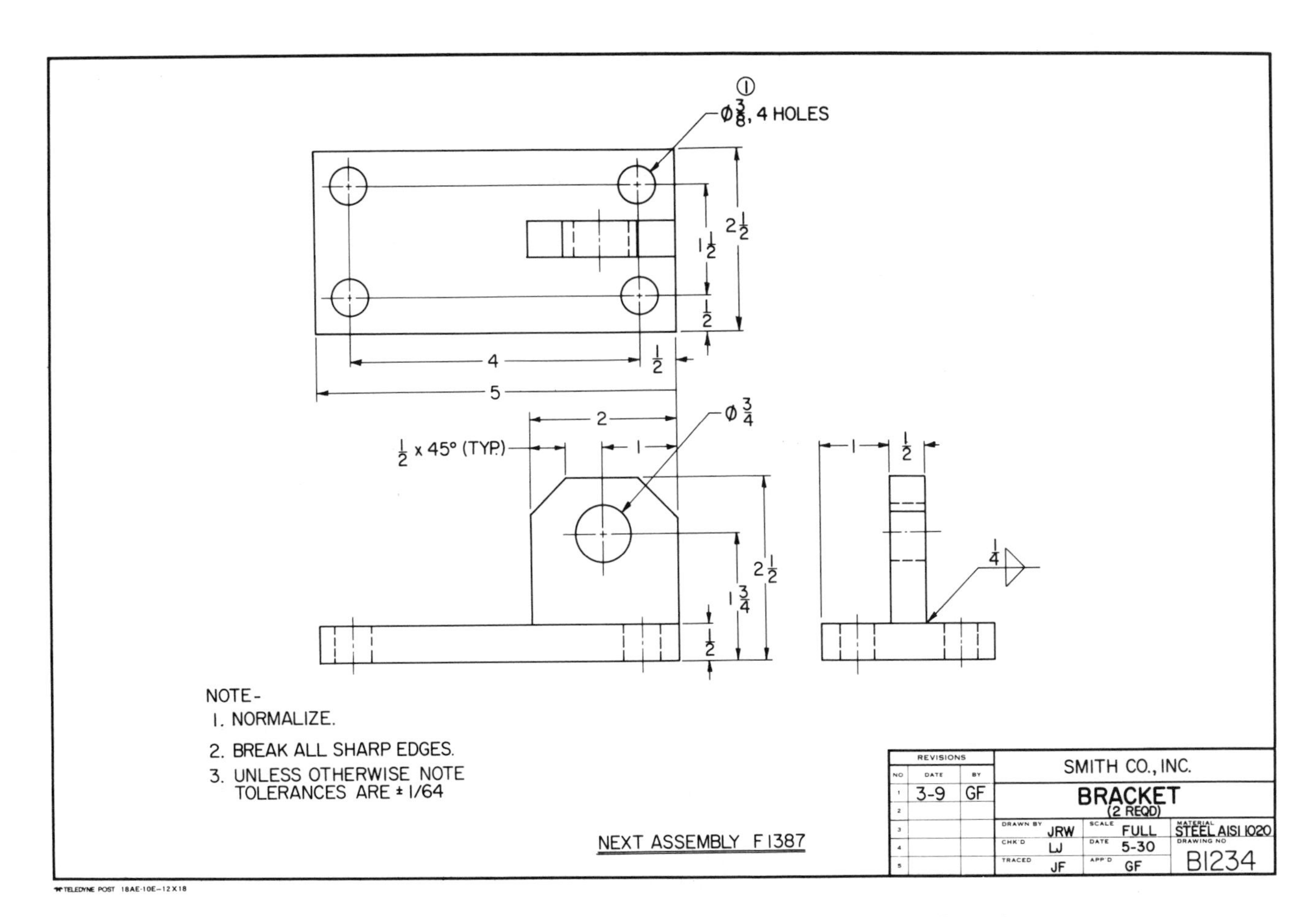

Fig. 9-1. This typical drawing is dimensioned in inches and fractions of an inch.

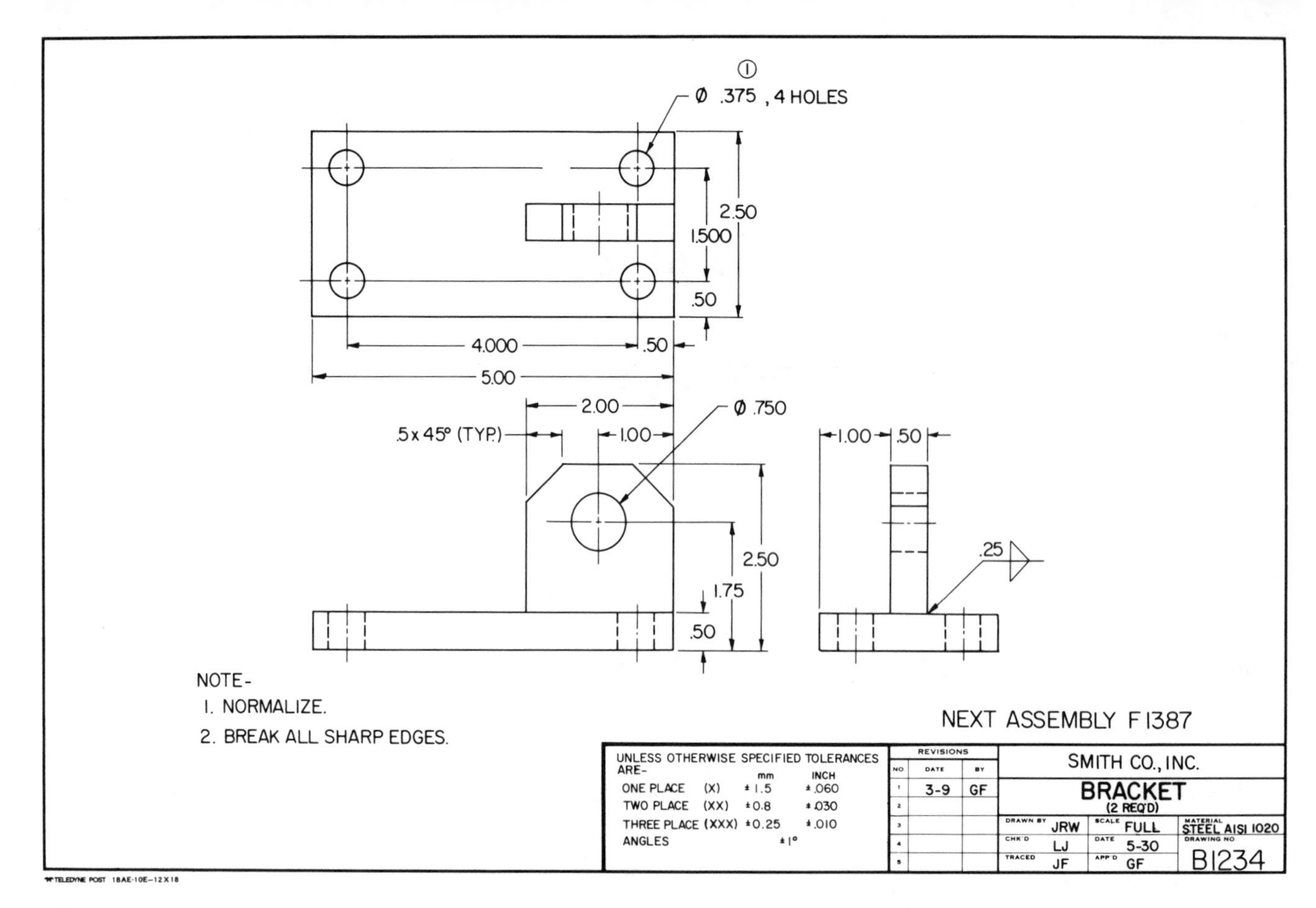

Fig. 9-2. Same drawing is dimensioned in decimal fractions.

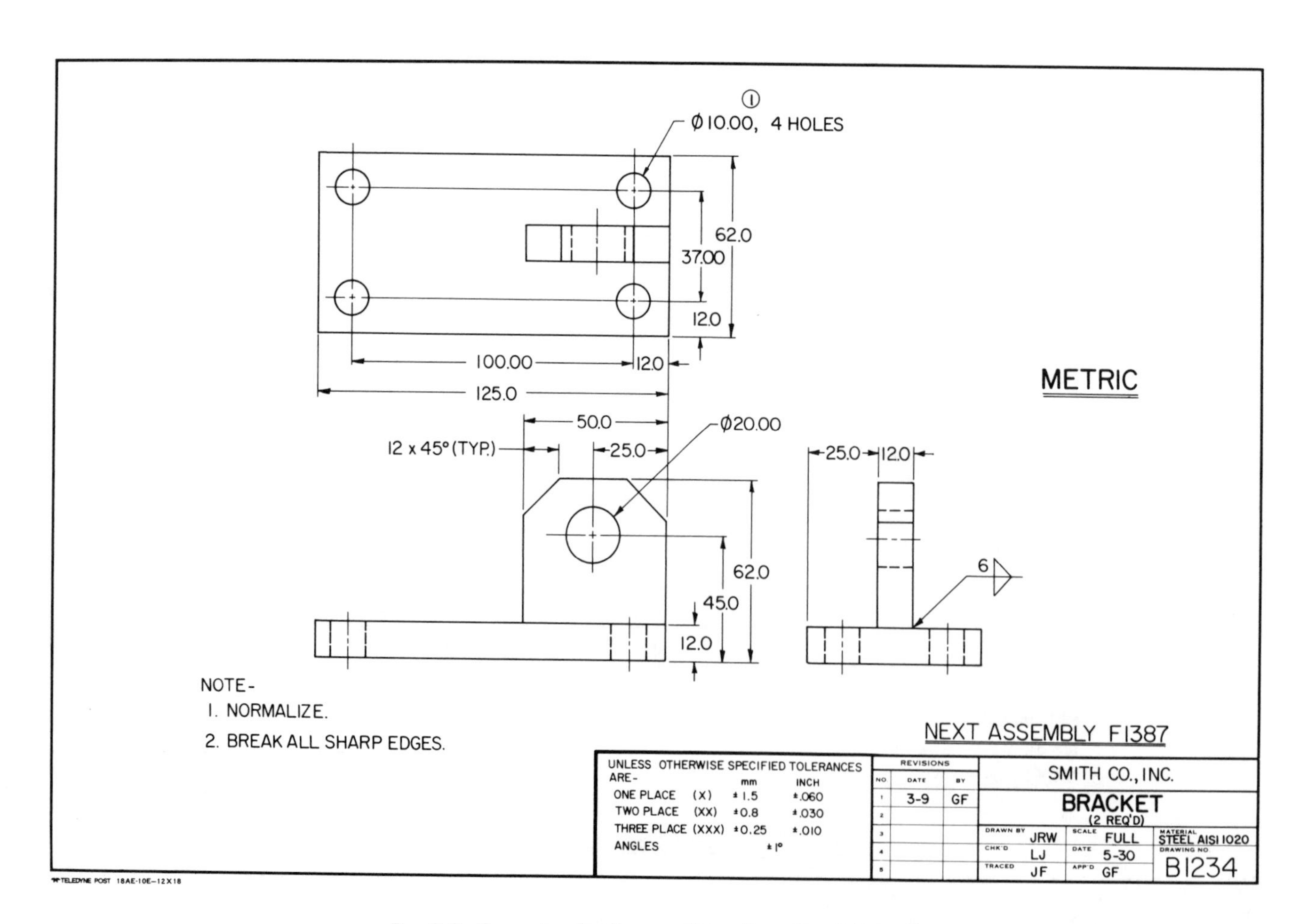

Fig. 9-3. Same drawing is now dimensioned in millimeters.

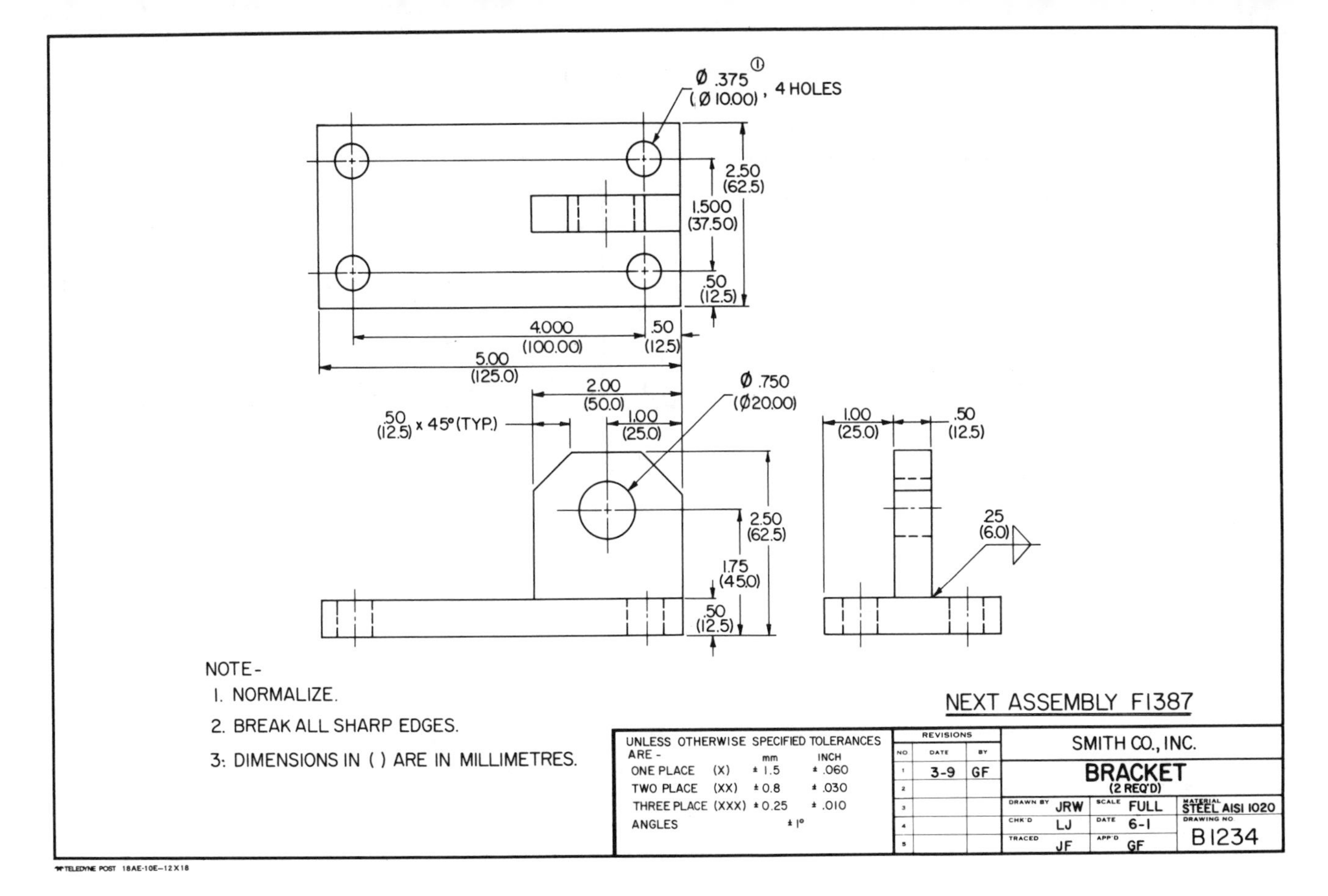

Fig. 9-4. Print uses dual dimensioning to describe size of part. How does a welder know which dimensions are in millimeters?

A *dual dimensioned print* has both English and metric values. Metric dimensions are usually in millimeters (mm). Fig. 9-5 shows the various dual dimensioning techniques, and how the dimensions are placed on a drawing.

PRINT TITLE BLOCK

Additional information about the product shown on the print can be found in the title block. Fig. 9-7 summarizes the information in a title block.

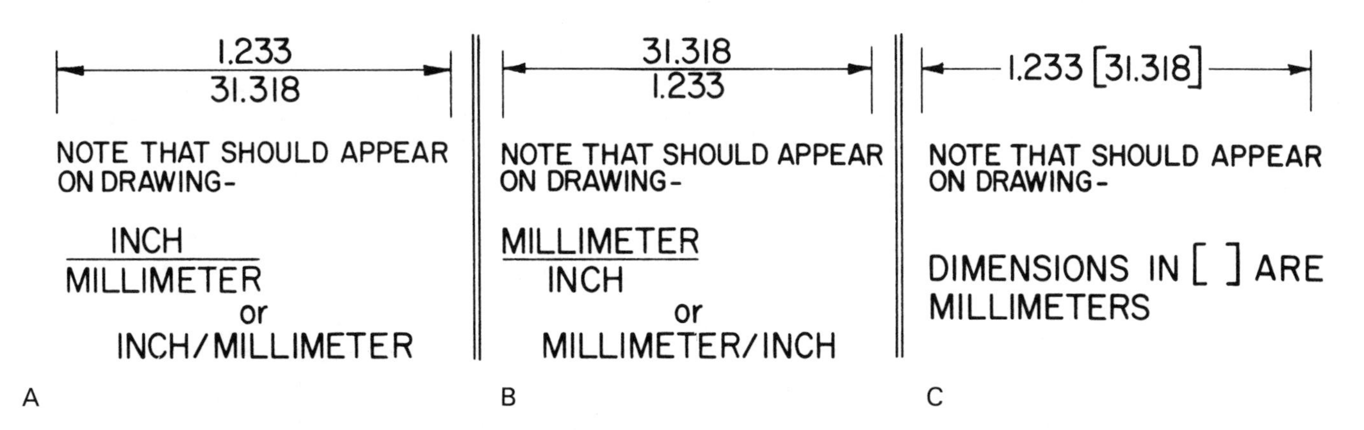

Fig. 9-5. Study how inch and metric dimenions are indicated on a dual dimensioned drawing. A—Method used when drawing is to be utilized in United States. B—Method used when drawing is to be utilized primarily in a metric country and United States. C—Brackets sometimes indicate metric equivalent on a drawing to be used in United States.

Drawings are usually dimensioned in inches and common fractions when the object does NOT require a high degree of accuracy. Greater precision is indicated when dimensions are in inches and decimal fractions, Fig. 9-6.

Material to be used

The *material specification* on the print gives the exact grade or type substance (1020 steel, aluminum, etc.)

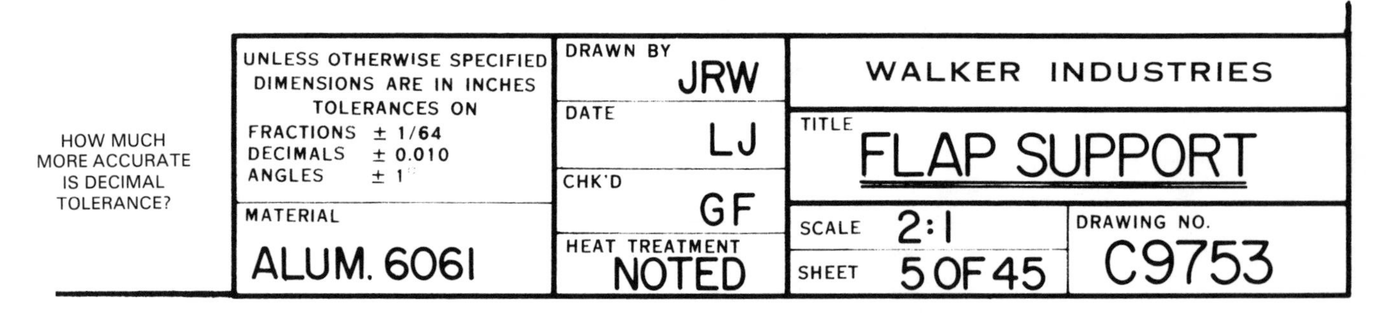

Fig. 9-6. When dimensions are in inches and decimal fractions, greater precision is normally required.

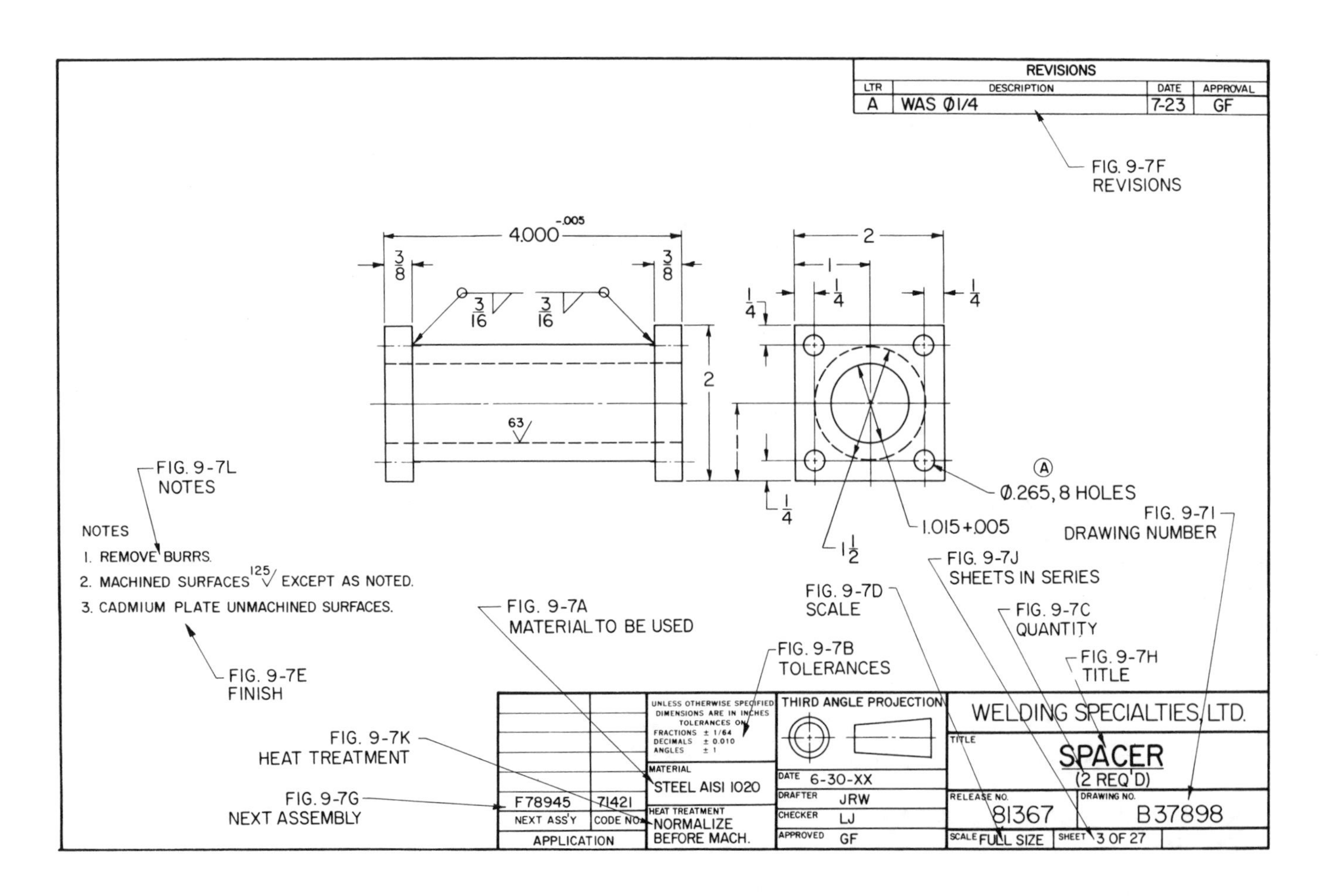

Fig. 9-7. Much information on the part is found in title block of drawing.

to be used in the weldment. It is usually included in a section of the title block. Refer to Fig. 9-7A.

Notes can also serve as material specifications NOT given in the title block. They are used at times and are given elsewhere on the print.

Remember! Under no condition should the welder substitute a different grade or type of material for the material specified on the print.

Tolerances

It would be costly to make everything to EXACT specified sizes. To keep costs within practical limits, tolerances or allowances are permitted.

A *tolerance* is a size allowance on a print. It indicates how much larger or smaller a portion of a product can be made and still be within specifications. Acceptable tolerances may be shown in the title block or on the drawing in several different ways, Figs. 9-7B and 9-8.

In general, when a dimension is given in inches and fractions of an inch, UNLESS OTHERWISE INDICATED ON THE DRAWING, permissible tolerances can be assumed to be ± 1/64 in.

The symbol ± means that the part can be made *plus* (larger) or *minus* (smaller) on that dimension by 1/64 in. and still be acceptable.

When the tolerance is plus AND minus, it is called a *bilateral tolerance.*

For example, if it is permissible to make the dimension larger, but NOT smaller, the dimension would read $2\ 1/2^{+1/64}$. If only a minus tolerance is permitted, the dimension would read $2\ 1/2^{-1/64}$.

When the tolerance permitted is plus OR minus (one direction), it is called a *unilateral tolerance.*

Dimensions shown as inches and decimals usually indicate that the work must be done more accurately. Two methods of showing these tolerances are in general use on modern prints.

A PLUS tolerance would be shown as $2.500^{+.010}$ or $\frac{2.510}{2.500}$. A MINUS tolerance would be shown as $2.500^{-.010}$ or $\frac{2.500}{2.490}$.

The dimensions show that the part can be used as long as that dimension measures within the indicated limits.

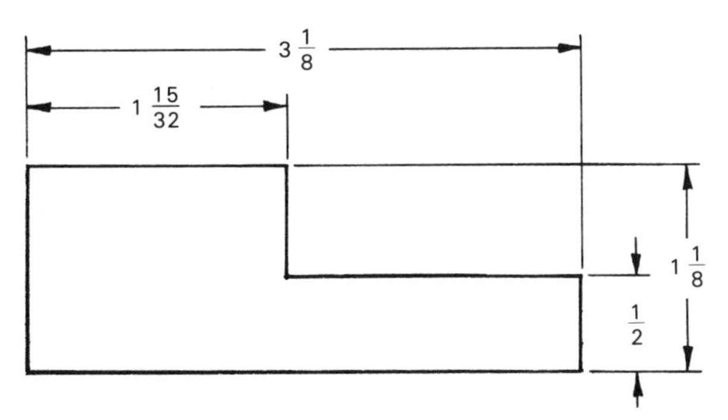

A—When fractional dimensions are used and no tolerances are shown with dimensions, tolerances are usually ± 1/64 in.

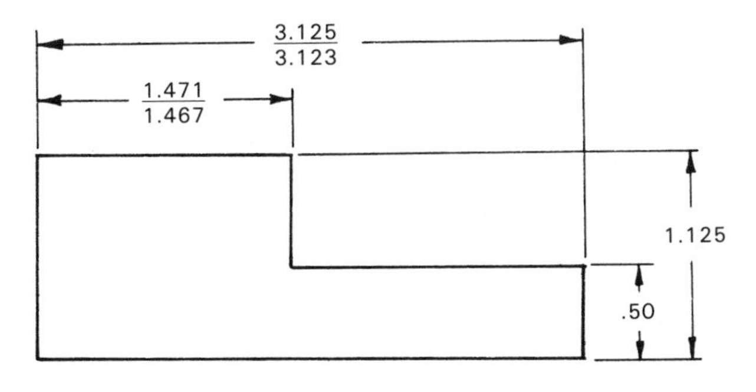

B—Decimal dimensions indicate tolerances of ±0.010 in., unless otherwise specified.

Fig. 9-8. Two ways of indicating tolerances on a drawing.

Quantity of units

The print should also show the number of units needed in each assembly, Fig. 9-7C. A WORK ORDER, included with each job received by the shop, will indicate the total number of units to be manufactured. This aids in ordering sufficient material, and will help determine the most economical method for manufacturing the part.

Scale of drawing

When views on the drawing are made other than actual size, they are called *scale drawings.* This occurs when the part drawn is too large or too small and it would not be practical to draw the views full size.

A print with views drawn ONE-HALF SIZE would be shown by the figures 1:2. The figure 2:1 would indicate that the drawing is made TWICE the actual size of the part. Drawings made ACTUAL SIZE would be listed as 1:1, *full,* or *full size,* Fig. 9-7D.

CAUTION! NEVER measure or scale the views on a print to get a needed dimension.

Finish

General *finish* requirements (sand blasting, plating, painting, etc.) may be specified in a section of the title block on some drawings. However, full finish specifications are usually listed elsewhere on the drawing along with the standards the finish must meet. See Fig. 9-7E.

Revisions

The *revision* block indicates what changes have been made to the original drawing, Fig. 9-7F.

Next assembly

Next assembly information is necessary to provide the next step in the manufacturing and assembly operations. The term *application* is sometimes used in place of next assembly. See Fig. 9-7G.

Drawing title

A section of the title block gives this information. The *drawing title* tells the welder the correct name of the part, Fig. 9-7H.

Print number

For your convenience in filing and locating drawings, industry provides each master tracing with a *print* or *identifying number,* Fig. 9-7I.

Each sheet in a series of drawings is also numbered to indicate the consecutive order and the total number of sheets in the series, Fig. 9-7J.

Heat treatment

Heat treatment includes a number of processes involving the controlled heating and cooling of a metal or alloy to obtain desirable changes in its physical characteristics. Heat treating can improve the metal's toughness, hardness, and resistance to shock.

Also, heat treatment can be used to *anneal* (soften) metals to make them easier to work. Annealing will also stress relieve metals that have been welded or machined.

The heat treatment requirements are shown in this section, Fig. 9-7K. The term *noted* is placed in the block when the heat treatment must conform to specific standards and is placed elsewhere on the print, Fig. 9-9.

If NO heat treatment is required, the word *none* or a *diagnoal line* is entered in the block.

PRINT SHEET SIZE

The bulk of the drawings used by industry are put on standard size drawing sheets. This makes them easier to file and identify. A *block* often shows sheet size. A *prefix letter* added to the print identification number may also be used to indicate print size.

A listing of standard sheet sizes is shown in Figs. 9-10 and 9-11.

NOTES

Information not included in the title block or dimensioned views, but pertinent to the manufacture or assembly of the part, is included under *notes.* Refer to Fig. 9-7L.

ZONING

Zoning is a technique used to aid in locating details on larger size drawings. The zones are indicated outside the border as letters and numbers. Fig. 9-12 illustrates how zoning is used on a print.

SECURITY CLASSIFICATION

Some drawings have a *security classification* of top secret, secret, confidential, or restricted. The type of classification will be noted on the top of the sheet and below the title block.

If your work involves using classified drawings, you will be instructed how to safeguard the security of these prints.

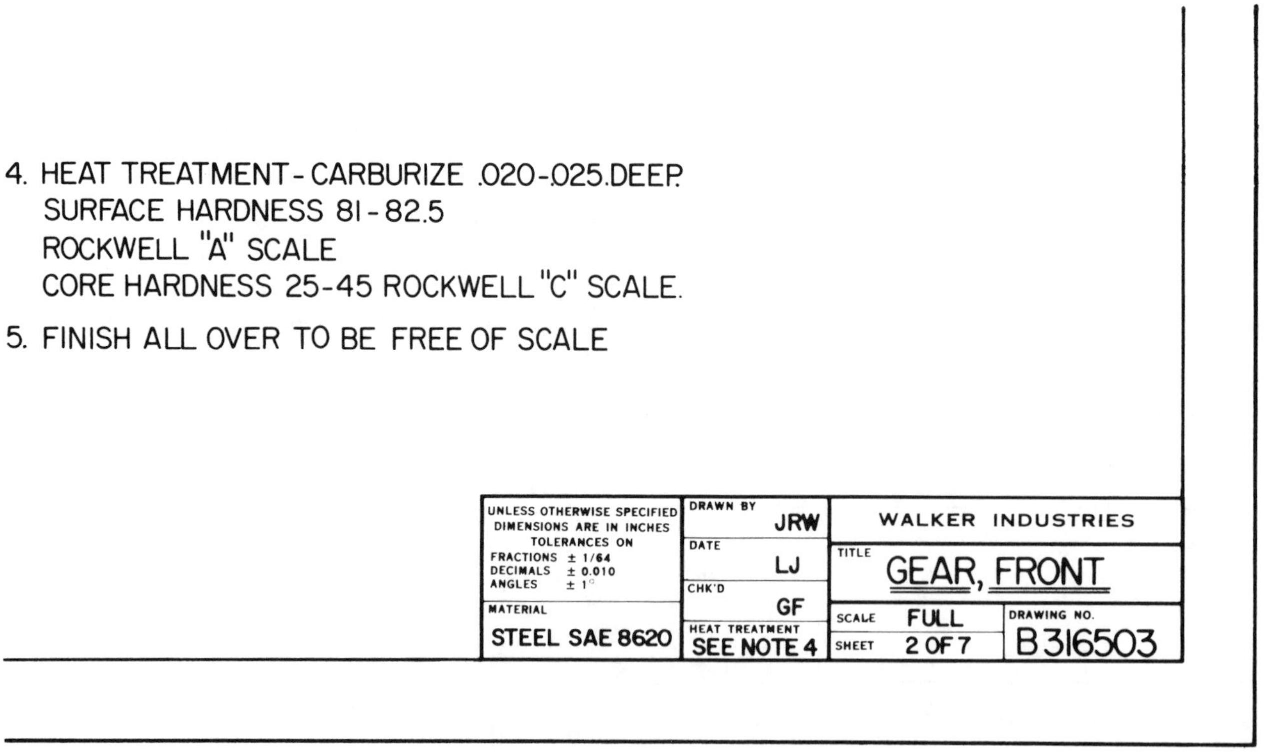

4. HEAT TREATMENT - CARBURIZE .020-.025 DEEP.
   SURFACE HARDNESS 81 - 82.5
   ROCKWELL "A" SCALE
   CORE HARDNESS 25-45 ROCKWELL "C" SCALE.
5. FINISH ALL OVER TO BE FREE OF SCALE

UNLESS OTHERWISE SPECIFIED DIMENSIONS ARE IN INCHES
TOLERANCES ON
FRACTIONS ± 1/64
DECIMALS ± 0.010
ANGLES ± 1°
MATERIAL STEEL SAE 8620
DRAWN BY JRW
DATE LJ
CHK'D GF
HEAT TREATMENT SEE NOTE 4
WALKER INDUSTRIES
TITLE GEAR, FRONT
SCALE FULL
SHEET 2 OF 7
DRAWING NO. B316503

Fig. 9-9. Heat treatment information is usually found in title block. However, the term "noted" is used in heat treatment block if heat treatment specifications are given elsewhere on drawing.

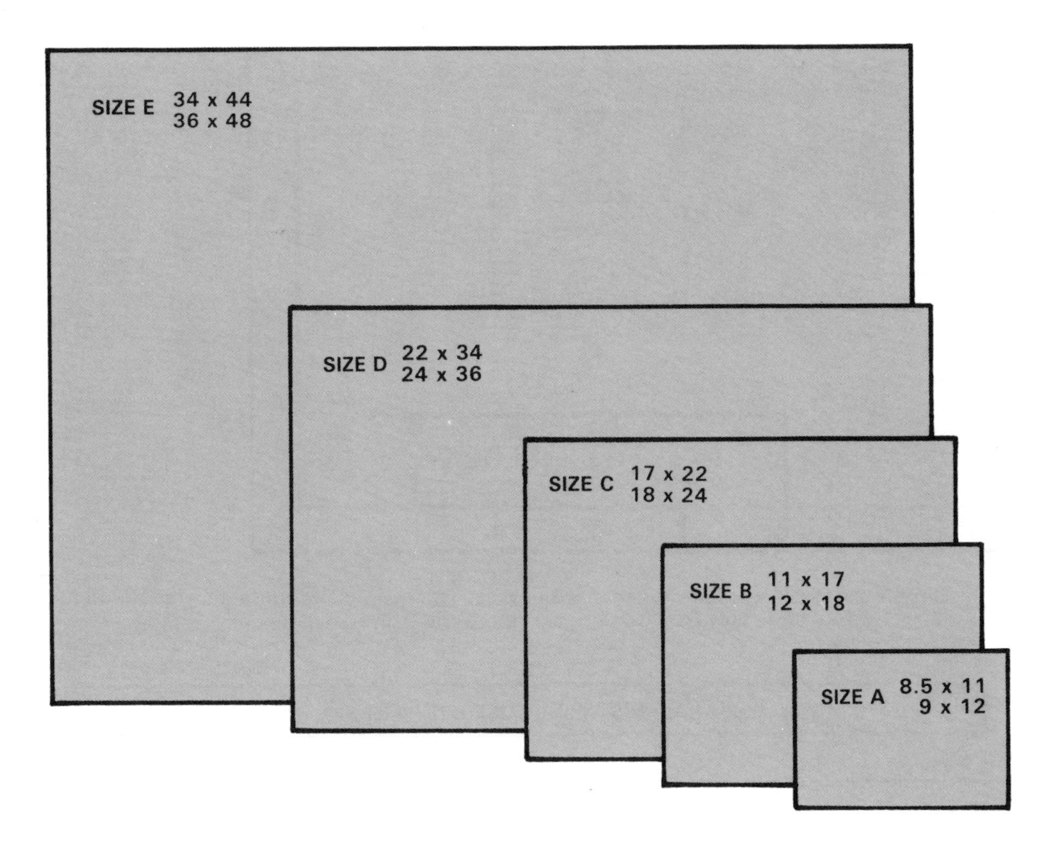

Fig. 9-10. Standard inch print sizes. Sheet sizes are given as "A", "B", etc.

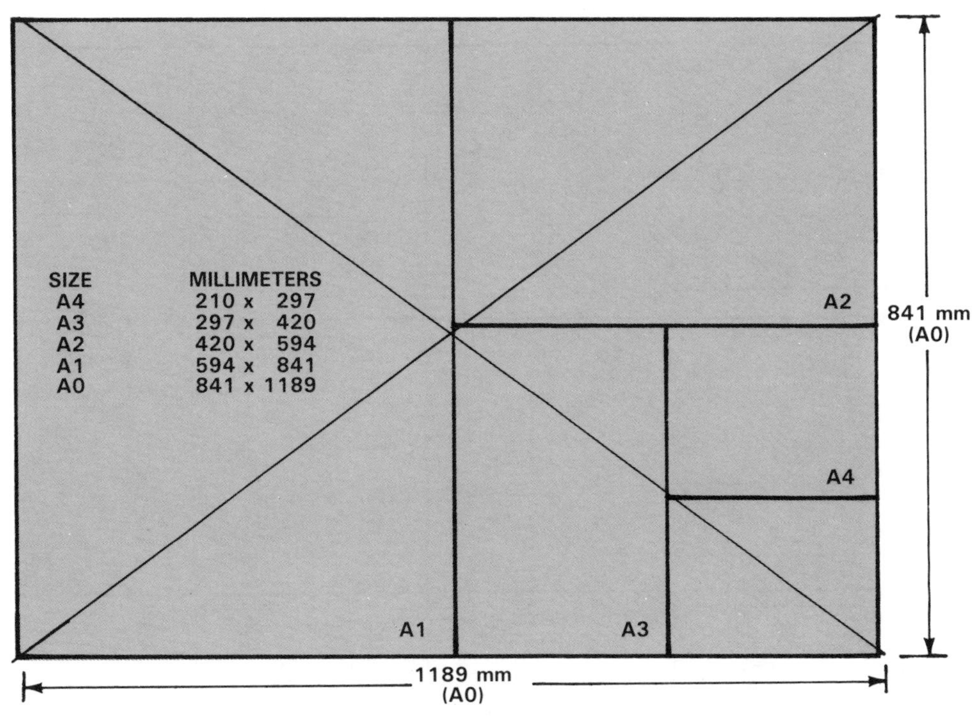

Fig. 9-11. Standard metric print sizes. Sheet sizes are given as A1, A2, etc.

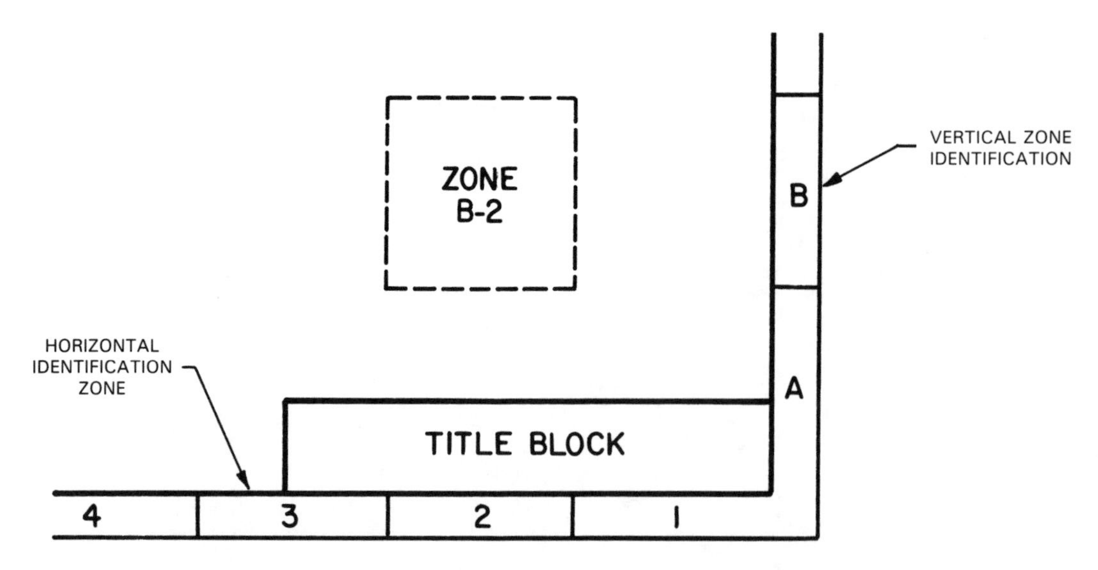

Fig. 9-12. Zoning is a technique used to locate details on large size drawings. A note will indicate location of specific information on a large print. Note how zone B-2 is adjacent to B and 2 on border of print.

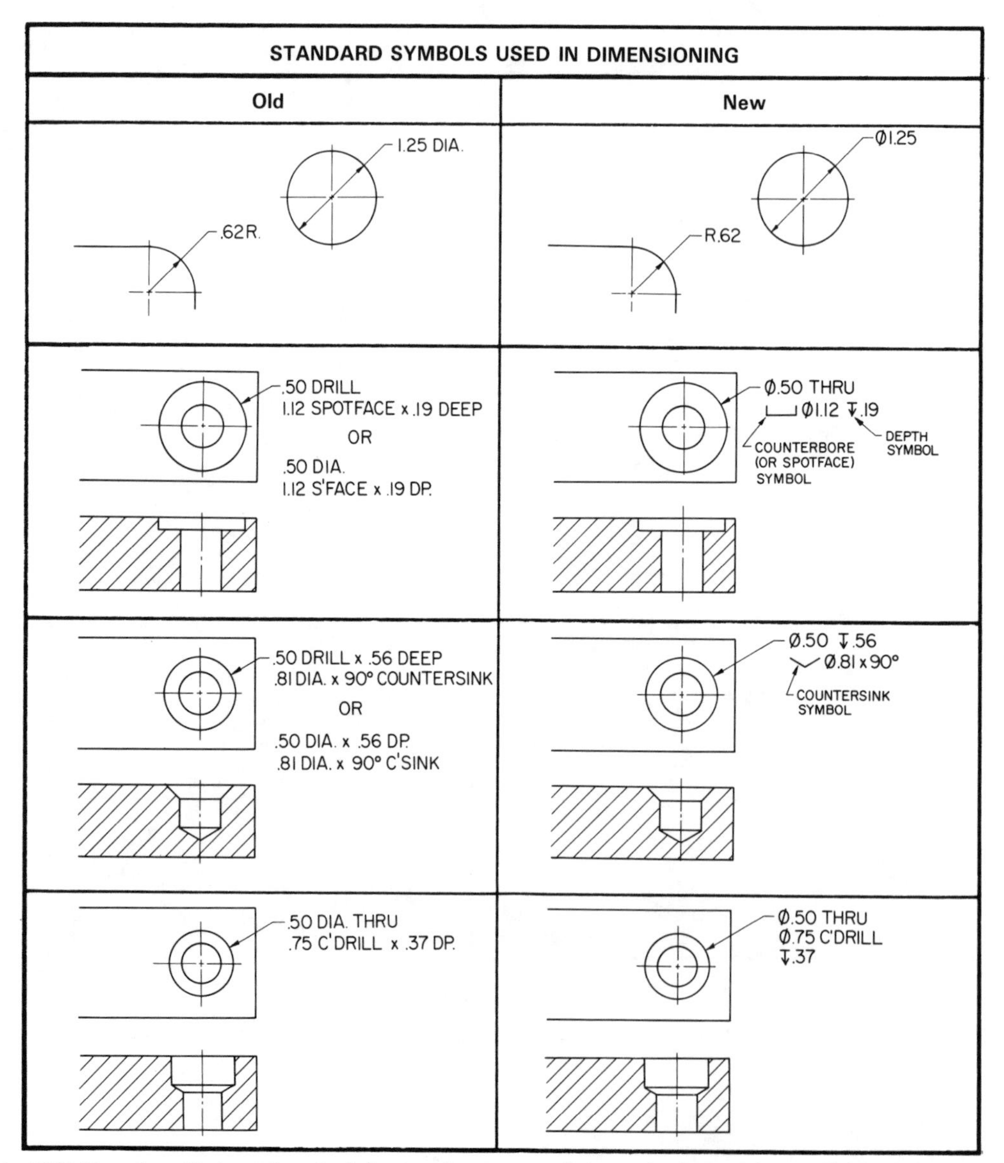

Recently ANSI (American National Standards Institute) recommended certain changes in specifying circles and holes. Since it will be many years before present standards will be found on ALL drawings and prints, both methods are shown in this text. Study and compare the examples shown above!

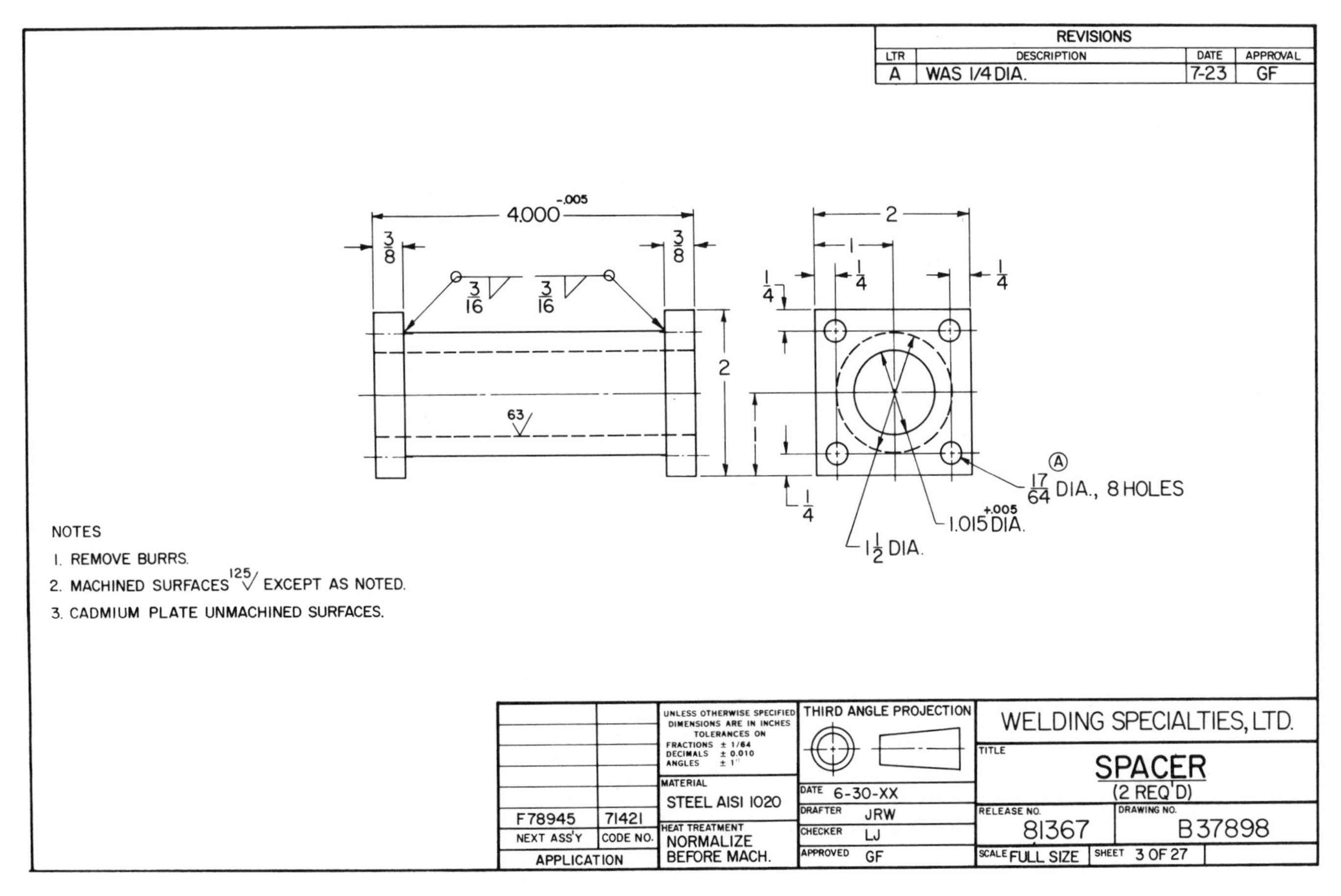

Use this drawing (B37898) to answer the questions in Unit 9—Test Your Knowledge.

UNIT 9—TEST YOUR KNOWLEDGE

Refer to the provided drawing (B 37898) and answer the following questions.

1. What is the name of the object shown on the drawing?

   ____

2. From what material is it made?

   ____

3. What is the next assembly?

   ____

4. How many drawings are used in the manufacture of this product?

   ____

5. This drawing is #________ of ________.

6. How is this drawing dimensioned?

   ____

7. The views of the object are drawn to what size?

   ____

8. List the drawing file number.

   ____

9. What revisions have been made to the drawing?

   ____

   ____

10. What tolerance is allowed on most dimensions?

    ____

    ____

11. What heat treatment is specified?

    ____

    ____

12. What finish technique is indicated?

    ____

    ____

Unit 10

UNDERSTANDING PRINTS

A print shows a series of views that give the welder an exact shape and size description of an object. See Fig. 10-1. Additional information necessary to make or assemble the product is also included on the print.

MULTIVIEWS

Multiviews are needed when more than one view is required to give an accurate shape description of the object. Most prints are in the form of multiviews.

The views are arranged in a systematic manner on the print. This makes it easier for the welder to merge the views in his or her mind and form a mental picture of the object. Refer to Fig. 10-2.

On most prints, the object is drawn in operating position.

In developing the needed views, the object is normally viewed from six directions, as in Fig. 10-3.

The various directions of sight will give the FRONT, TOP, RIGHT SIDE, LEFT SIDE, REAR, and BOTTOM VIEWS. These are shown in Fig. 10-4.

To obtain the views, think of the object as being enclosed in a hinged glass box. Study Fig. 10-5 carefully. Imagine that the views are projected on the sides

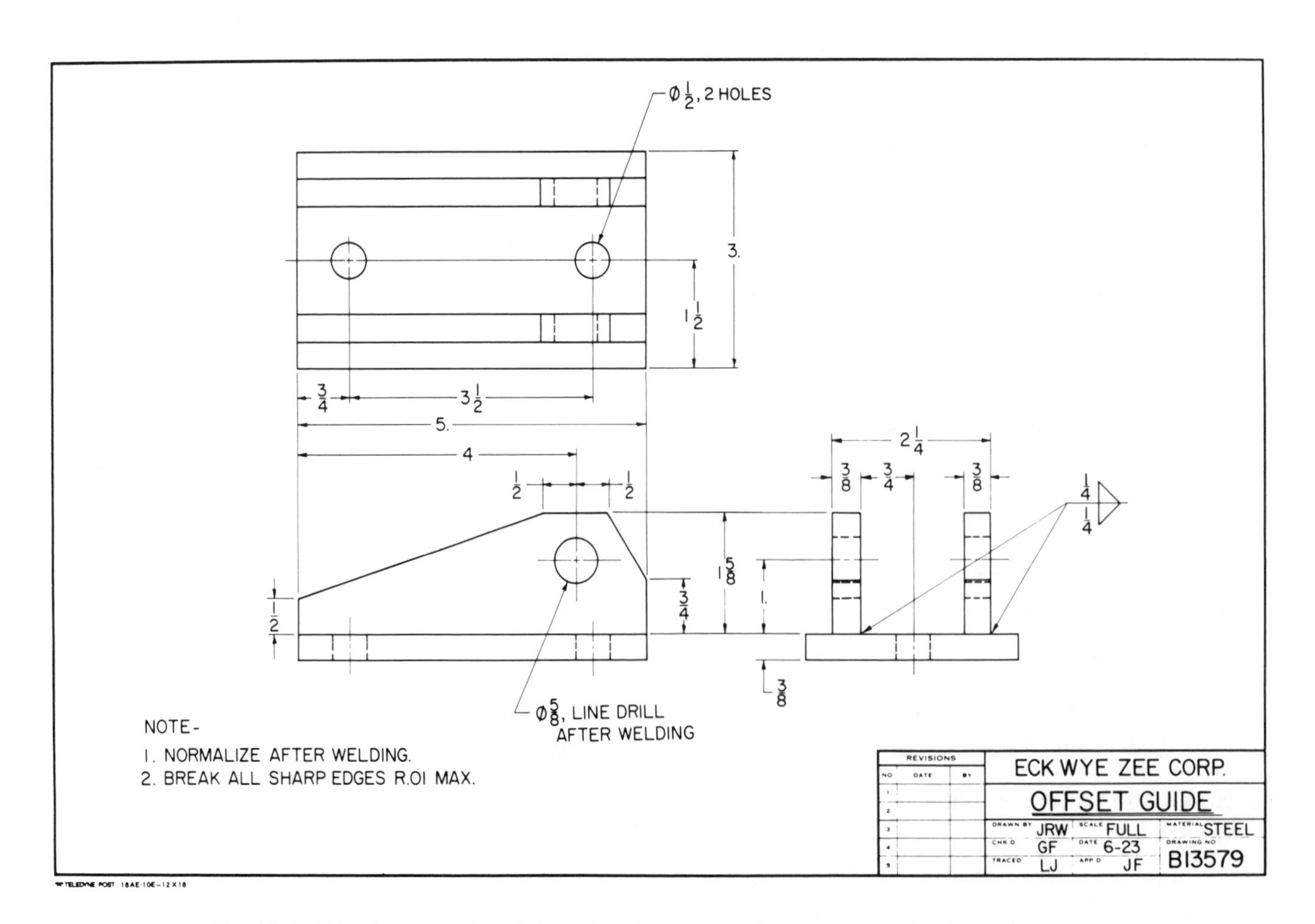

Fig. 10-1. Print shows a series of views that give an exact shape and size description of product.

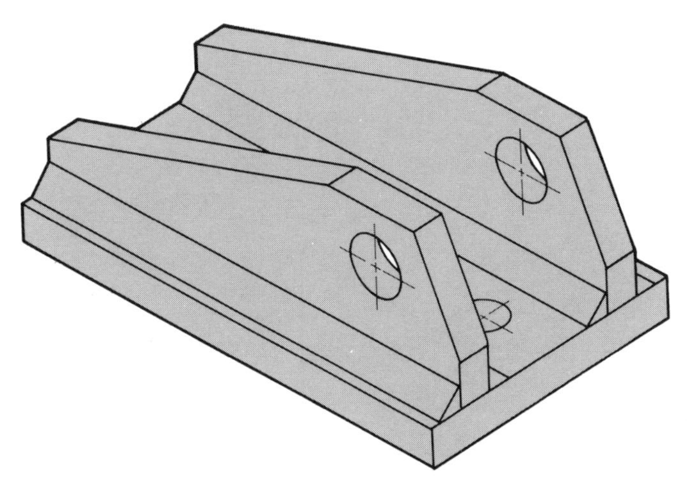

Fig. 10-2. Compare this drawing with views on print in Fig. 10-1. Views are arranged in a systematic manner and make it easier to form mental image of weldment.

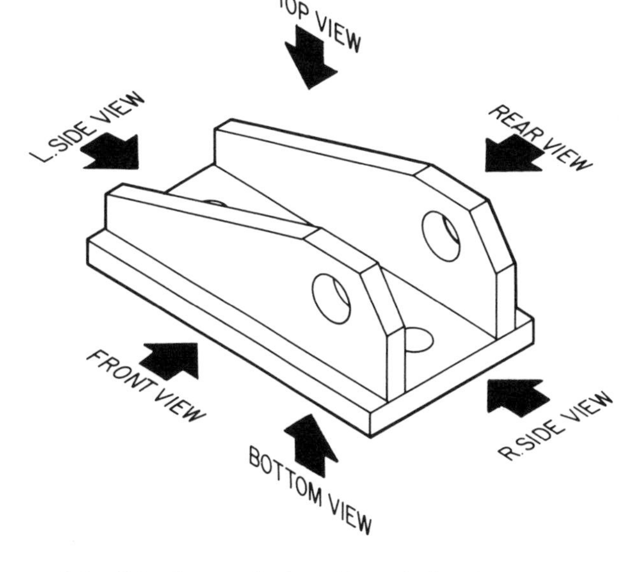

Fig. 10-3. Object is normally viewed from six directions when developing views on a print.

of the box. The top view of the object is seen on the top of the box, the front view on the front of the box, and so on for the remaining views.

This technique is called *orthographic projection.* It permits a three-dimensional object to be described on a flat sheet of paper having only two dimensions.

As can be seen, at least six views will be developed. Not all of them, however, are needed. Only those views required to give an accurate shape description of the object are included on the print. A view that repeats the same shape description as another view is NOT used, Fig. 10-6.

In the United States and Canada, all engineering

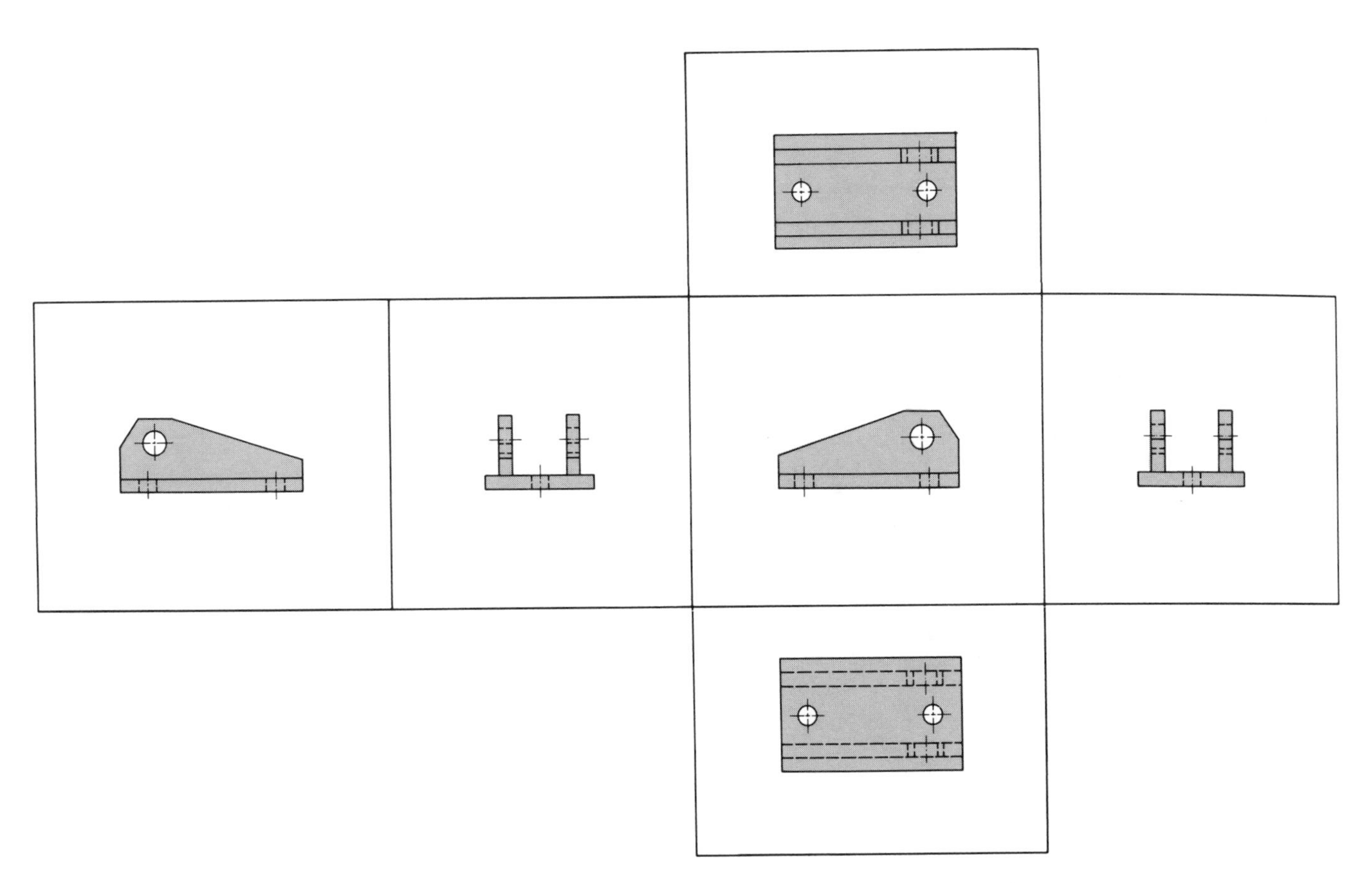

Fig. 10-4. Views given by six directions of sight.

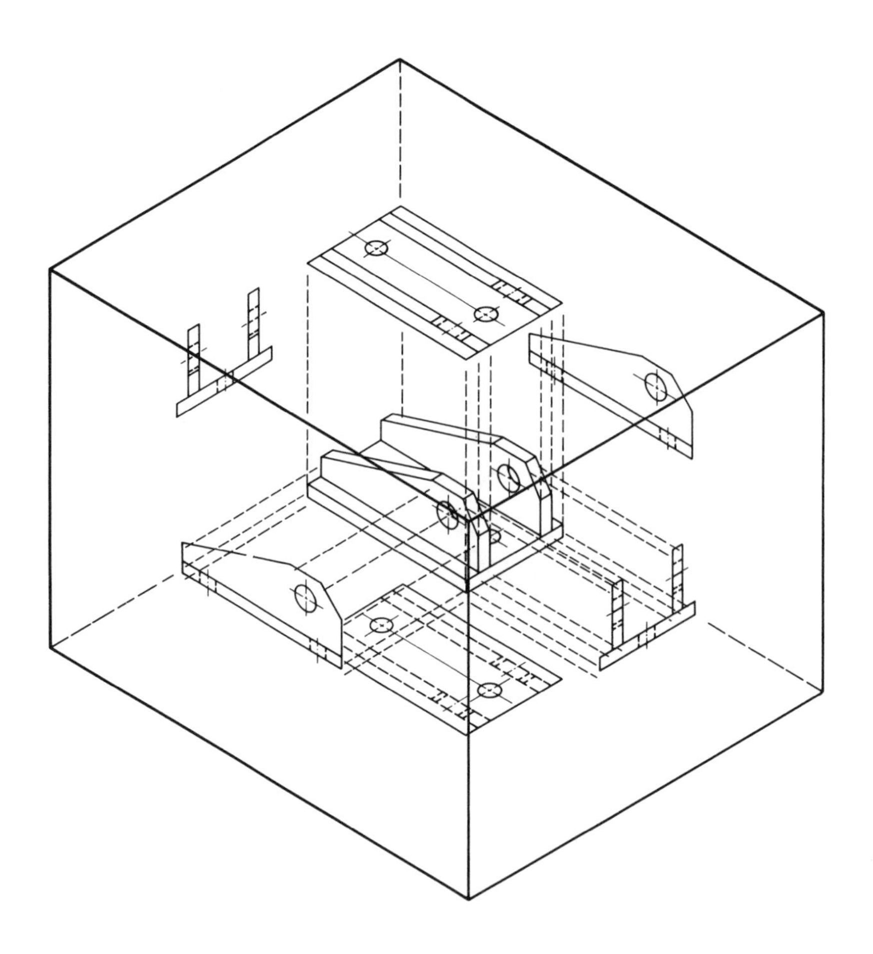

Fig. 10-5. Think of object as being enclosed in a hinged glass box to obtain views. Note relationships.

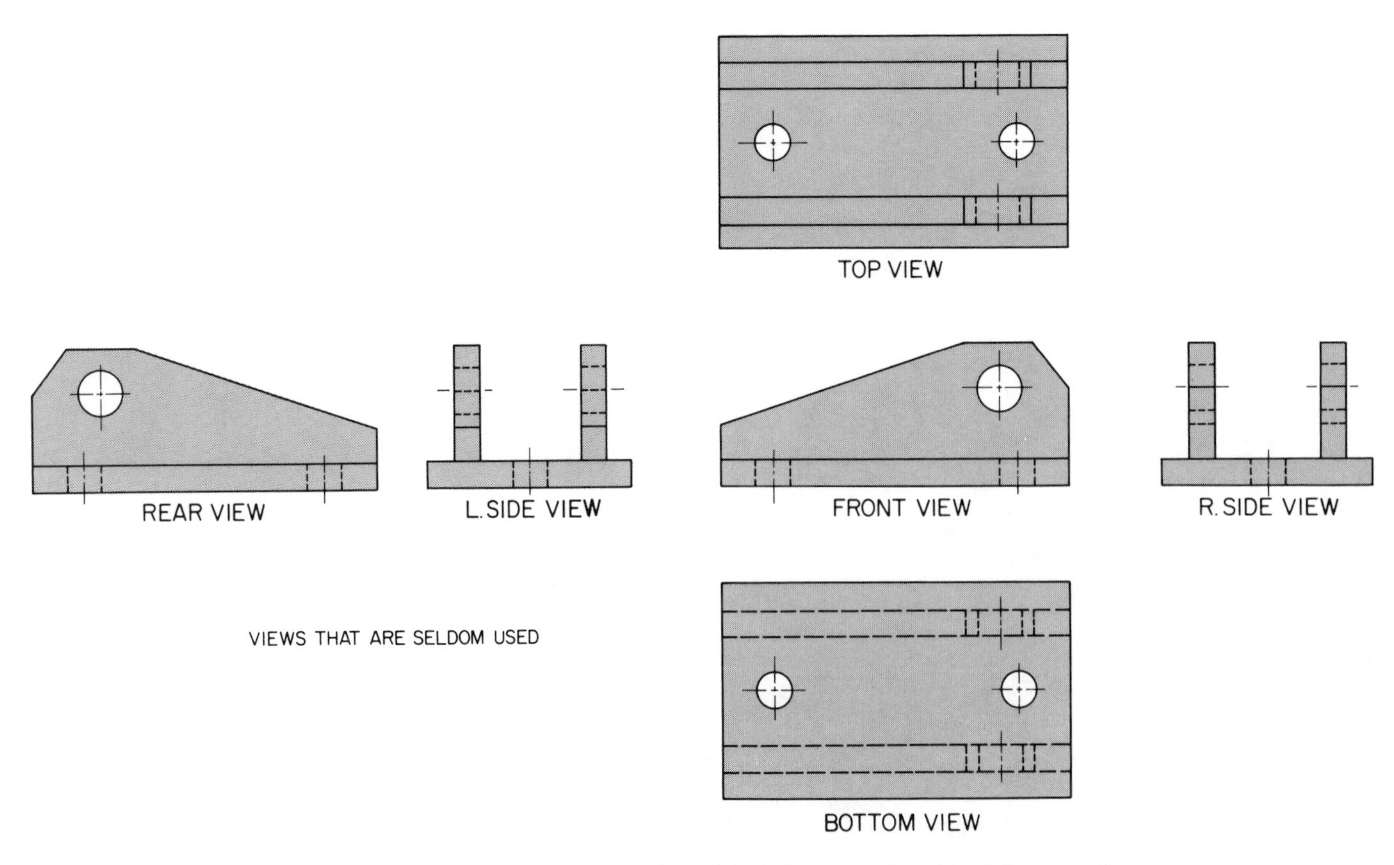

Fig. 10-6. When a view repeats same shape description as another view, it is not used.

drawings are *third angle projection,* with the views projected forward. The top view is always directly ABOVE the front view. The right side view is to the RIGHT of the front view and in line with it.

Drawings used in European countries are drawn in *first angle projection,* with the views projected rearward. Fig. 10-7 shows both third angle and first angle projection. Compare the two.

A block similar to Fig. 10-8 is usually included on prints to identify the type of projection angle.

top and side views but neither view shows its true length.

An additional or *auxiliary view* is used to show the true shape and size of the angular surface. Fig. 10-10 gives an example.

The auxiliary view is always projected at right angles (90 °) from the regular view on which the angular surface appears as a line.

Quite often with auxiliary views, it is possible to eliminate one of the regular views, as in Fig. 10-11.

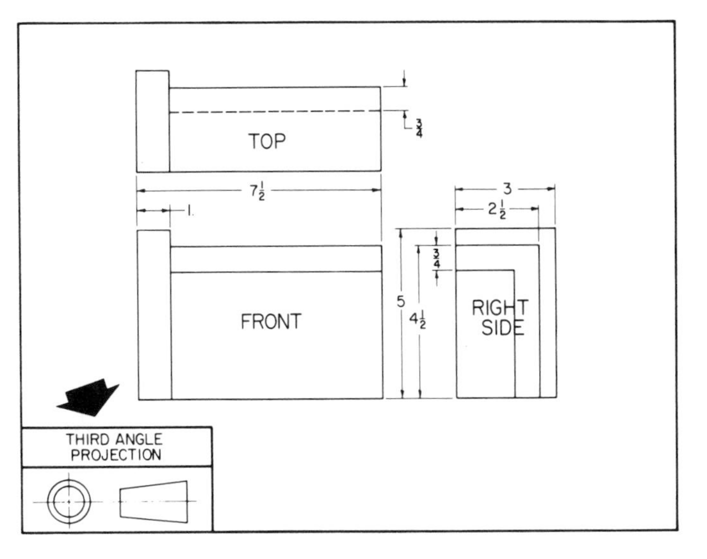

A – This is called third angle projection. Drawings made in the United States use this type which has drawings arranged as shown.

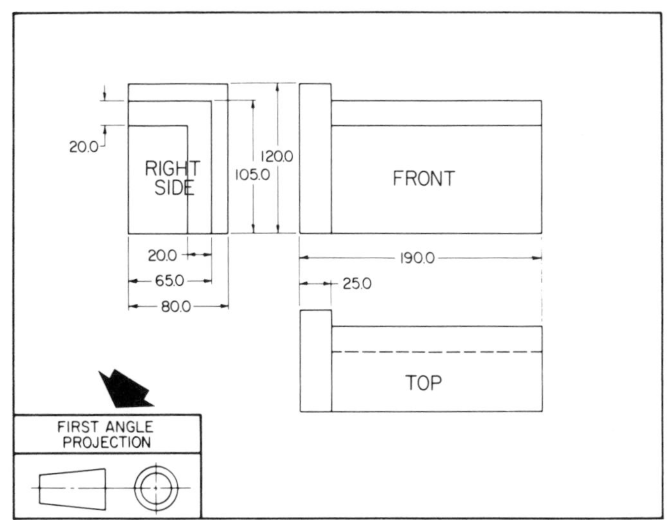

B – This is termed first angle projection. Drawings used in European countries are drawn like this.

Fig. 10-7. Study third and first angle projection.

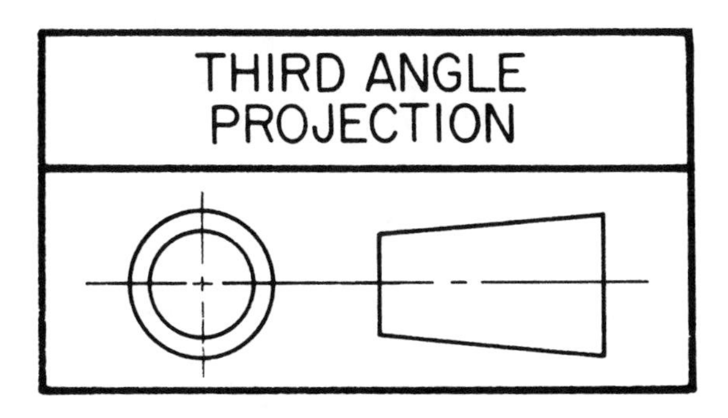

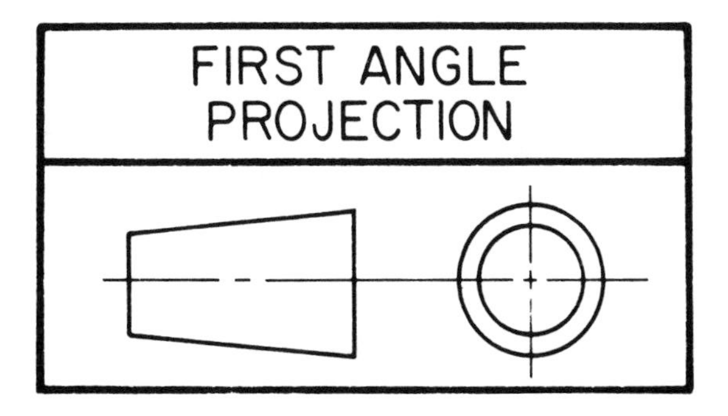

Fig. 10-8. A block like one of these is normally on print to identify angle of projection.

AUXILIARY VIEWS

The true shape and size of objects having angular or slanted surfaces cannot be shown using the regular (top, front, side) views. This is illustrated in Fig. 10-9.

The true length of the angular surface is shown on the front view, but this view does not show its width. The true width of the angular surface is shown on the

SECTIONAL VIEWS

When an object is relatively simple in design, its shape can be described on a drawing without difficulty, Fig. 10-12. For a complex object with many features obscured from view, however, it is often not easy to show its internal structure without a "jumble" of hidden lines, Fig. 10-13. The drawing would be hard to

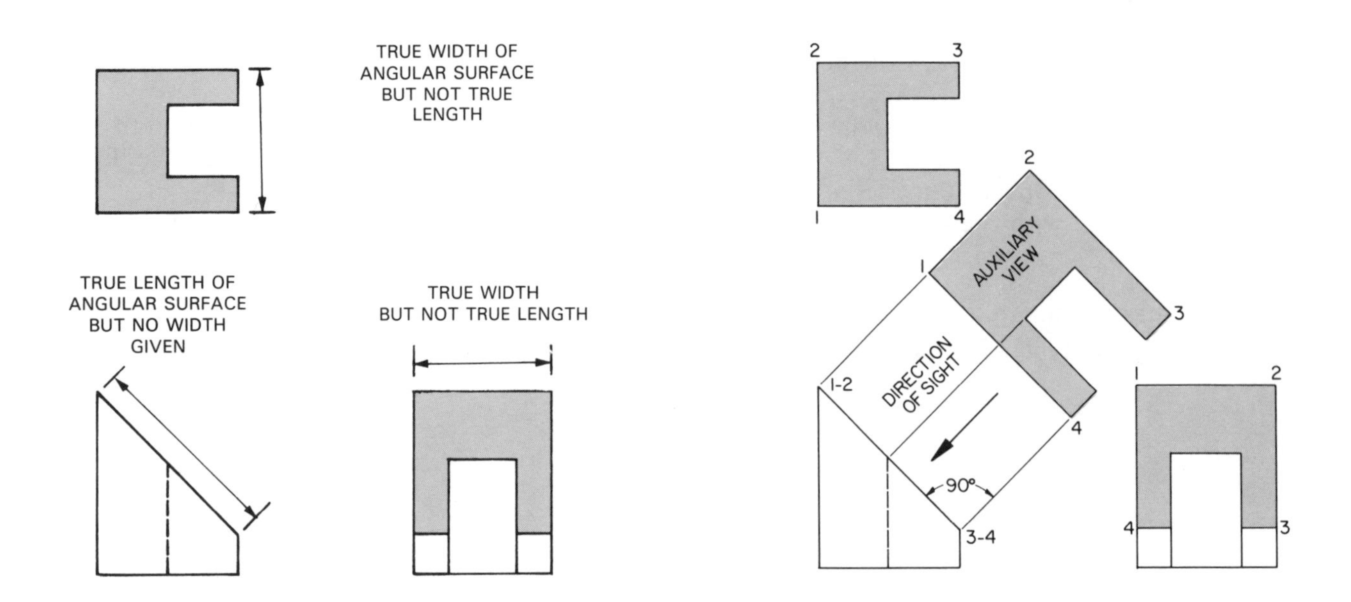

Fig. 10-9. The regular top, front, and side views do not show true shape and size of angular or slanted surfaces.

Fig. 10-10. Dimensions of angular surface are given by extra auxiliary view.

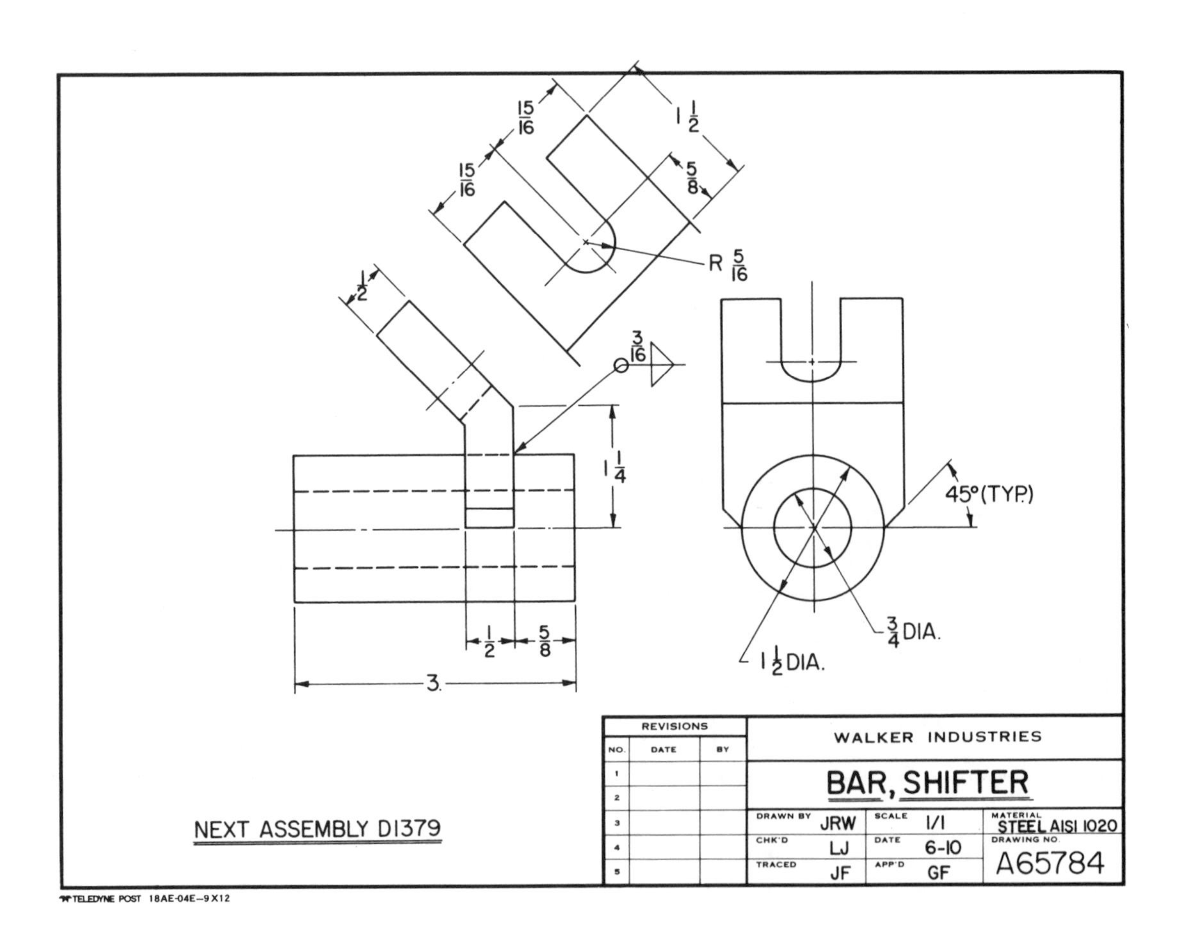

Fig. 10-11. Note how print may not have one of the regular views when an auxiliary view is used.

understand and interpret.

Sections or *sectional views* permit the true internal shape of a complex object to be shown without the confusion caused by a myriad of hidden lines.

A sectional view shows how the object would appear if an imaginary cut (called *cutting plane*) were made through the object perpendicular to the direction of sight. Shown in Fig. 10-14, the section or portion of the object between the eye and the cutting plane is removed or broken away to reveal the interior features of the object. This makes the shape of the object more understandable.

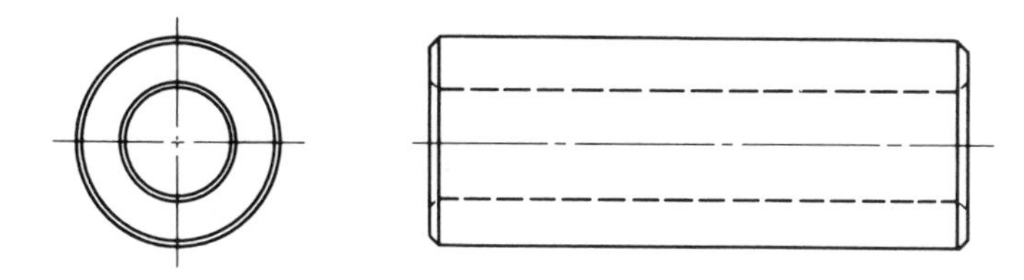

Fig. 10-12. Simple objects, like this piston pin, can be fully illustrated on a drawing with little difficulty. Only two views are needed.

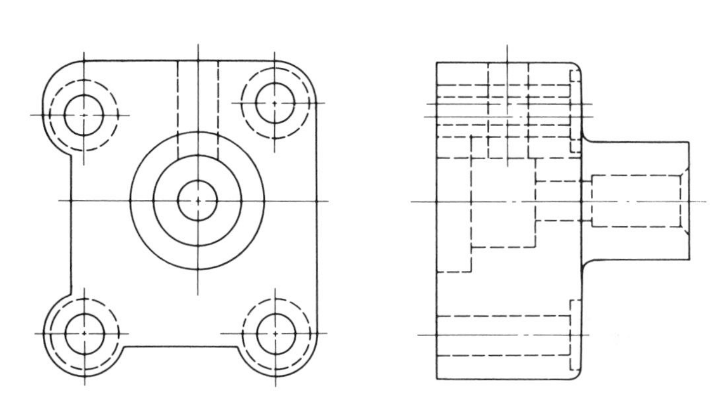

Fig. 10-13. Complex objects with many interior features, when described graphically, can often result in a drawing with a maze of hidden lines. These lines can make drawing confusing.

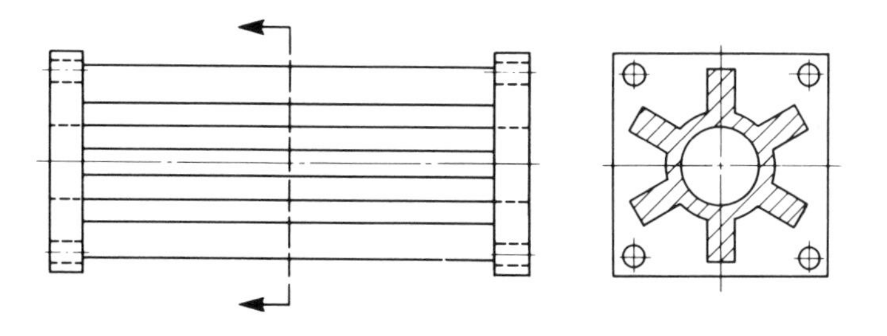

Fig. 10-14. A sectional view shows how object would appear if an imaginary cut were made through object perpendicular to direction of sight. This permits interior features of object to be seen without confusion of many hidden object lines.

Cutting plane line

The *cutting plane line* indicates the point from which the section is taken from the part. An example is given in Fig. 10-15. The arrows at the end of the cutting plane line show the direction of sight for viewing the section.

Two forms of cutting plane lines are accepted for general use. These are given in Fig. 10-16.

Sections are usually identified with bold capital letters (A-A, B-B, etc.).

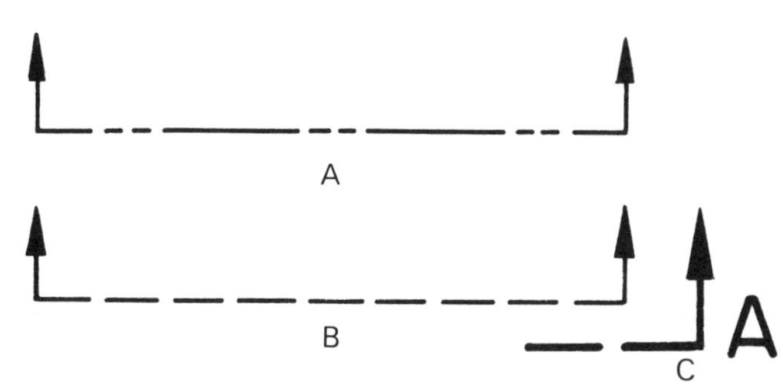

Fig. 10-16. Compare two forms of cutting plane lines. Either of them may be found on a drawing. A—Cutting plane line having long dashes and pairs of short dashes. B—Cutting plane line with equal length short dashes. C—Letters can be used to denote cut surface in diferent locations on print. Direction of arrows show direction of sight.

Section lining

The type of exposed cut surface of a section is represented by *section lining,* which is also called *"cross hatching."*

The American National Standards Institute (ANSI) has recommended the symbols for section lining shown in Fig. 10-17. They depict the different types of materials.

General purpose section lining (cast iron) is usually used on drawings when exact material specifications are located elsewhere on the print.

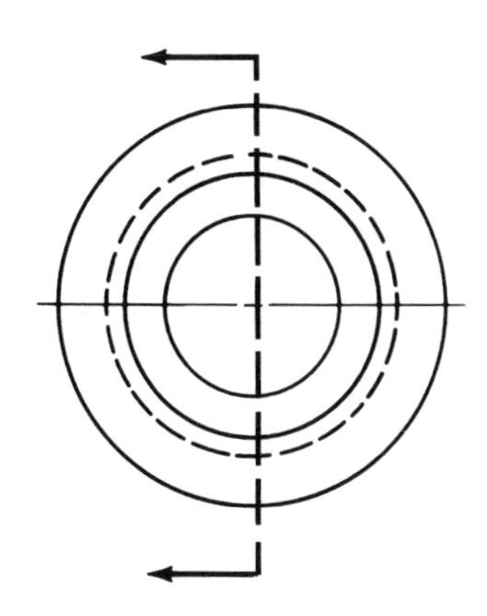

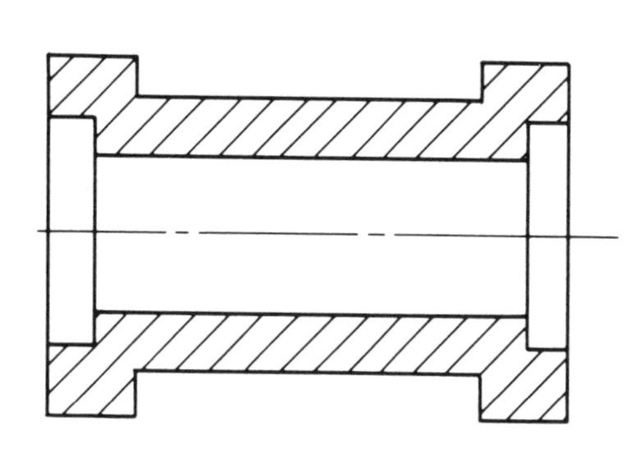

Fig. 10-15. Cutting plane line shows point from which section is removed from part.

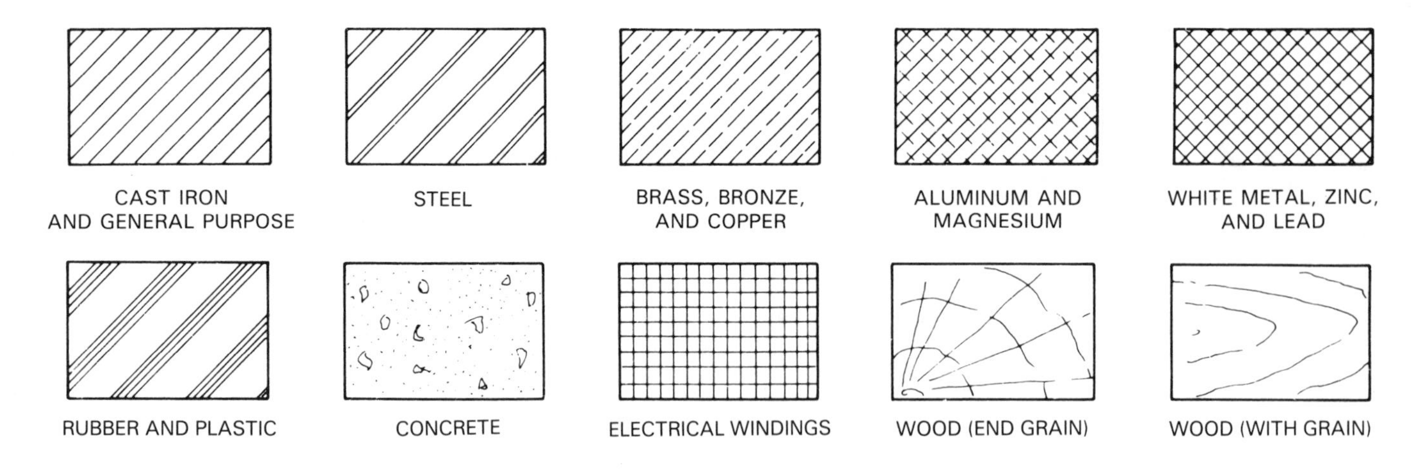

Fig. 10-17. Study standard code symbols for various materials in section.

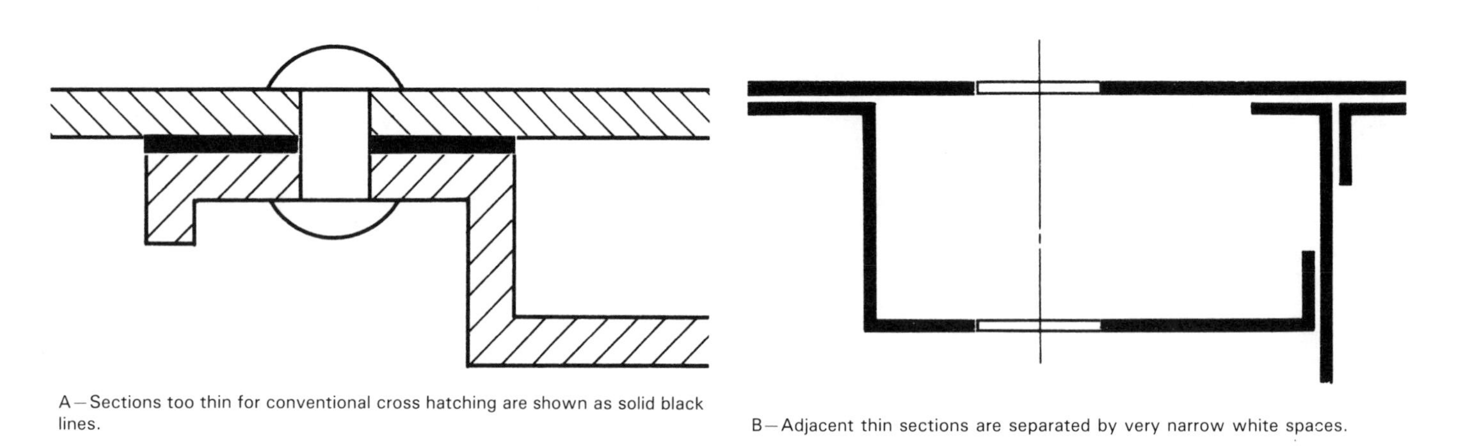

A—Sections too thin for conventional cross hatching are shown as solid black lines.

B—Adjacent thin sections are separated by very narrow white spaces.

Fig. 10-18. Denoting thin section.

Thin sections

Thin sections or sections not thick enough for conventional cross hatching (sheet metal, gaskets, etc.) are shown as solid black lines. See Fig. 10-18.

Full sections

A *full section* is shown when the cutting plane line passes entirely through the object. Fig. 10-19 gives an example.

Half section

Half sections are primarily limited to symmetrical objects. The shape of one-half of the interior features and one-half of the exterior features of the object are shown in the *half section.* Refer to Fig. 10-20.

Revolved section

Revolved sections rotate or turn the cut section 90 degrees. They are primarily used to show the shape of such things as spokes, ribs, and stock metal shapes. See Fig. 10-21.

Removed section

A *removed section* is used when it is not possible to show the sectional views on one of the regular views. This is illustrated in Fig. 10-22.

Offset section

The *offset section* is employed when the needed information cannot be shown by using a single, straight cutting plane. See Fig. 10-23. In this case, the cutting plane line is stepped or offset to pass through these features.

Broken out section

The *broken out section* is utilized when a small portion of a sectional view will provide the necessary information. One is shown in Fig. 10-24.

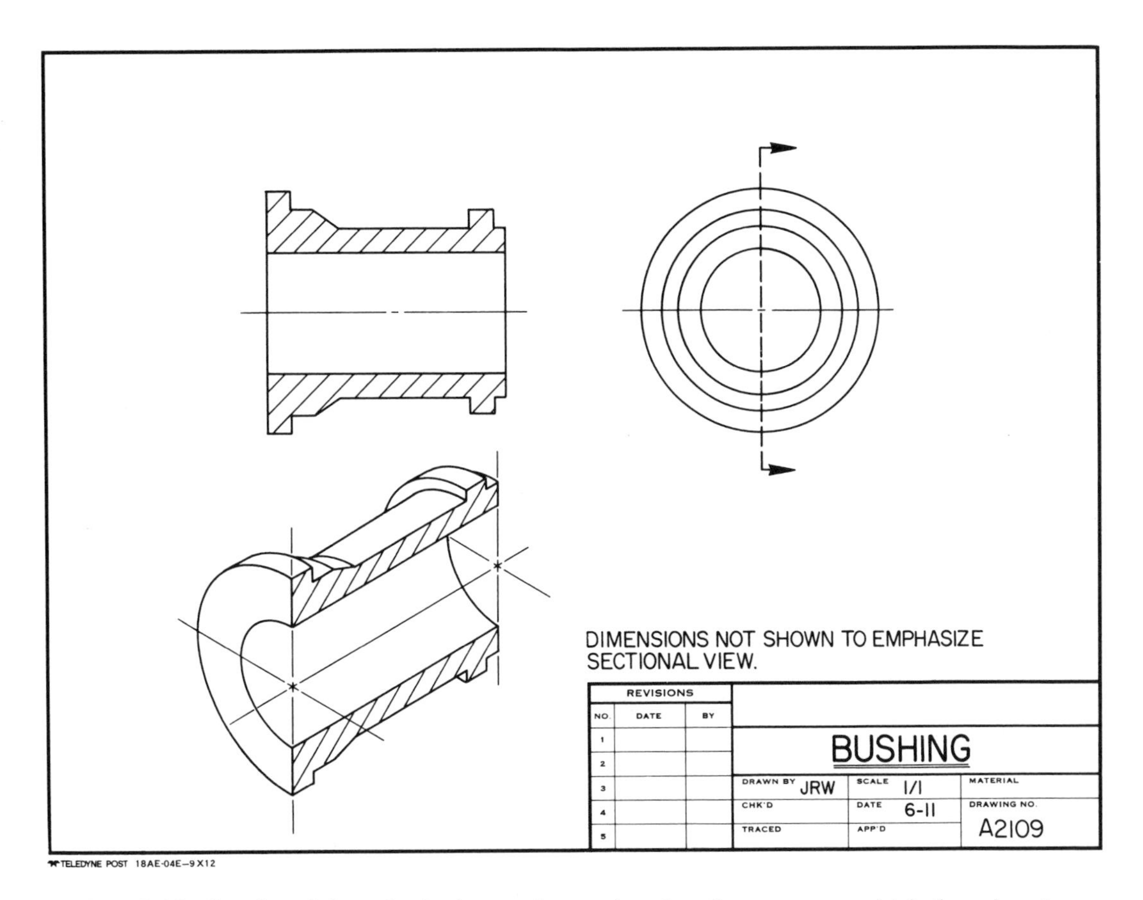

Fig. 10-19. Note how full section is shown when cutting plane line passes completely through part.

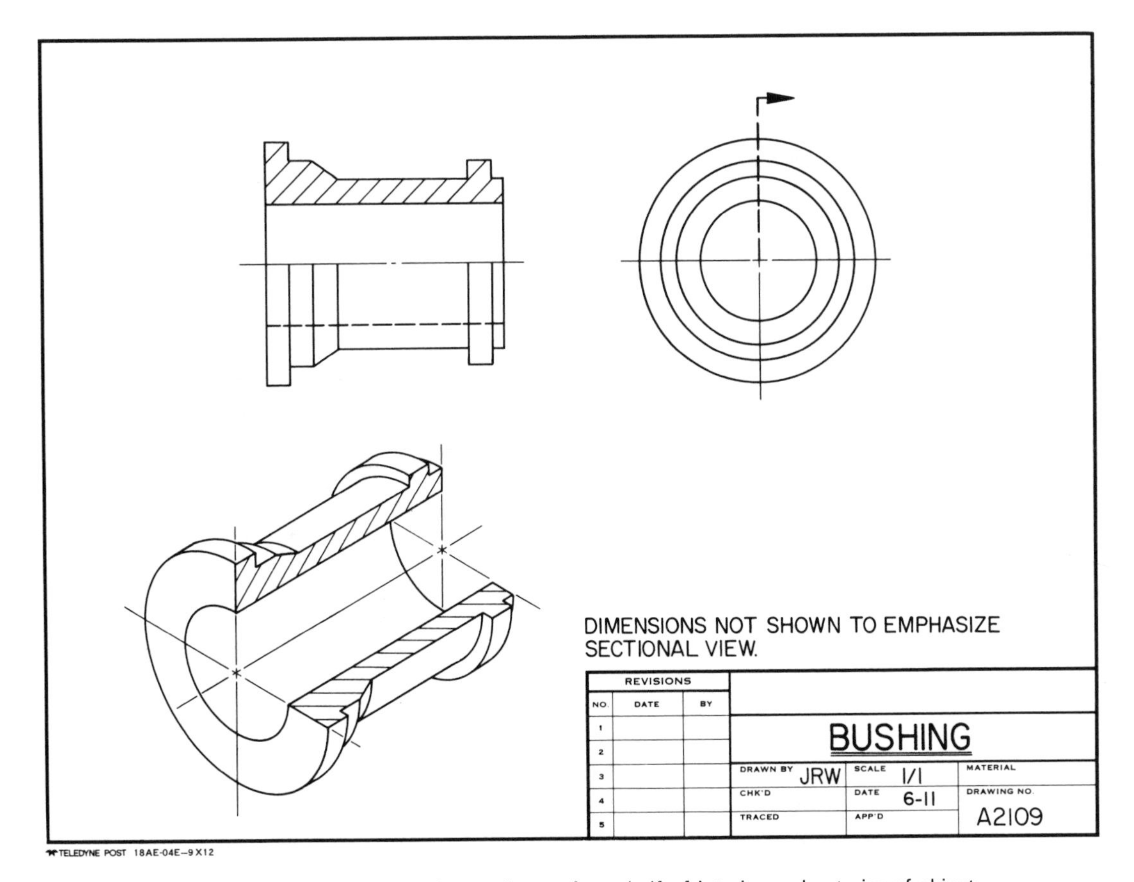

Fig. 10-20. A half section shows shape of one-half of interior and exterior of object.

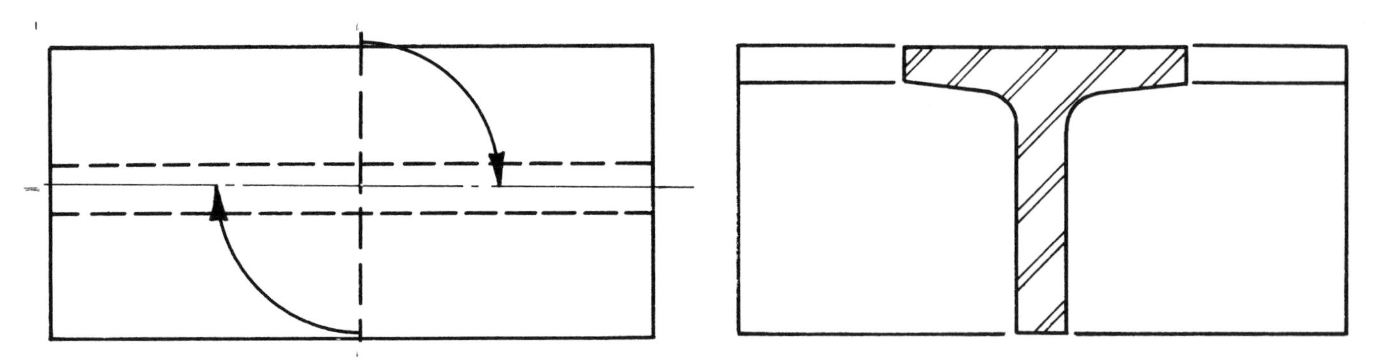

Fig. 10-21. Revolved section helps show shape of such things as spokes, ribs, and stock metal shapes.

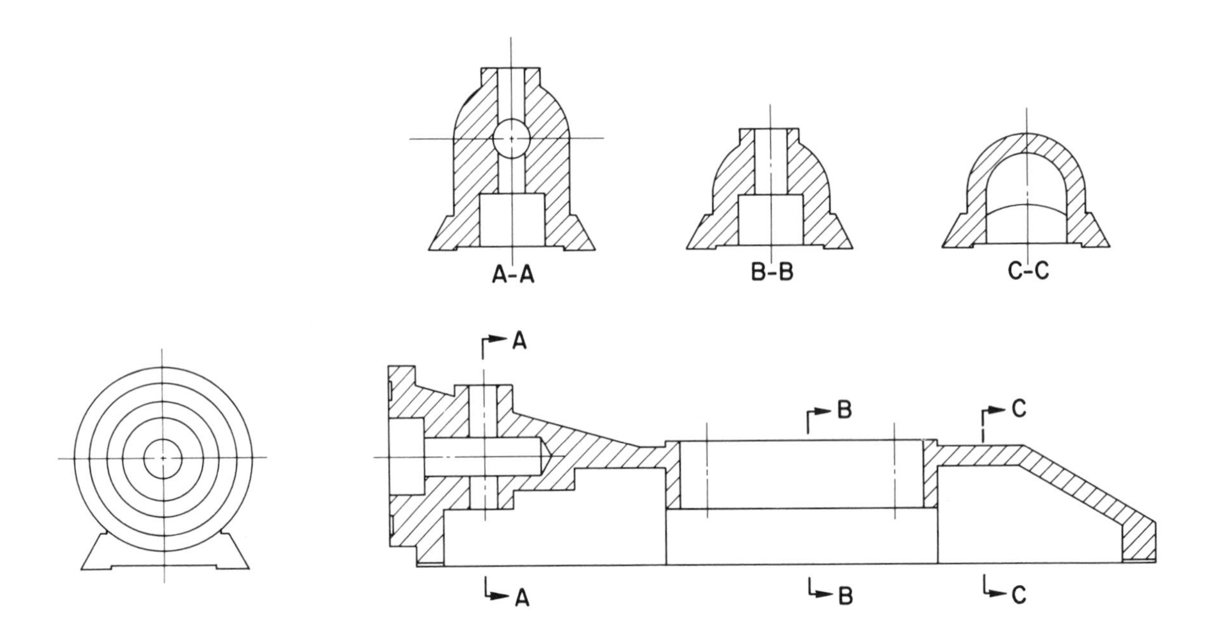

Fig. 10-22. A removed section shows a sectional view taken from object and shows it on another part of print.

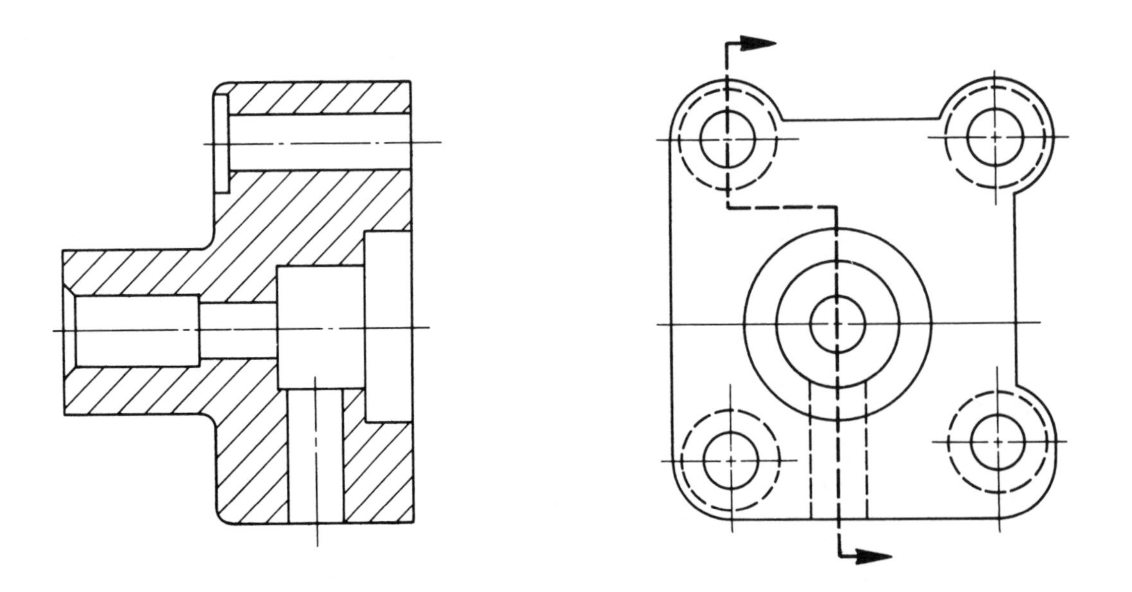

Fig. 10-23. Now how offset section is employed. It is needed when information cannot be shown by single, straight cutting plane.

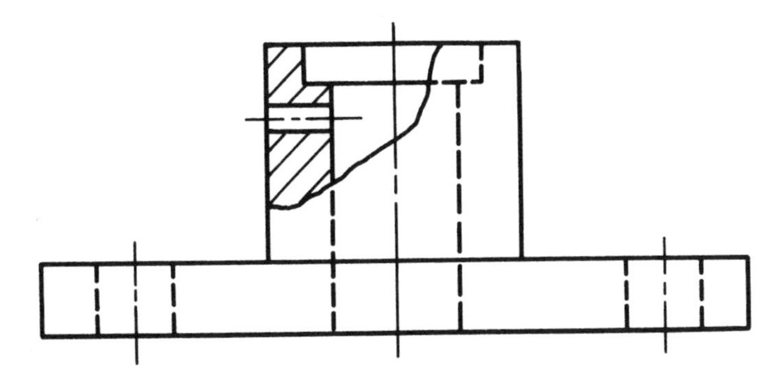

Fig. 10-24. With broken out section, only a small portion of a sectional view will adequately show essential information.

Conventional breaks

Long, uniformly shaped objects are sometimes difficult to present in a scale large enough to show its details clearly. *Conventional breaks* permit enlongated objects to be shortened so that a large enough scale can be employed to present details with clarity. Refer to Fig. 10-25.

Sections through webs and ribs

Webs and ribs are added to some objects for increased strength and rigidity. Fig. 10-26 shows how ribs and webs are represented in sectional view.

RULES ON READING PRINTS

There is no one best way to read a print or drawing. Most welders come up with their own method. The following rules are suggested to help YOU get started. Eventually, you will develop a method best suited to your own needs.

1. Carefully review the print.
2. Study one view at a time. Identify surface limits and lines that describe the intersection of surfaces. This will help you to visualize the shape of the object.
3. Establish sizes from the dimensions.
4. Review the other information (notes, title block, revisions, etc.) on the print.
5. Determine what your responsibilities will be in producing the object and the sequence you will follow in performing the operations.

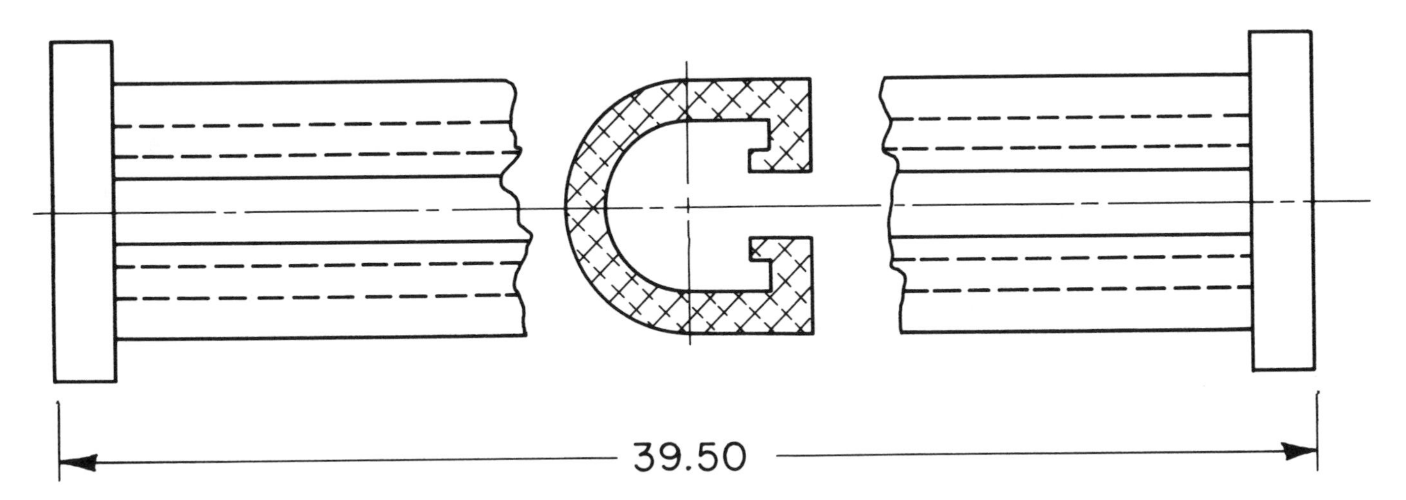

Fig. 10-25. A conventional break allows elongated objects to be shortened so that a larger scale can be used to show details more clearly.

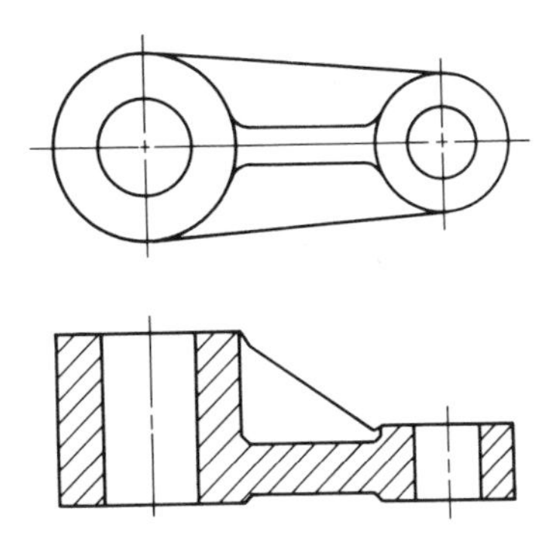

Fig. 10-26. Note this section through a web or rib. Section lines are not drawn through webbed or ribbed area.

6. Do not be afraid to ask for help if you do not understand something on a print. A mistake caused by not understanding something can be very costly to your employer and may cause injury to someone using the product at a later date.
7. Practice print reading until it becomes second nature to you.

UNIT 10—TEST YOUR KNOWLEDGE

Study the drawing (A 4678) and answer the questions in Parts I, II, and III.

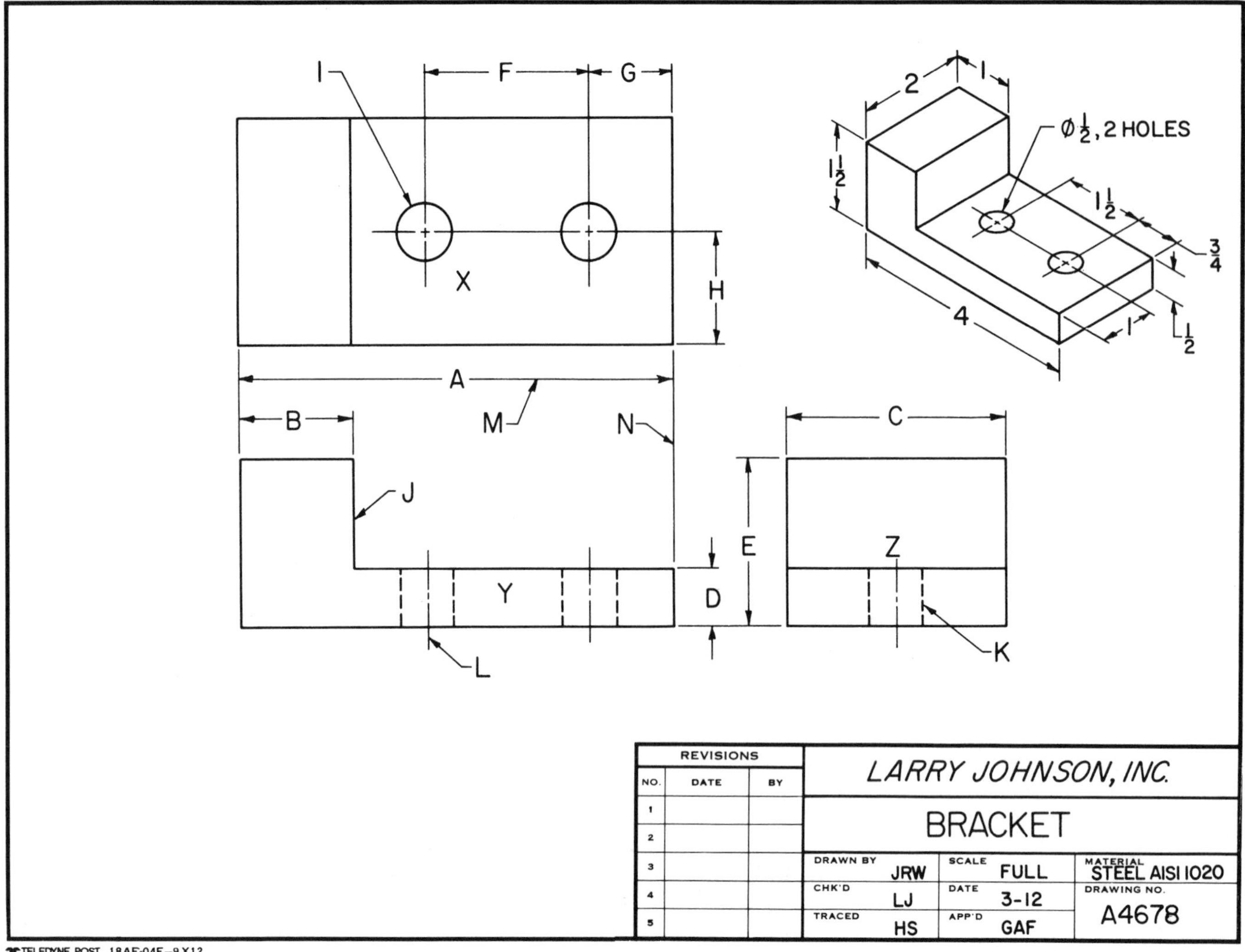

Use this print (A4678) to answer questions in Parts I, II, and III of Unit 10—Test Your Knowledge.

Part I

Refer to the pictorial view of the bracket and write in the dimensions indicated by the following letters.

1. A ____________________
2. B ____________________
3. C ____________________
4. D ____________________
5. E ____________________
6. F ____________________
7. G ____________________
8. H ____________________
9. I ____________________
10. How many holes are specified? __________

Part II

What type of lines are indicated by the following letters?

11. J ____________________
12. K ____________________
13. L ____________________
14. M ____________________
15. N ____________________

Part III

Answer the following questions.

16. Which view is indicated by surface X? __________

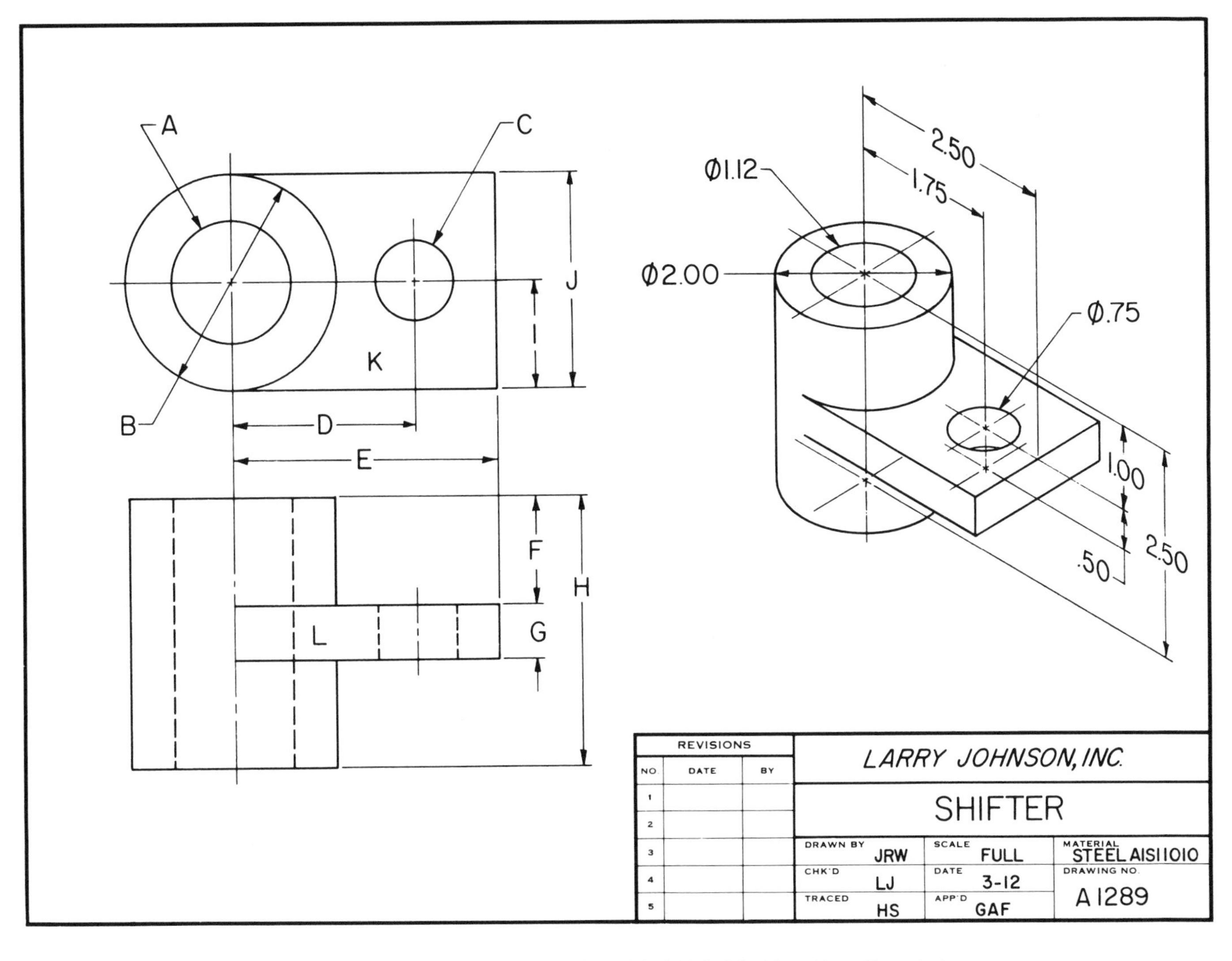

Use this drawing (A1289) to do Part IV of Unit 10—Test Your Knowledge.

17. Surface Y indicates which view? ____________

18. The __________ view of the bracket is indicated by surface Z.

19. What is the print number? ________________

20. The material specified is ________________.

Part IV

Study the pictorial drawing of the shifter (A 1289) and write in the dimensions indicated by the following letters on the orthographic views:

1. A ________________
2. B ________________
3. C ________________
4. D ________________
5. E ________________
6. F ________________
7. G ________________
8. H ________________
9. I ________________
10. J ________________

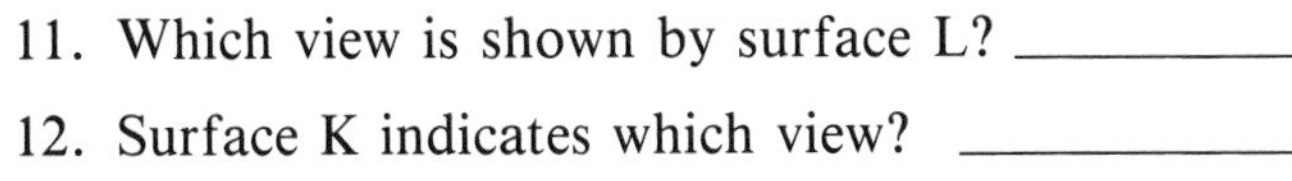

11. Which view is shown by surface L? __________
12. Surface K indicates which view? __________
13. What is the scale of the original drawing? ______
14. What is the drawing number? __________
15. The material specified is __________.

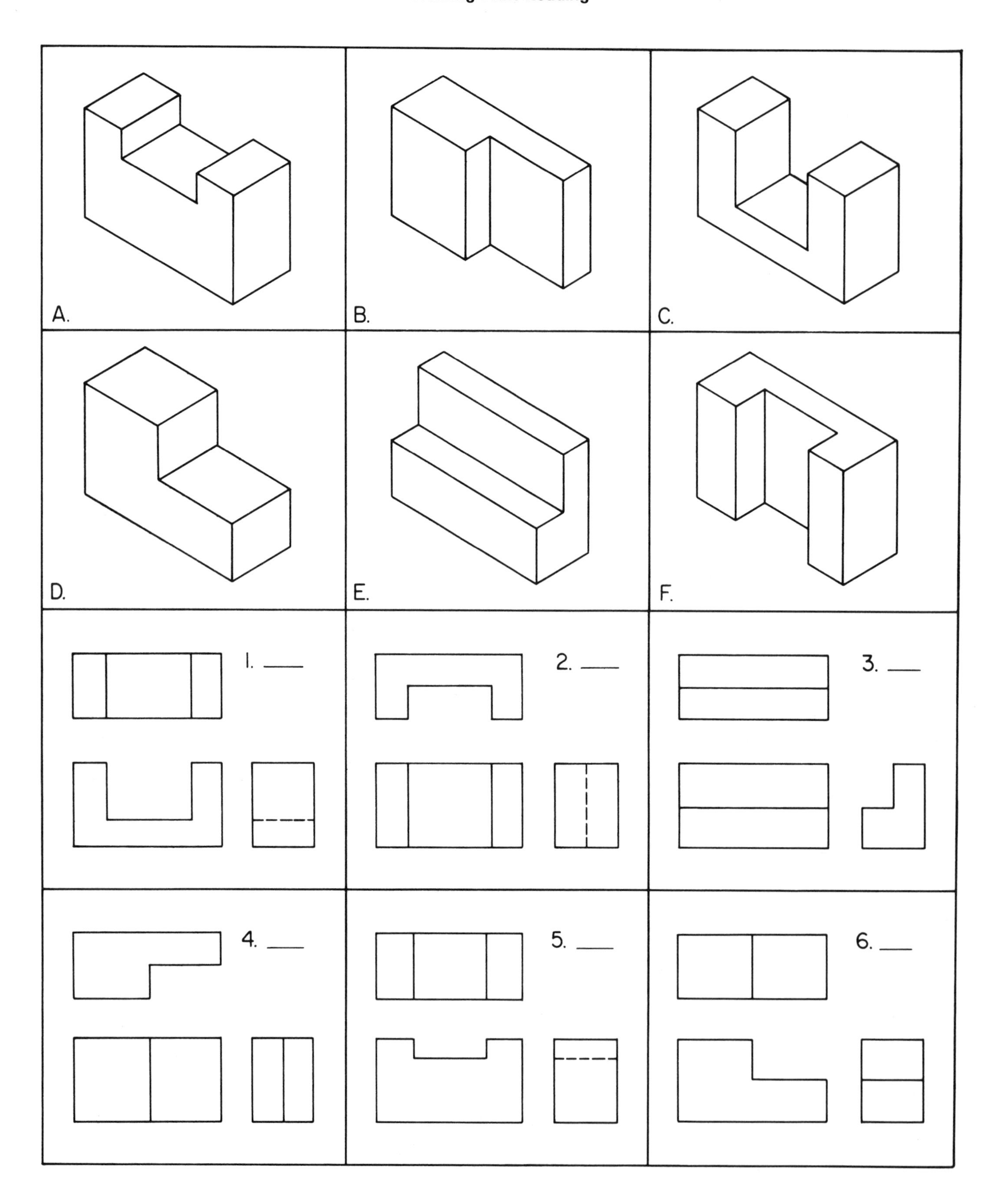

Part V

Study the pictorial views and match each orthographic drawing with its pictorial drawing by inserting the correct letter in the space provided.

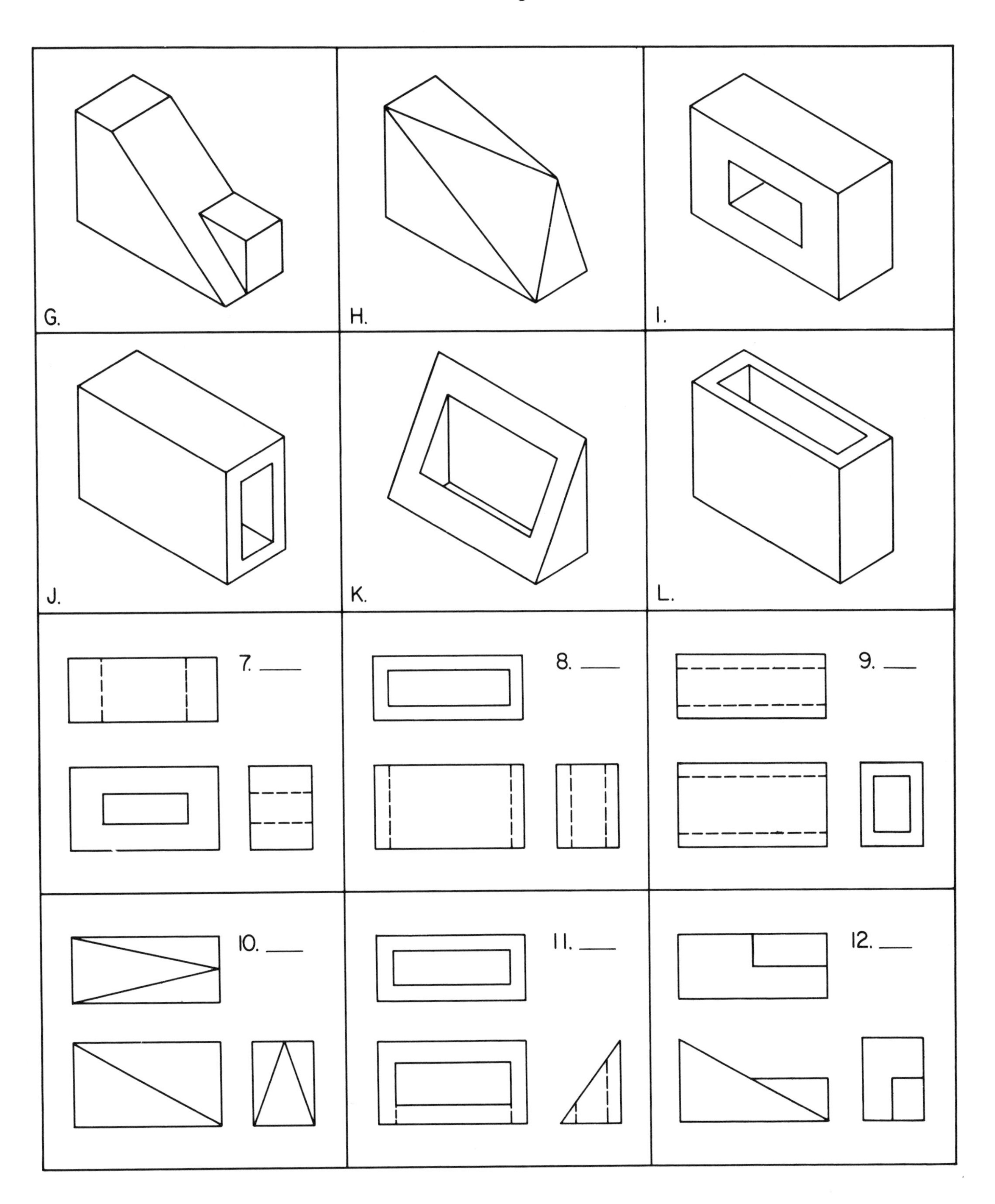
G.
H.
I.
J.
K.
L.
7. ___
8. ___
9. ___
10. ___
11. ___
12. ___

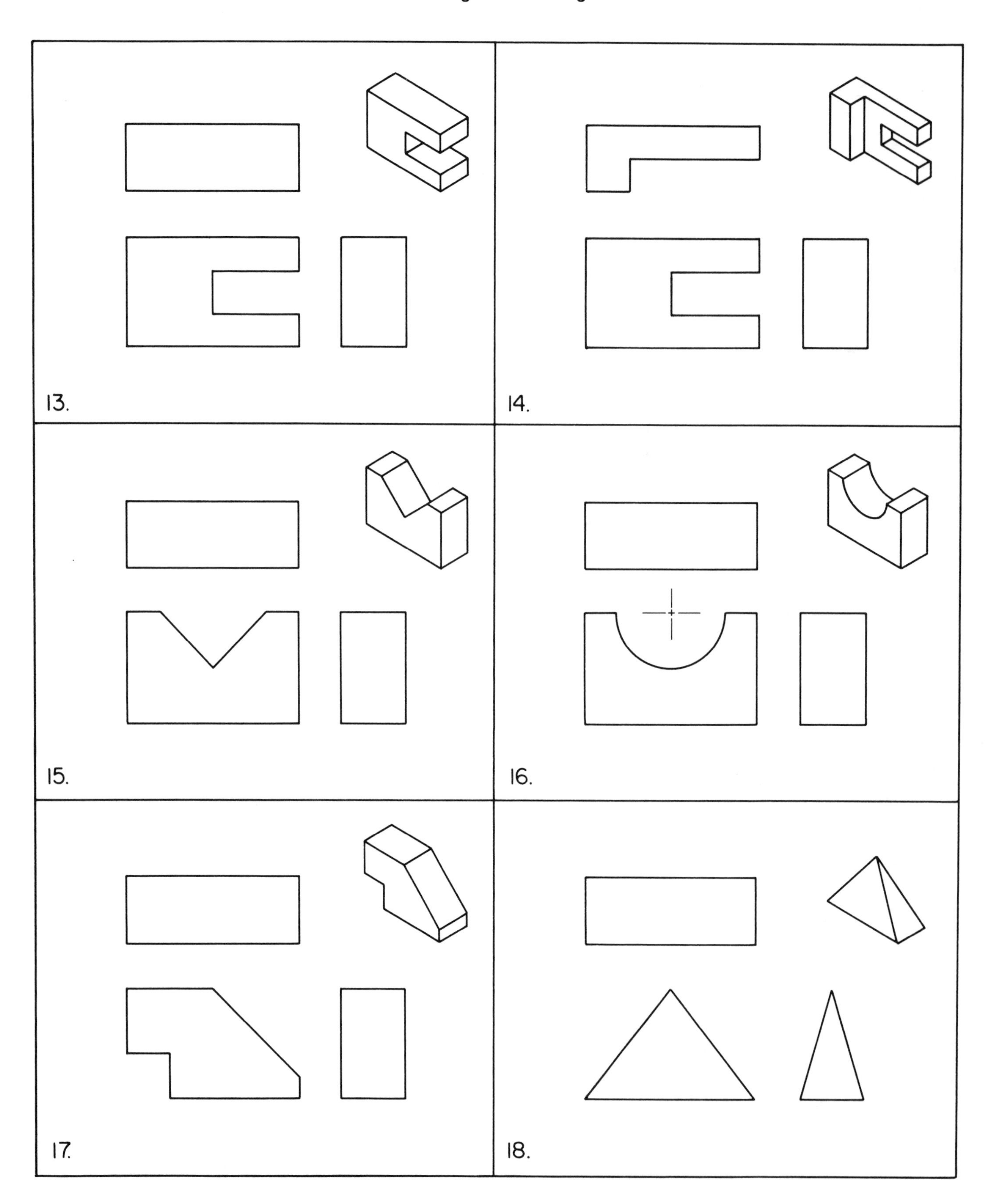

Part VI

Sketch in the correct missing lines in the orthographic projection drawings.

Unit 11

COMMON TYPES OF WELDS AND JOINTS

This chapter reviews basic types of welds and joints. It is important for you to be able to visualize weld and joint types before studying weld symbols in the next chapter.

WELD TYPES

Each welding job will require one or more of the welds shown in Fig. 11-1. Study each carefully.

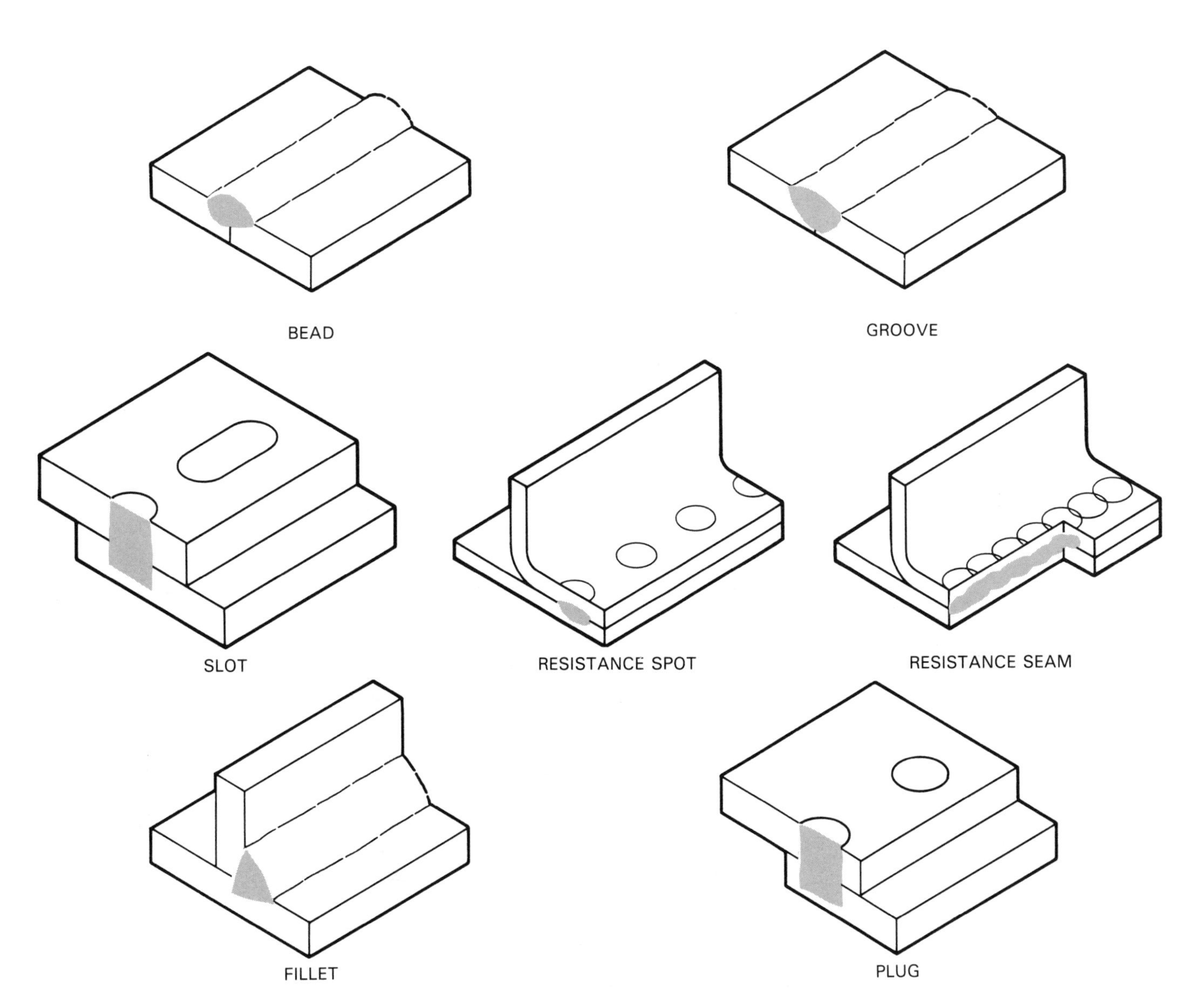

Fig. 11-1. One or more of these welds will be used on a weldment.

The *bead weld* is composed of a narrow layer or layers of metal deposited in an unbroken puddle on the surface of the metal.

A *groove weld* is a weld made in a groove on one or both surfaces to be joined.

The *fillet weld* is approximately triangular in shape and is used when joining two surfaces at an angle.

Plug and *slot welds* are welds made THROUGH one piece of metal to join it to another piece of metal. The opening may be partially or completely filled with weld.

A *spot weld* is an individually formed weld where the shape and size of the weld nugget is limited by the size and contour of the welding electrodes.

The *seam weld* is a series of overlapping spot welds made progressively along the joint by rotating electrodes.

JOINT TYPES

A *basic joint* is a way of arranging metal pieces in relation to one another so they can be welded. Common

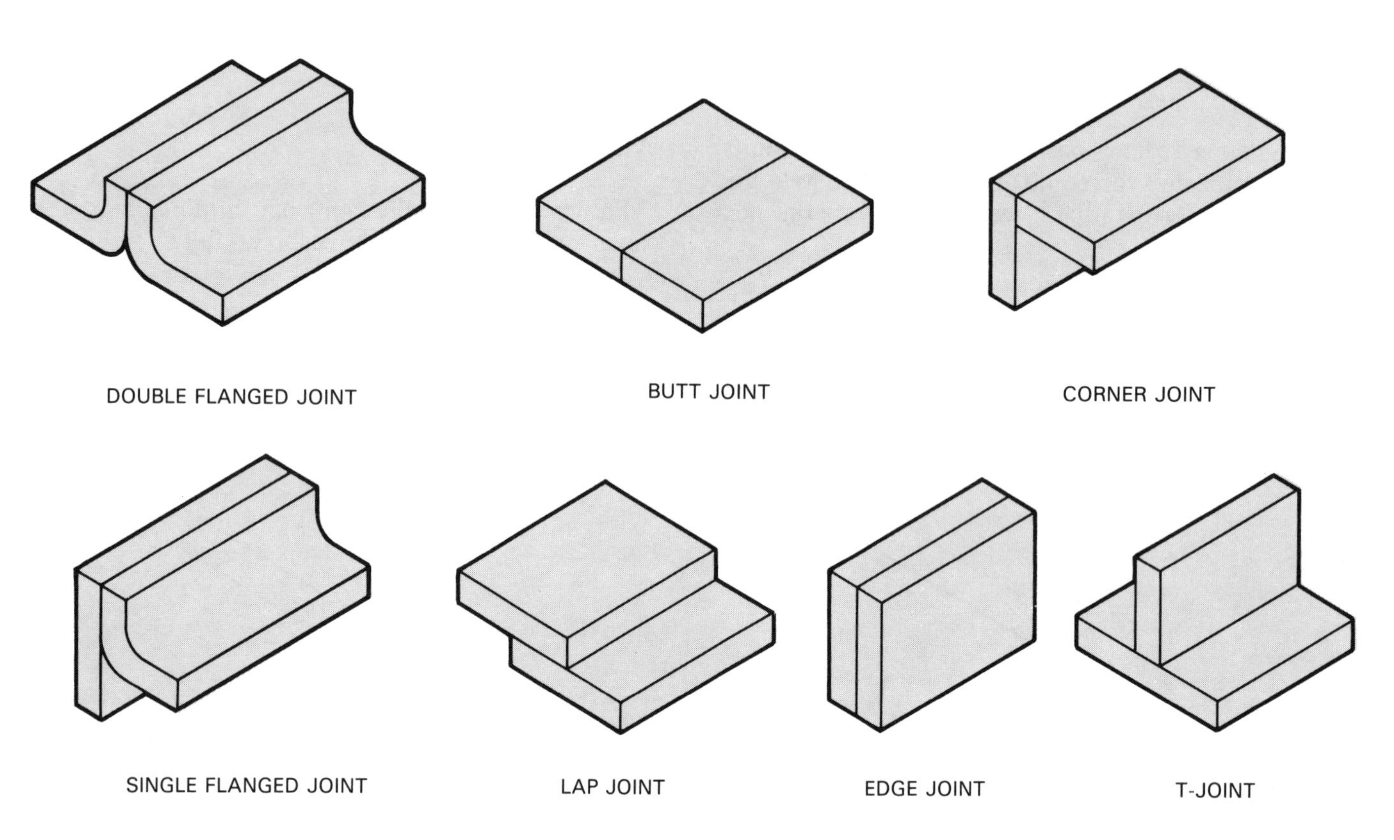

Fig. 11-2. Study names and shapes of basic joints used in welding.

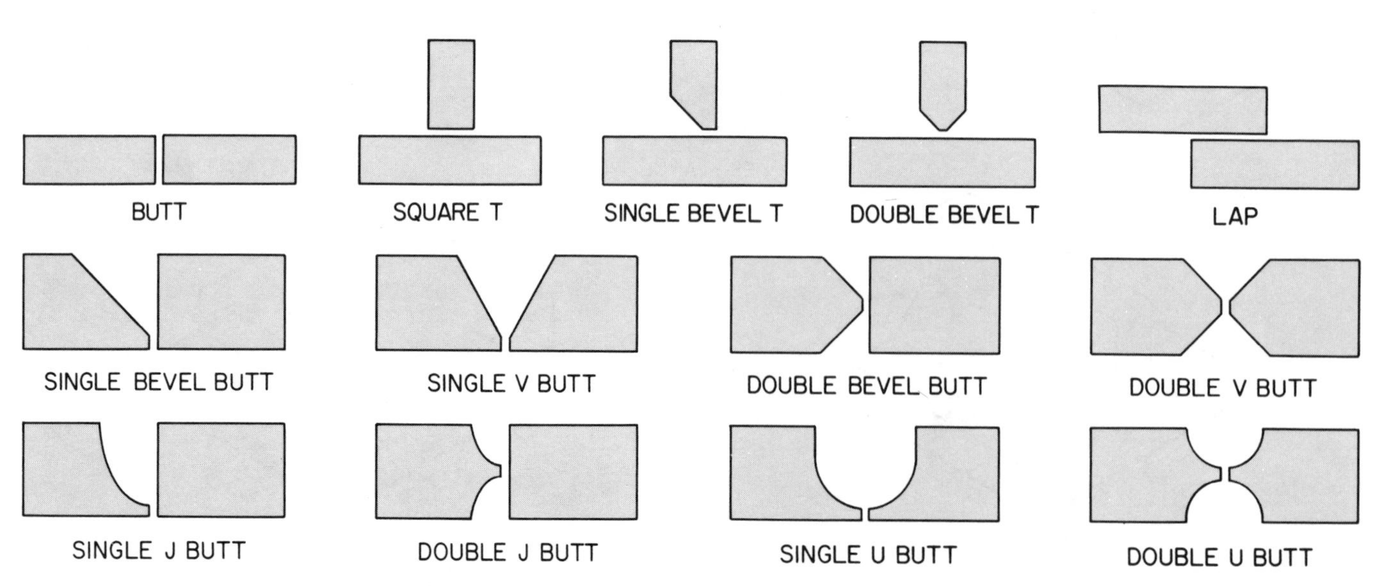

Fig. 11-3. You must often combine joint with one of these groove styles.

joints are shown in Fig. 11-2. Other joints can be made by combinations and variations of the basic joint designs.

GROOVES

To assure a solid weld, it is frequently necessary to combine the joint with one of the groove styles shown in Fig. 11-3. The *groove* is the opening provided between the two meteal pieces being joined by a groove weld.

UNIT 11—TEST YOUR KNOWLEDGE

Identify the following types of welds and joints.

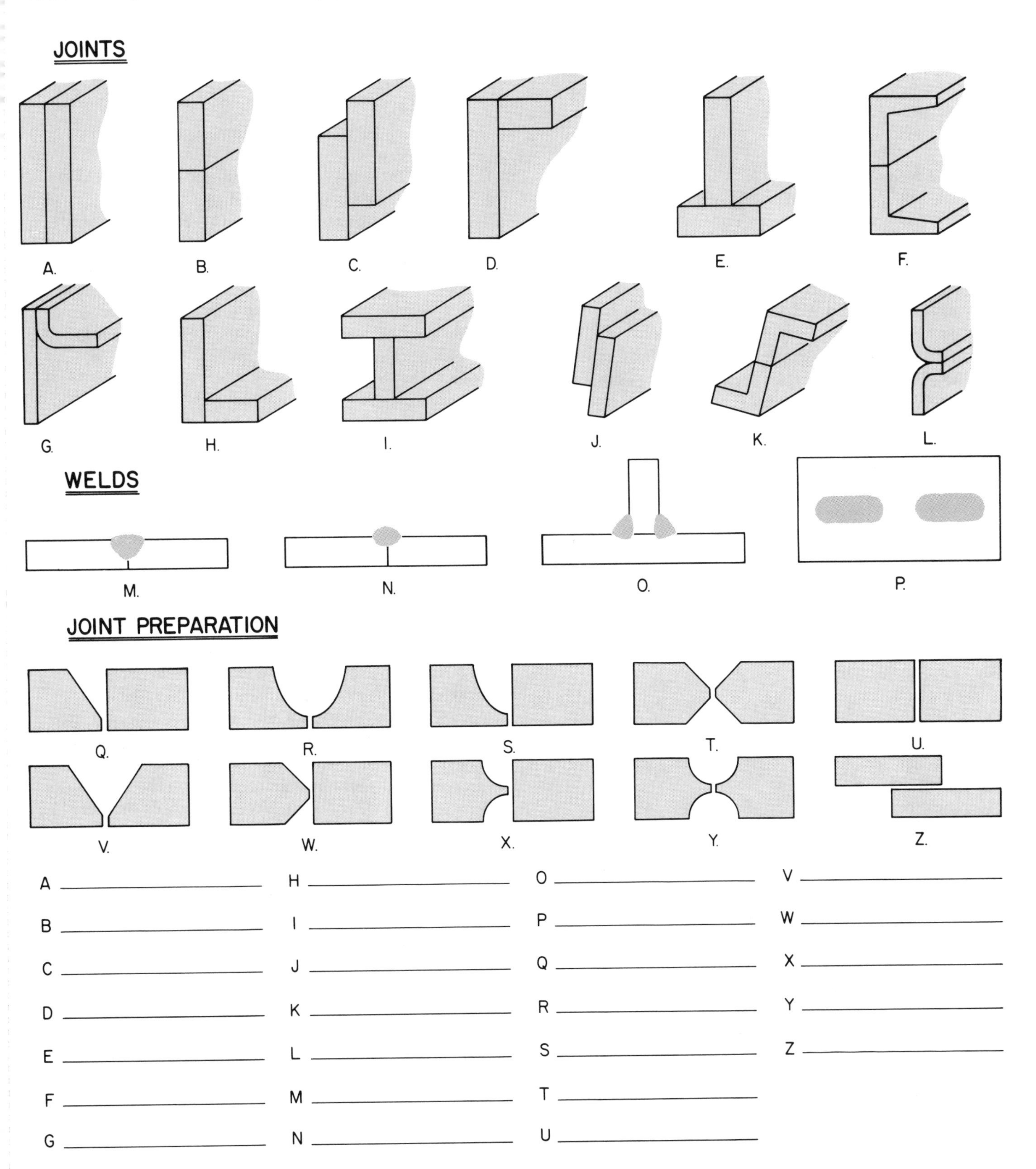

Unit 12

WELDING SYMBOLS

To make welds indicated on a drawing, the welder must be able to interpret welding symbols. Symbols are used to condense a large quantity of information about the weld into a small amount of space. This is shown in Fig. 12-1.

A *welding symbol* gives all of the important information for making a weld. Welding symbols simplify communications between engineers who design the product and shop personnel who must fabricate the product. Used on drawings, they contain the data needed to "tell" a welder the exact type of weld wanted by the designer or engineer.

ELEMENTS OF A WELDING SYMBOL

A welding symbol is a graphic assembly of the elements needed to fully specify weld requirements. See Fig. 12-2. Unless needed for clarity, all of the elements do NOT have to be used on every welding symbol.

The eight possible elements of a welding symbol include:

A. Reference line.
B. Basic weld symbol.
C. Arrow.
D. Tail.
E. Supplementary symbols.
F. Dimensions.
G. Finish symbols.
H. Notations pretaining to the process, filler metal, and any related standards.

Reference line

The *reference line* is the "backbone" (required central element) of the welding symbol. It is always shown in a horizontal position. Other elements describing weld requirements are located on, above, below, and/or at either end of the reference line.

Basic weld symbol

The basic WELD SYMBOL should be differentiated from the WELDING SYMBOL. The *basic weld symbol* depicts the cross-sectional shape of the weld or joint. It is one part of the welding symbol.

Basic weld symbols are shown in Fig. 12-3. Study them carefully!

Either a basic weld symbol or notations in the tail are mandatory on a welding symbol. In some situations, however, both elements may be included on a welding symbol to furnish complete weld specifications.

Arrow

An *arrow* connects the reference line of a welding symbol to one side of the joint to be welded. The shape and location of the arrow are quite important. The use of the arrow is mandatory.

The placement of FILLET, GROOVE, and FLANGE welds is indicated by running the arrow from the reference line to one side of the required weld.

The *arrow side* is on the lower side of the reference line. It indicates the same or near side of the joint. See Fig. 12-4.

The *other side* is on the upper side of the reference line. It is opposite the arrow side. The other side indicates the far side of the joint. See Fig. 12-4.

A weld symbol BELOW the reference line signifies that the weld should be on the arrow side (same side) of the joint. When the weld symbol is ABOVE the reference line, a weld in only required on the other side (opposite side) of the joint. Weld symbols on BOTH sides of the reference line indicate welds are needed on both the arrow side and other side of the joint. Refer to Fig. 12-5.

The arrow for BEVEL and J-GROOVE welds uses a *bent leader* and the arrow head points to the particular section to be machined, Fig. 12-6.

When PLUG, SLOT, SPOT, SEAM, and PROJECTION welds are required, the arrow points to the outer surface of one of the joint members at the center of the desired weld. This member is considered the *arrow side member*. The other section of the joint is the *other side member*. See Fig. 12-7.

Resistance welds have NO arrow side or other side significance because the weld is made at the interface

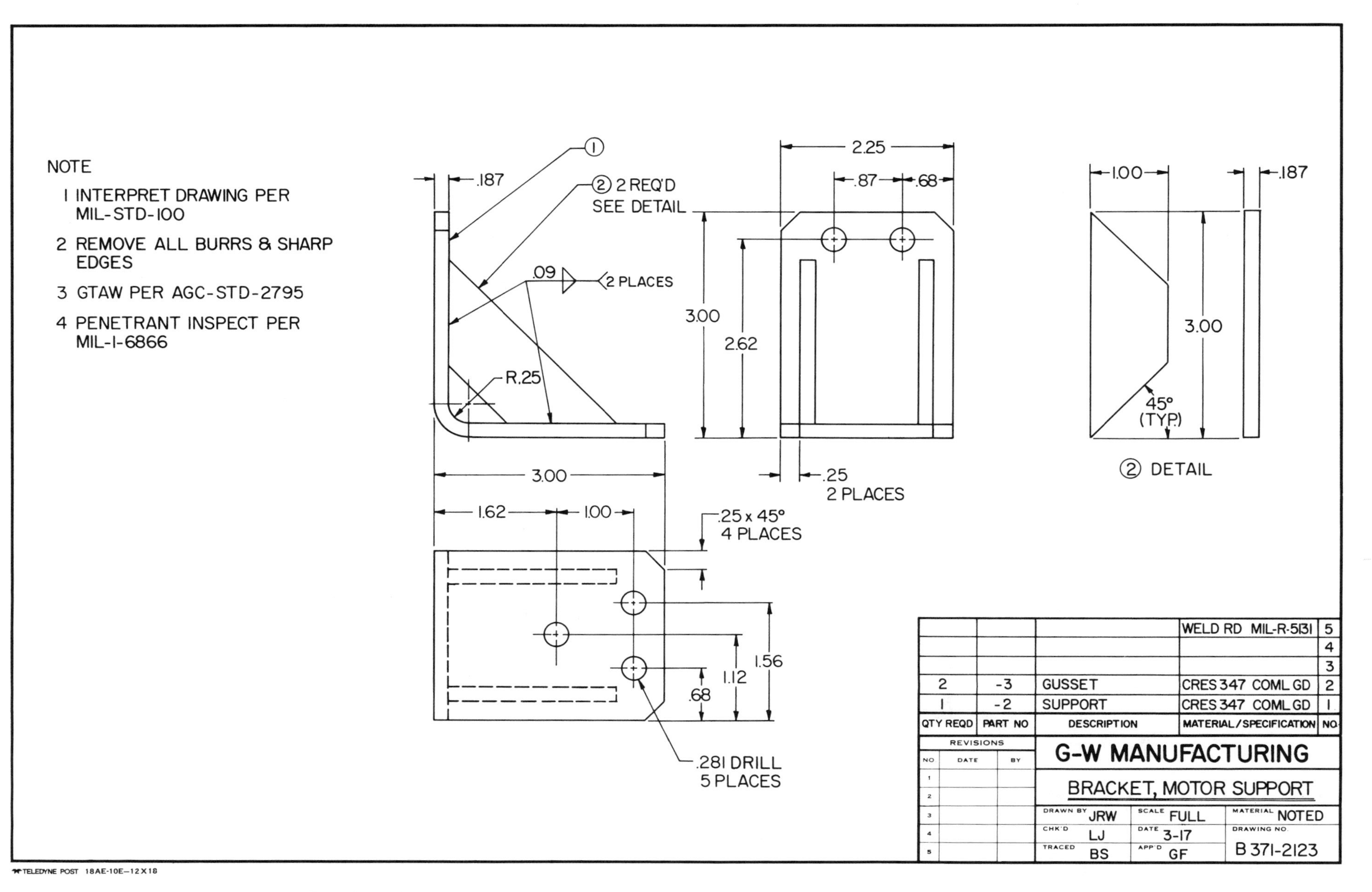

Fig. 12-1. To make welds indicated on this drawing, welder must be able to "read" symbols that furnish weld specifications.

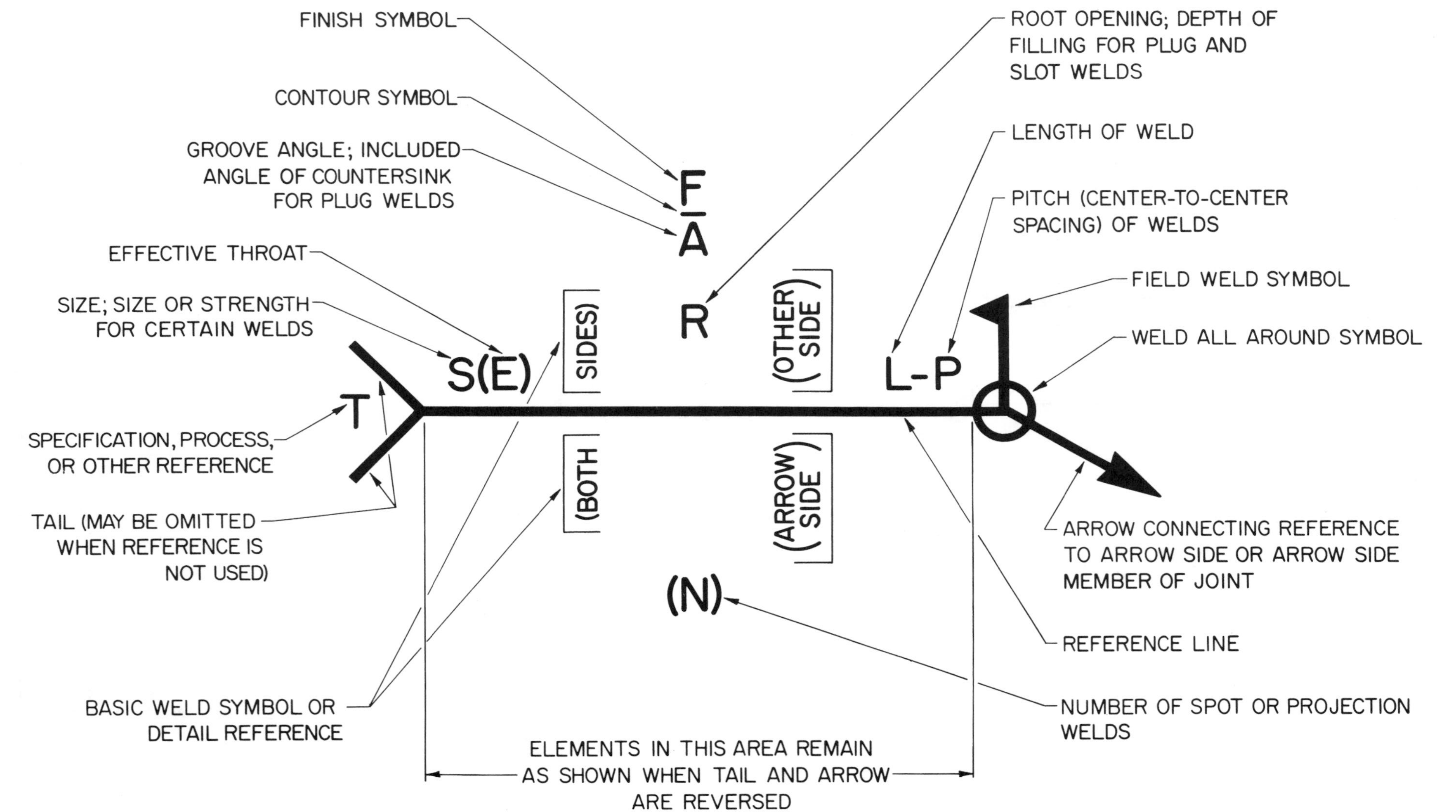

Fig. 12-2. Study elements of welding symbol carefully. It is a graphic explanation needed to fully specify weld requirements.

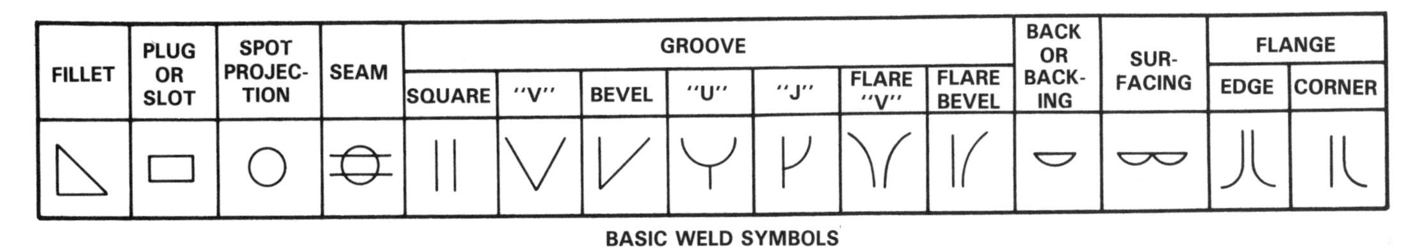

BASIC WELD SYMBOLS

Fig. 12-3. Memorize names and shapes of basic weld symbols.

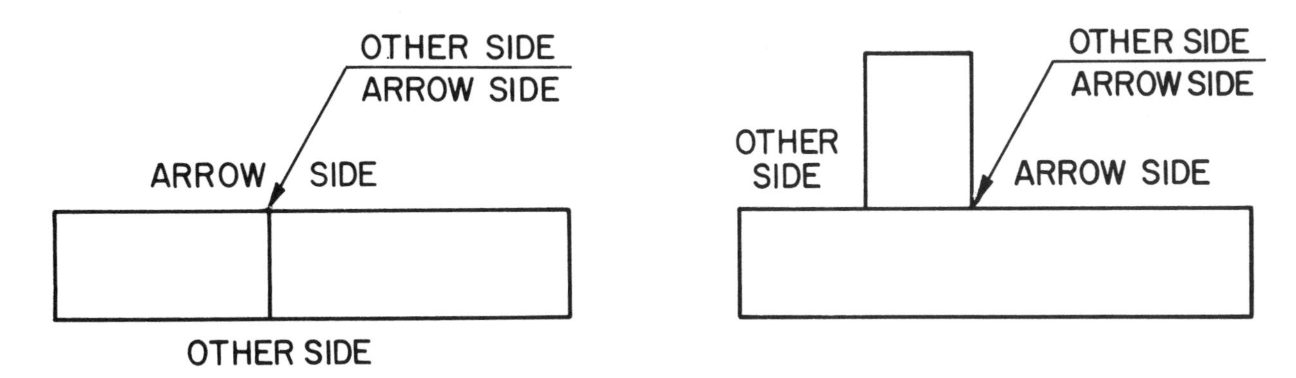

Fig. 12-4. Study how arrow side and other side welds are specified.

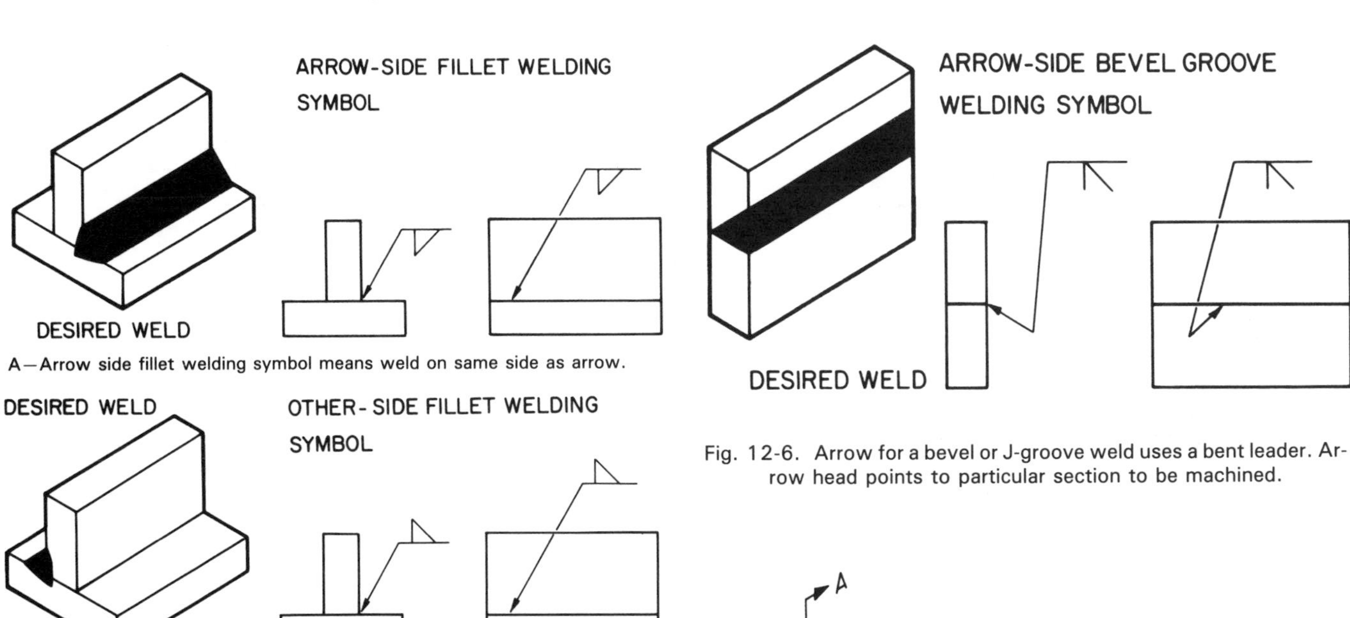

A—Arrow side fillet welding symbol means weld on same side as arrow.

B—Other side fillet welding symbol means weld on opposite side of arrow.

Fig. 12-6. Arrow for a bevel or J-groove weld uses a bent leader. Arrow head points to particular section to be machined.

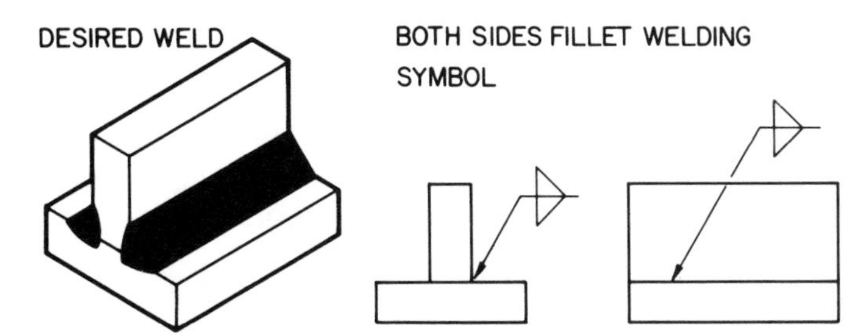

C—This welding symbol means weld both sides.

Fig. 12-5. Note how locations of weld symbol above or below reference line show where to weld.

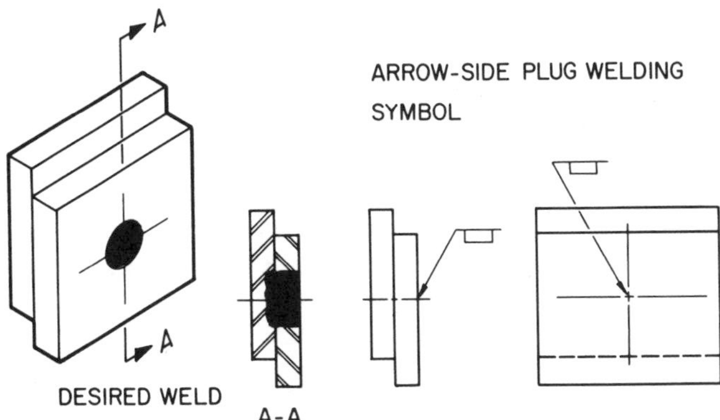

Fig. 12-7. On plug, slot, spot, seam, and projection welds, arrow points to outer surface of one of joint members at center of weld. This member is considered arrow side member.

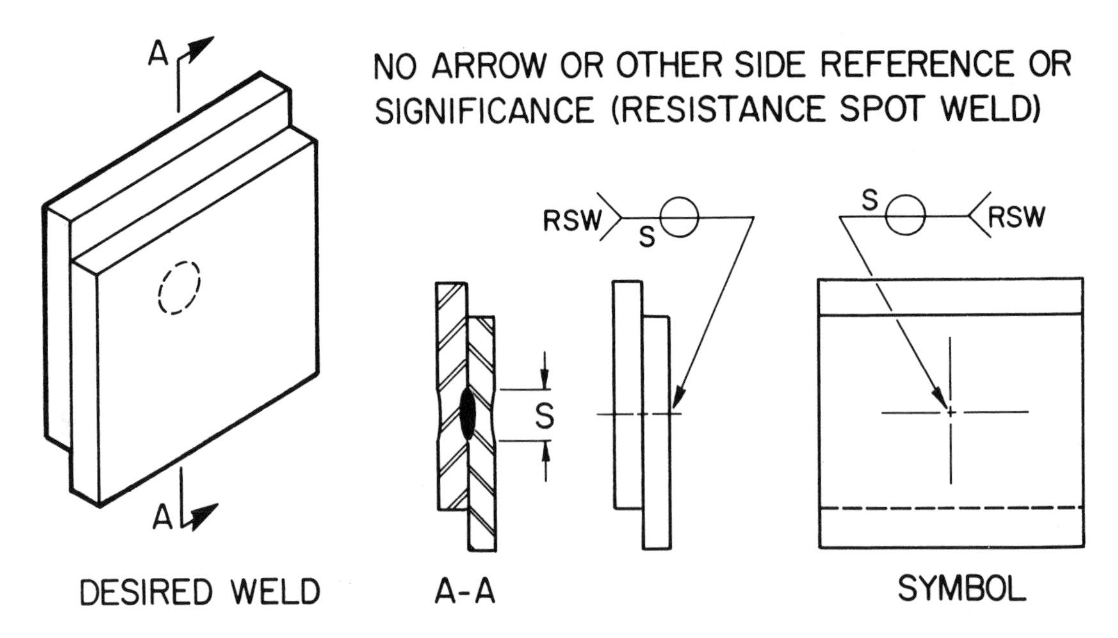

Fig. 12-8. Resistance weld is made at interface of members. It has no arrow side or other side significance.

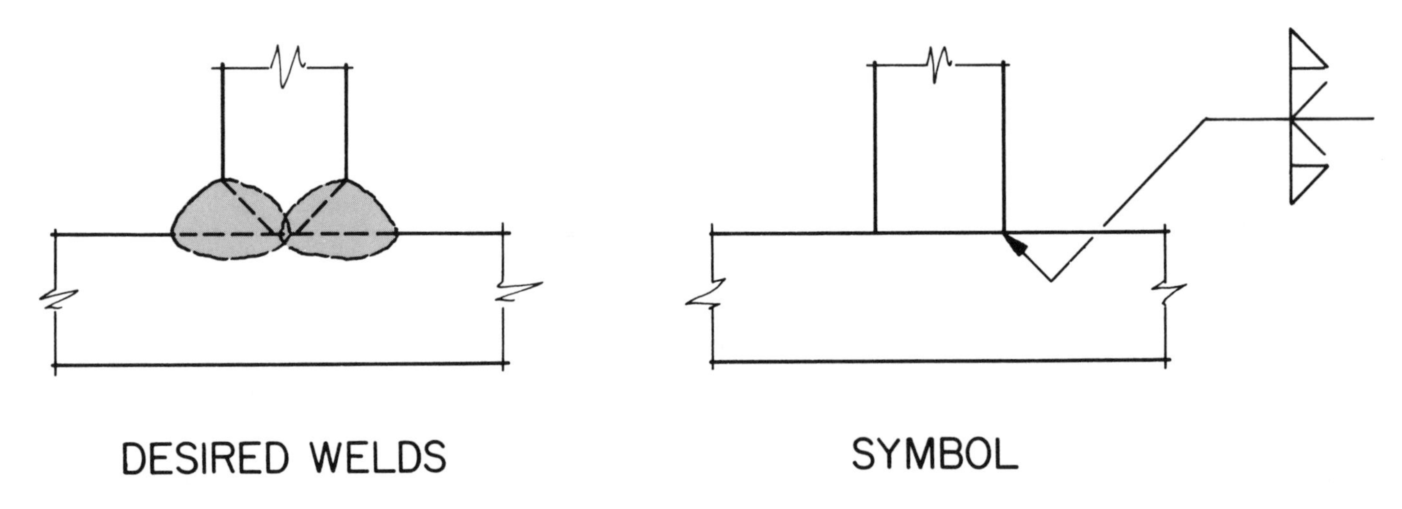

Fig. 12-9. Joint with more than one type weld will have a weld symbol for each weld. In this example, there is a bevel weld and a fillet weld on each side of joint.

of the members. Refer to Fig. 12-8.

Joints having more than one type weld will have a weld symbol for each weld. See Fig. 12-9.

A chart showing basic welding symbols and their location significance is given in Fig. 12-10. Learn to name each weld symbol and explain its location on the joint.

Tail

The *tail* contains notes pertaining to the process, filler metal, and any related standards needed to establish specific weld requirements. See Fig. 12-11. If notations are NOT used, the tail element may be omitted.

SUPPLEMENTARY SYMBOLS

Supplementary symbols are often included with basic weld symbols to provide more specific weld data not provided by other elements in the welding symbol. Supplementary weld symbols are shown in Fig. 12-12.

The *weld-all-around symbol* signifies that the weld is to be made completely around the joint without interruption. See Fig. 12-13.

A *field weld symbol,* Fig. 12-14, indicates the weld is NOT done where the unit is initially made, but when in the field.

The *backing symbol* indicates a bead type backing weld on the opposite side of the regular weld. This is

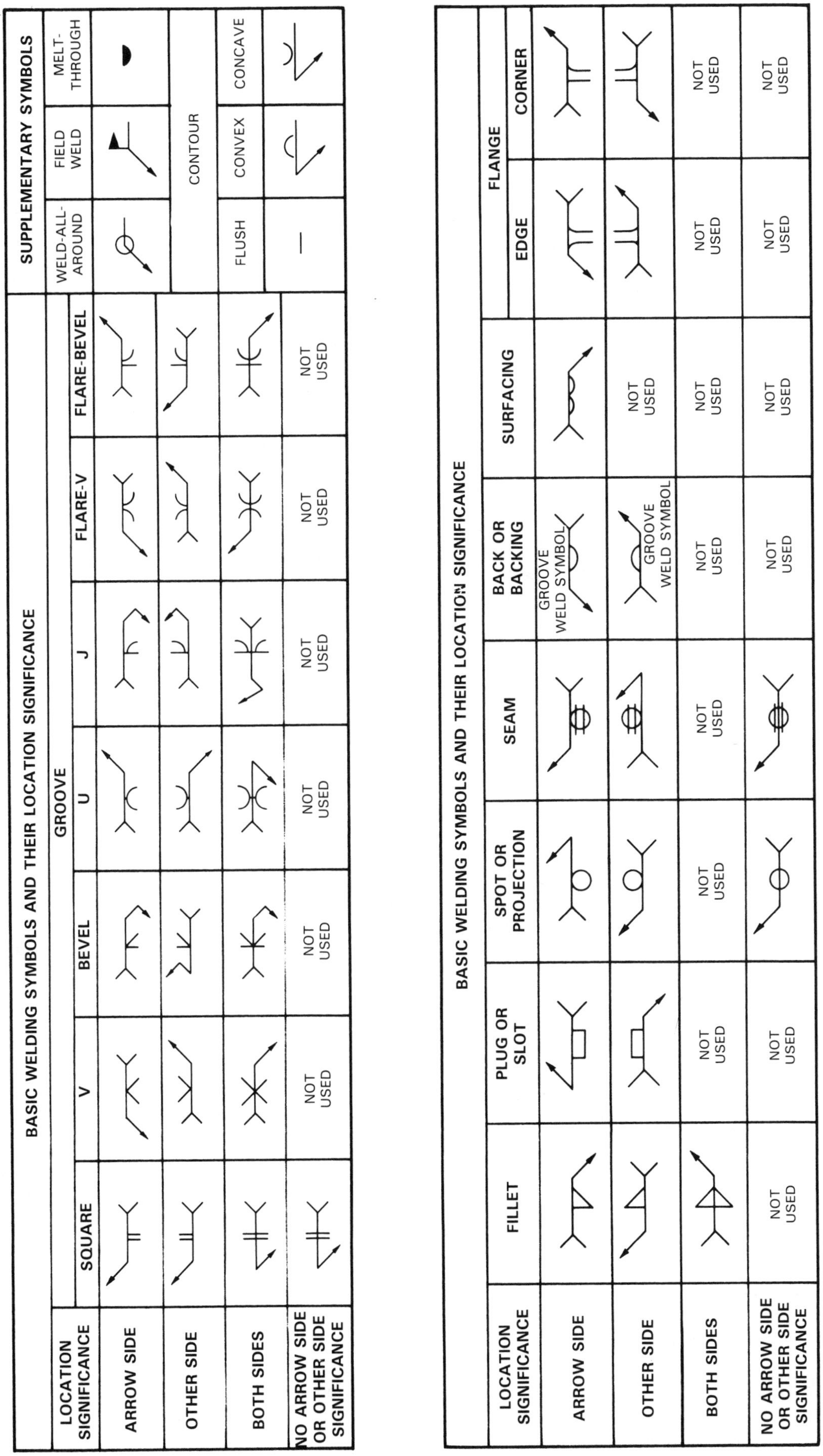

BASIC WELDING SYMBOLS AND THEIR LOCATION SIGNIFICANCE

LOCATION SIGNIFICANCE	GROOVE						
	SQUARE	V	BEVEL	U	J	FLARE-V	FLARE-BEVEL
ARROW SIDE							
OTHER SIDE							
BOTH SIDES							
NO ARROW SIDE OR OTHER SIDE SIGNIFICANCE		NOT USED	NOT USED	NOT USED	NOT USED	NOT USED	NOT USED

SUPPLEMENTARY SYMBOLS

WELD-ALL-AROUND	FIELD WELD	MELT-THROUGH
CONTOUR		
FLUSH	CONVEX	CONCAVE

BASIC WELDING SYMBOLS AND THEIR LOCATION SIGNIFICANCE

LOCATION SIGNIFICANCE	FILLET	PLUG OR SLOT	SPOT OR PROJECTION	SEAM	BACK OR BACKING	SURFACING	FLANGE	
							EDGE	CORNER
ARROW SIDE					GROOVE WELD SYMBOL			
OTHER SIDE					GROOVE WELD SYMBOL	NOT USED		
BOTH SIDES		NOT USED	NOT USED	NOT USED	NOT USED	NOT USED	NOT USED	NOT USED
NO ARROW SIDE OR OTHER SIDE SIGNIFICANCE	NOT USED	NOT USED			NOT USED	NOT USED	NOT USED	NOT USED

Fig. 12-10. Study basic welding symbols and their location significance carefully.

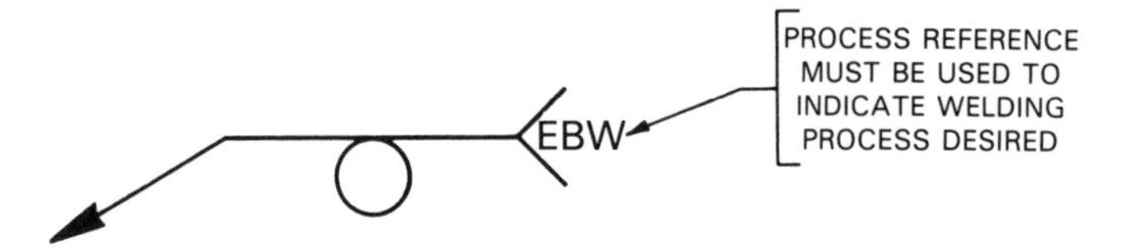

Fig. 12-11. Tail of welding symbol may contain notes on process, filler metal, and/or any related standards for weld requirements. This one denotes electron beam welding.

WELD-ALL-AROUND	FIELD WELD	MELT-THROUGH	CONTOUR		
			FLUSH	CONVEX	CONCAVE
			—		

Fig. 12-12. Study supplementary weld symbols.

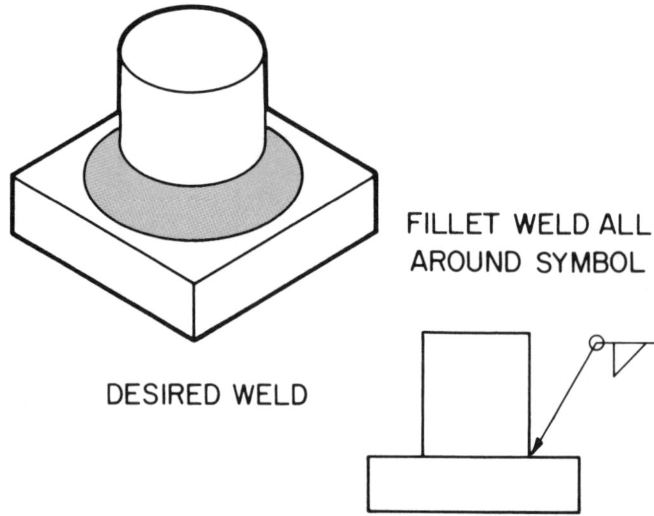

Fig. 12-13. Note fillet weld-all-around symbol.

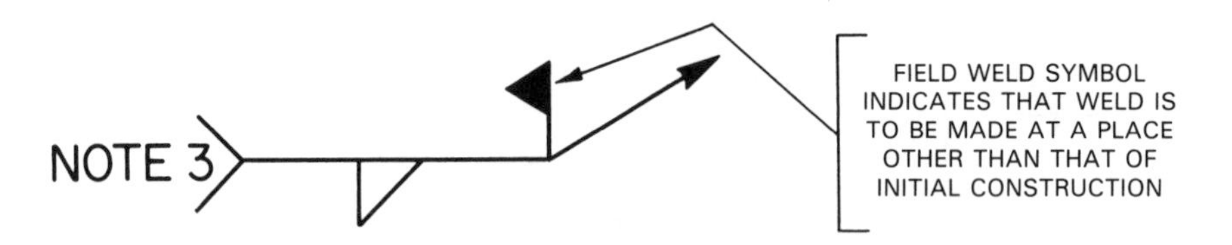

Fig. 12-14. Welding symbol denoting that weld must be made in field.

pictured in Fig. 12-15.

The *melt-through symbol* is used when complete joint penetration is required in a weld made from only one side. This is illustrated in Fig. 12-16. It insures 100 percent weld penetration.

A *surfacing symbol* indicates the surface is to be built up by single or multiple pass welding, as in Fig. 12-17. The symbol is always shown BELOW the reference line. The height of the built up surface is indicated to the LEFT of the surfacing symbol.

DIMENSIONS

Weld dimensions may be indicated in inches/fractions of an inch, or in millimeters (mm). Angles are specified in degree.

Weld size is placed on the LEFT SIDE of the weld symbol. *Weld length* is shown on the RIGHT SIDE of the weld symbol. NO length dimension is given when the weld is to be made the FULL LENGTH of the joint. See Fig. 12-18.

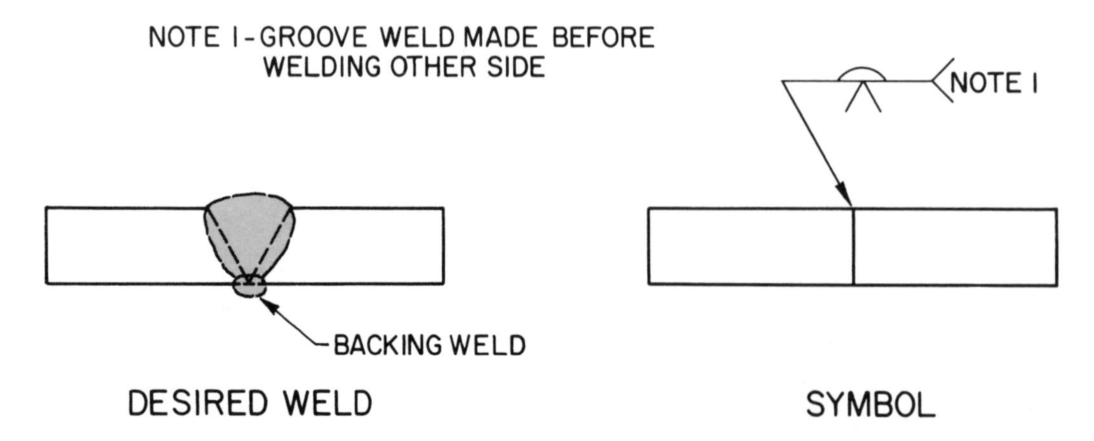

Fig. 12-15. Backing weld symbol means a bead type backing weld is to be made on opposite side of regular weld.

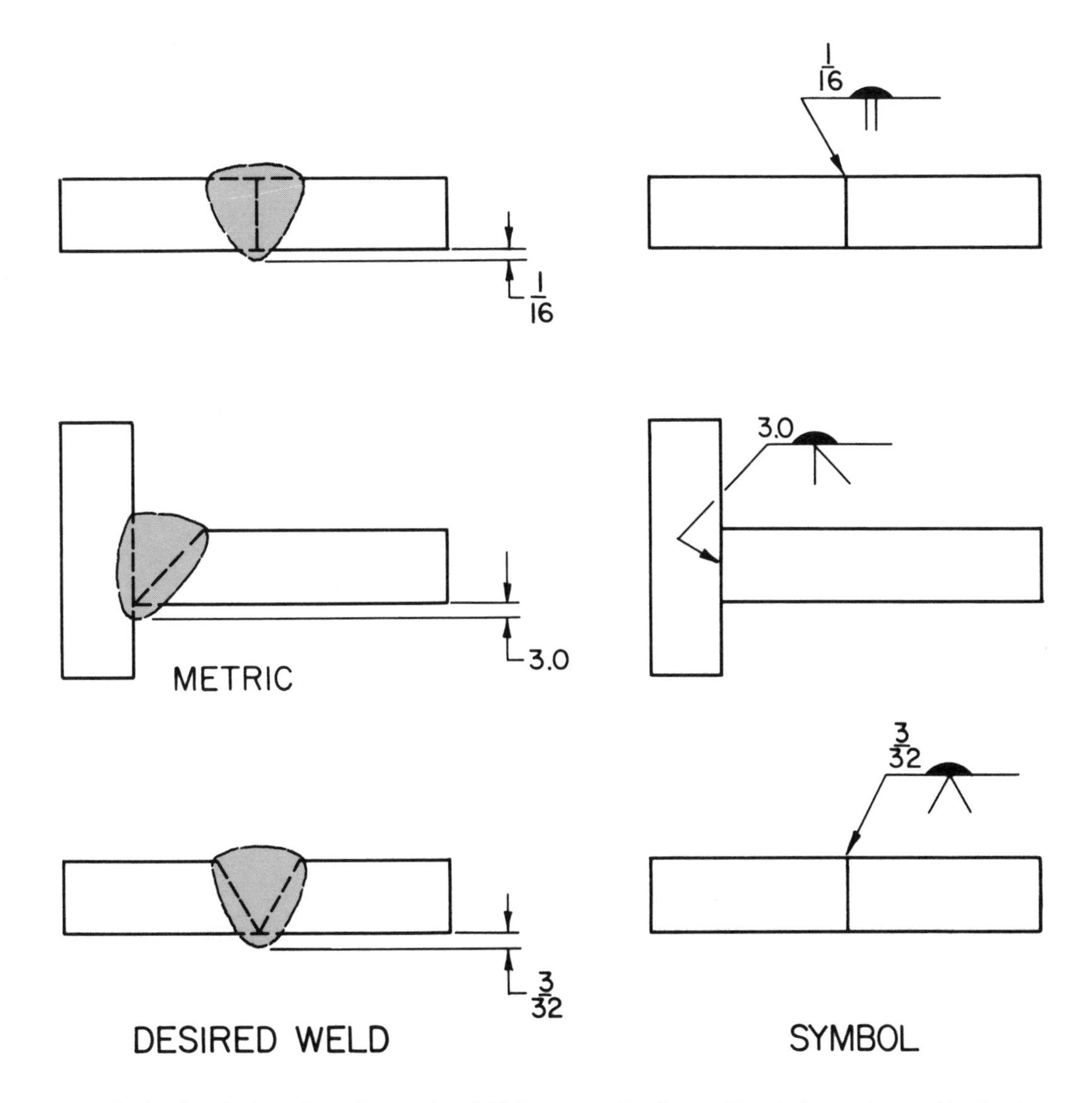

Fig. 12-16. A melt-through symbol requires full joint penetration in a weld made from only one side. Number denotes amount of penetration.

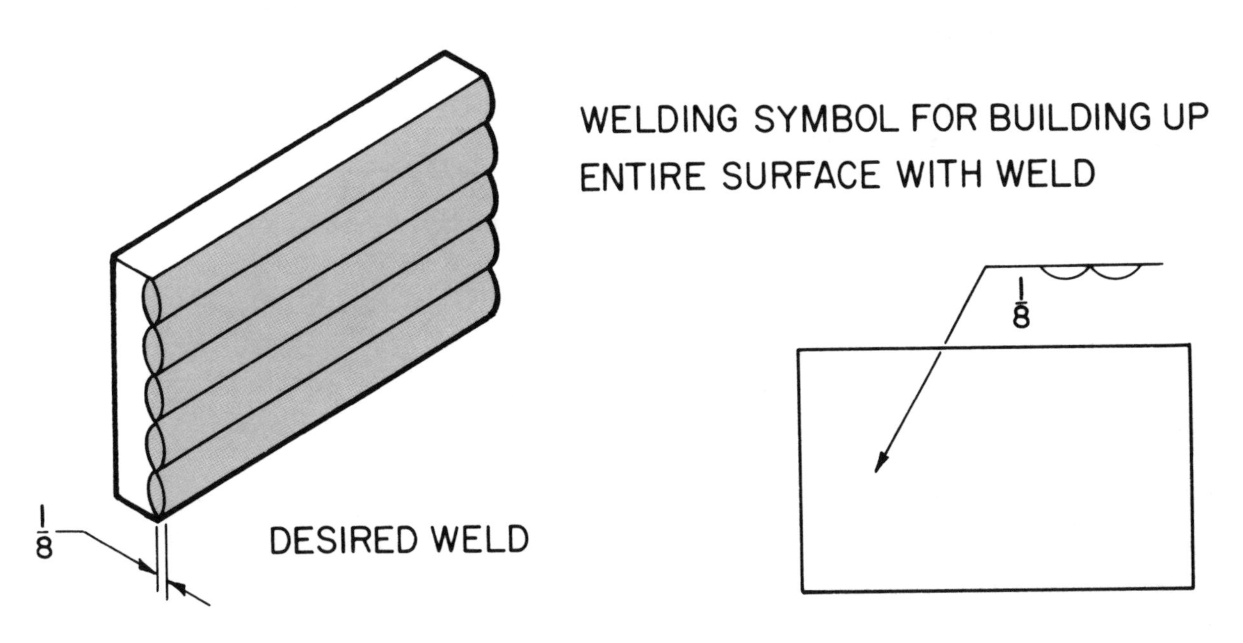

Fig. 12-17. Surfacing symbol means cover entire surface with weld as shown.

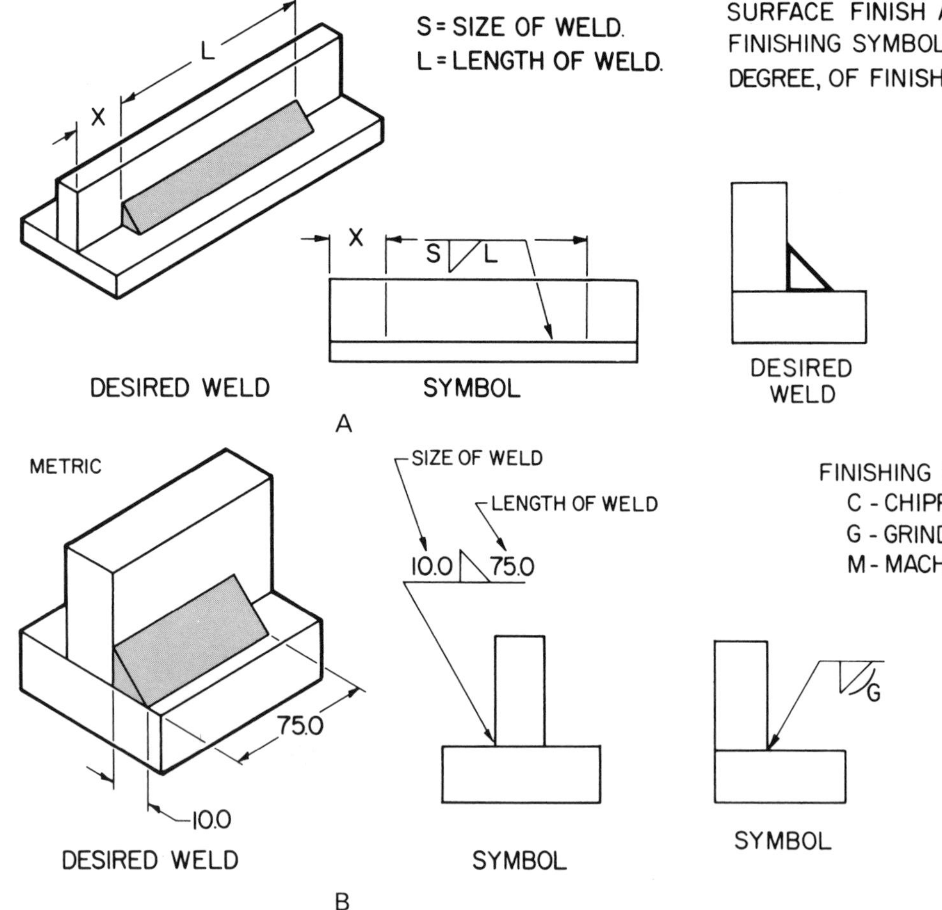

Fig. 12-18. Weld size is placed on left side of weld symbol. Weld length is shown on right side of weld symbol.

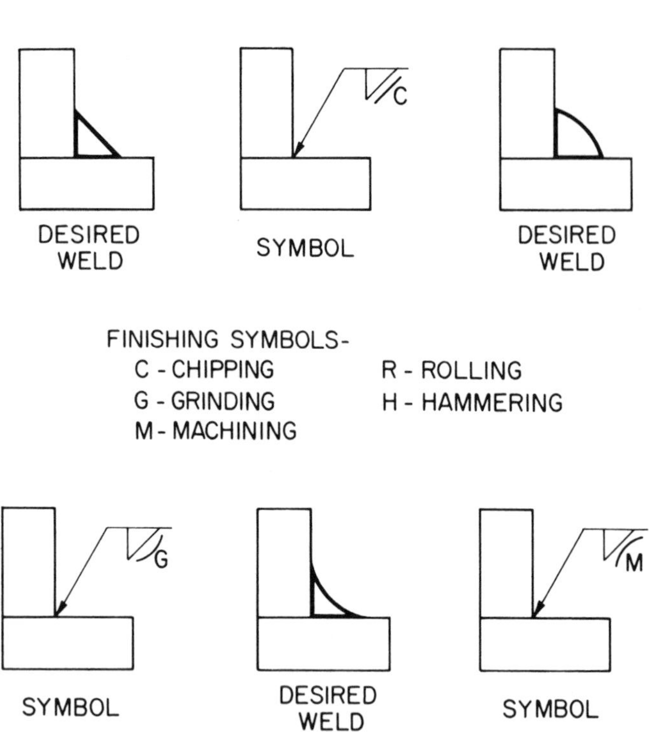

Fig. 12-19. Study contour and method of finish symbols.

Information on weld dimensions, as they refer to specific types of welds, is included in the textbook units that follow.

CONTOUR AND FINISH SYMBOLS

A *contour symbol* is used with the weld symbol when the finished shape of the weld is important. If a weld is to be contoured or finished (other than being cleaned), a *method of finish symbol* will be included with the contour symbol. See Fig. 12-19.

NOTATIONS

A *notation* is information called out in the tail of a welding symbol. It is often in the form of abbreviations or notes. If there is insufficient room in the tail for all of the notation, reference is made in the tail to where a note covering this information will be found elsewhere on the print. Fig. 12-20 gives an example of a notation.

For the table DESIGNATION OF WELDING AND RELATED PROCESSES BY LETTERS, refer to Fig. 12-21. Study this table carefully!

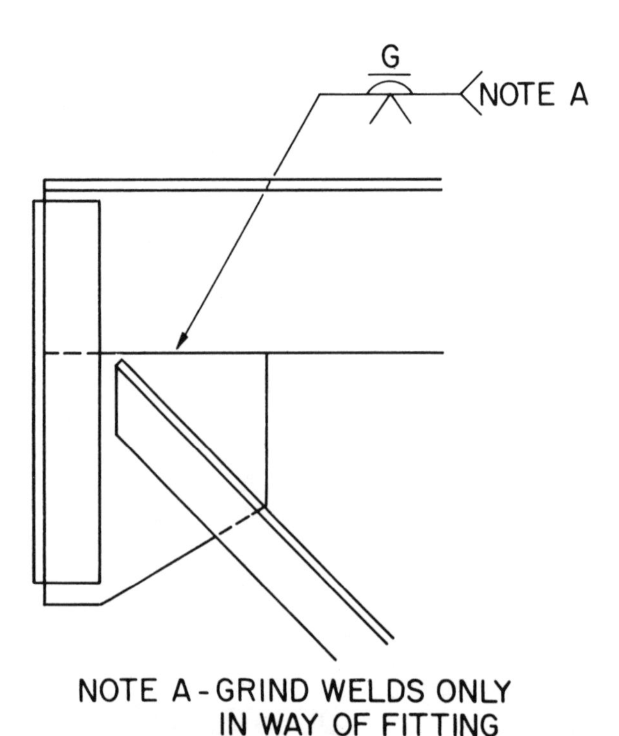

Fig. 12-20. Note used with welding symbol.

SUFFIXES FOR OPTIONAL USE IN APPLYING WELDING AND ALLIED PROCESSES

Automatic	AU	Manual	MA
Machine	ME	Semiautomatic	SA

DESIGNATION OF WELDING AND ALLIED PROCESSES BY LETTERS

LETTER DESIGNATION	WELDING AND ALLIED PROCESSES	LETTER DESIGNATION	WELDING AND ALLIED PROCESSES
AAC	air carbon arc cutting	GTAW	gas tungsten arc welding
AAW	air acetylene welding	GTAW-P	gas tungsten arc welding — pulsed arc
ABD	adhesive bonding	HFRW	high frequency resistance welding
AB	arc brazing	HPW	hot pressure welding
AC	arc cutting	IB	induction brazing
AHW	atomic hydrogen welding	INS	iron soldering
AOC	oxygen arc cutting	IRB	infrared brazing
AW	arc welding	IRS	infrared soldering
B	brazing	IS	induction soldering
BB	block brazing	IW	induction welding
BMAW	bare metal arc welding	LBC	laser beam cutting
CAC	carbon arc cutting	LBW	laser beam welding
CAW	carbon arc welding	LOC	oxygen lance cutting
CAW-G	gas carbon arc welding	MAC	metal arc cutting
CAW-S	shielded carbon arc welding	OAW	oxyacetylene welding
CAW-T	twin carbon arc welding	OC	oxygen cutting
CW	cold welding	OFC	oxyfuel gas cutting
DB	dip brazing	OFC-A	oxyacetylene cutting
DFB	diffusion brazing	OFC-H	oxyhydrogen cutting
DFW	diffusion welding	OFC-N	oxynatural gas cutting
DS	dip soldering	OFC-P	oxypropane cutting
EASP	electric arc spraying	OFW	oxyfuel gas welding
EBC	electron beam cutting	OHW	oxyhydrogen welding
EBW	electron beam welding	PAC	plasma arc cutting
ESW	electroslag welding	PAW	plasma arc welding
EXW	explosion welding	PEW	percussion welding
FB	furnace brazing	PGW	pressure gas welding
FCAW	flux cored arc welding	POC	metal powder cutting
FCAW-EG	flux cored arc welding — electrogas	PSP	plasma spraying
FLB	flow brazing	RB	resistance brazing
FLOW	flow welding	RPW	projection welding
FLSP	flame spraying	RS	resistance soldering
FOC	chemical flux cutting	RSEW	resistance seam welding
FOW	forge welding	RSW	resistance spot welding
FRW	friction welding	ROW	roll welding
FS	furnace soldering	RW	resistance welding
FW	flash welding	S	soldering
GMAC	gas metal arc cutting	SAW	submerged arc welding
GMAW	gas metal arc welding	SAW-S	series submerged arc welding
GMAW-EG	gas metal arc welding — electrogas	SMAC	shielded metal arc cutting
GMAW-P	gas metal arc welding — pulsed arc	SMAW	shielded metal arc welding
GMAW-S	gas metal arc welding — short circuiting arc	SSW	solid state welding
		SW	stud arc welding
GTAC	gas tungsten arc cutting	TB	torch brazing

Fig. 12-21. Study designation of welding and related processes by letter.

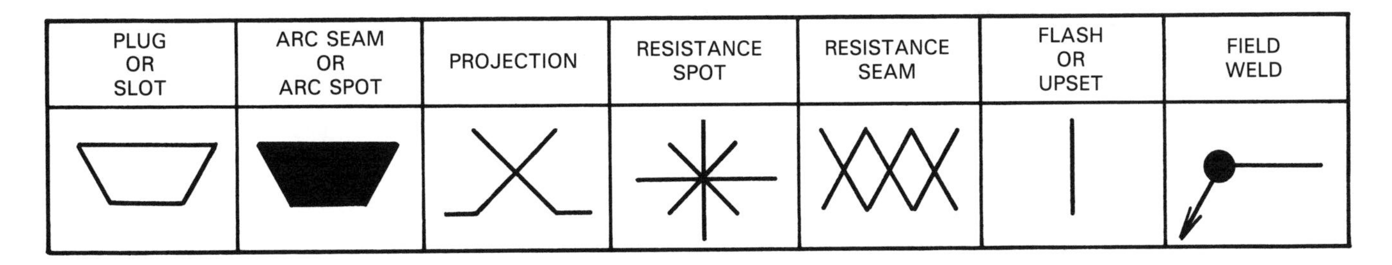

Fig.12-22. Nonpreferred weld symbols may be found on old prints.

NONPREFERRED WELD SYMBOLS

The *nonpreferred weld symbols* have been replaced by new symbols. They are NOT used on contemporary prints (prints drawn since new symbols were introduced). Nonpreferred weld symbols, however, may still be found on older prints and are included in Fig. 12-22 for reference purposes.

UNIT 12—TEST YOUR KNOWLEDGE

Part I

Draw the correct symbols for the following welds.

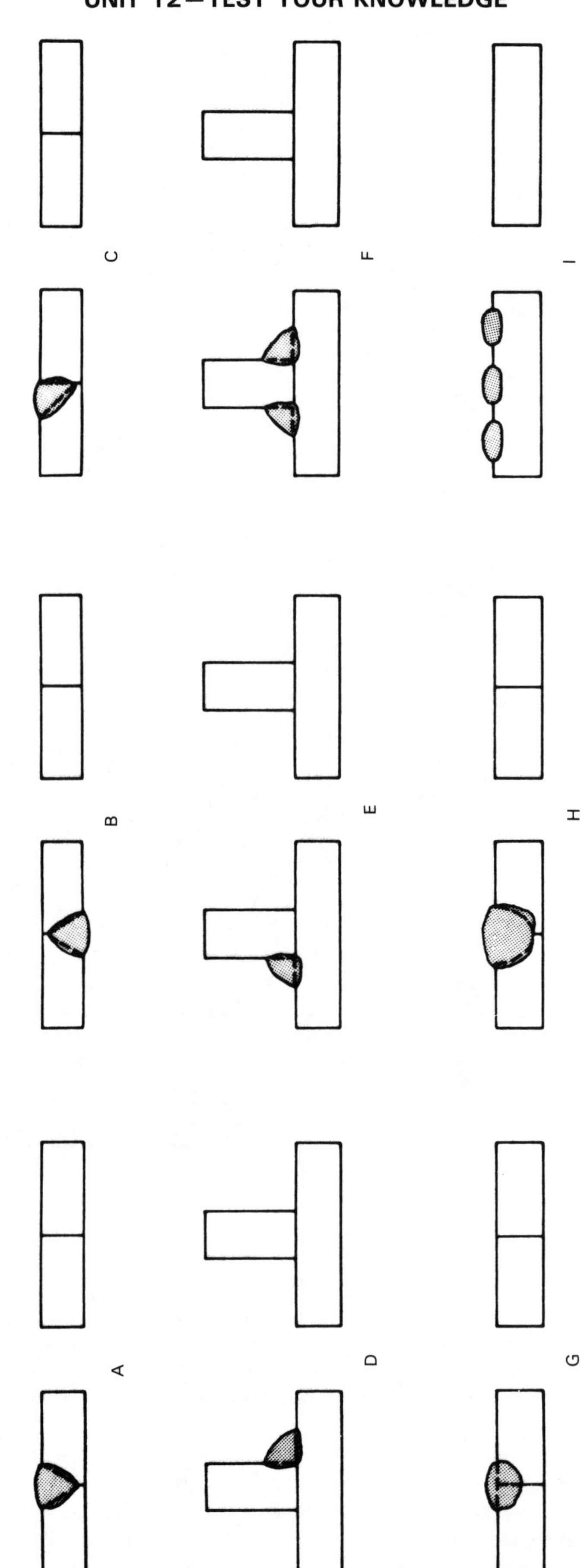

Part II

Draw the correct weld as indicated by the symbol.

J

K

L

M

N

O

P

Q

R

$\frac{1}{8}$

S

Unit 13

FILLET WELDS

A *fillet weld* is approximately triangular in shape. It is usually used when joining two surfaces at an angle. An example is given in Fig. 13-1.

The welding symbol for a fillet weld is shown in Fig. 13-2. The *dimension(s)* of this type of weld is placed on the SAME SIDE of the reference line as the weld symbol. *Weld size* is shown on the LEFT SIDE of the weld symbol.

Unless otherwise noted on the print, the *minimum size* of a fillet weld is the specified dimension.

When a fillet welding symbol does NOT include weld size, a *general note* governing weld size will be found elsewhere on the print. See Fig. 13-3.

At times, fillet welds with unequal legs will be specified. *Weld orientation* is NOT indicated by the symbol but will be shown on the print. This is il-

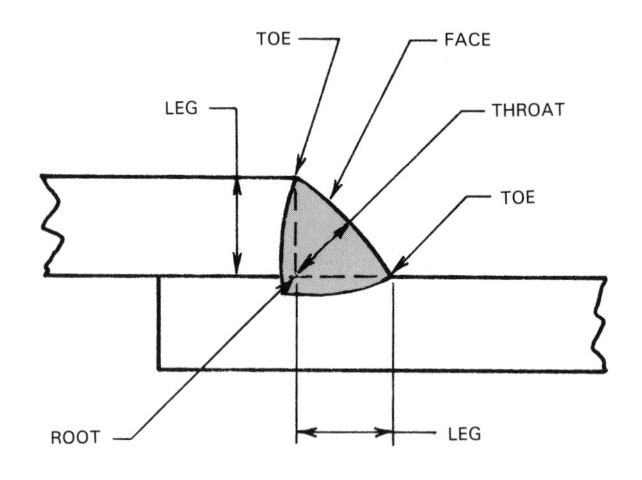

Fig. 13-1. Study parts of this fillet weld.

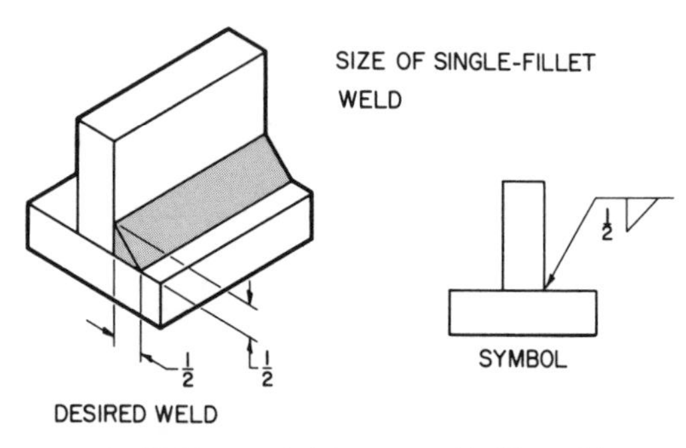

A—Welding symbol for arrow side fillet weld.

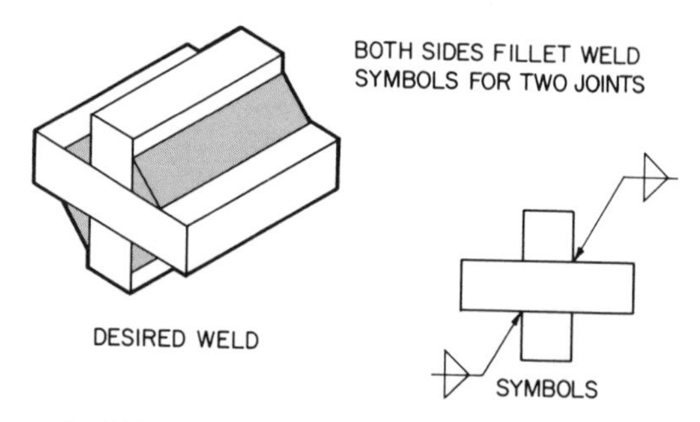

C—Welding symbol for unequal double-fillet welds.

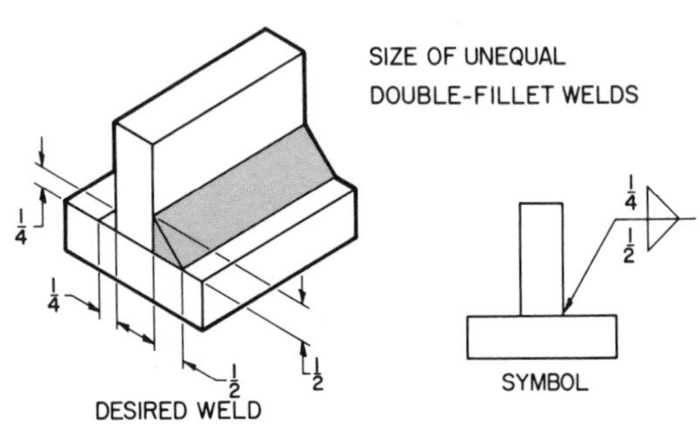

B—Welding symbol for arrow side and other side fillet weld.

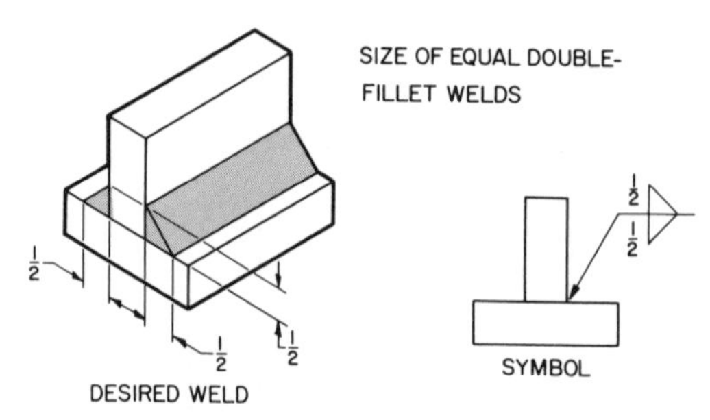

D—Welding symbol for both sides fillet welds for two joints.

Fig. 13-2. Note variations of fillet weld symbol.

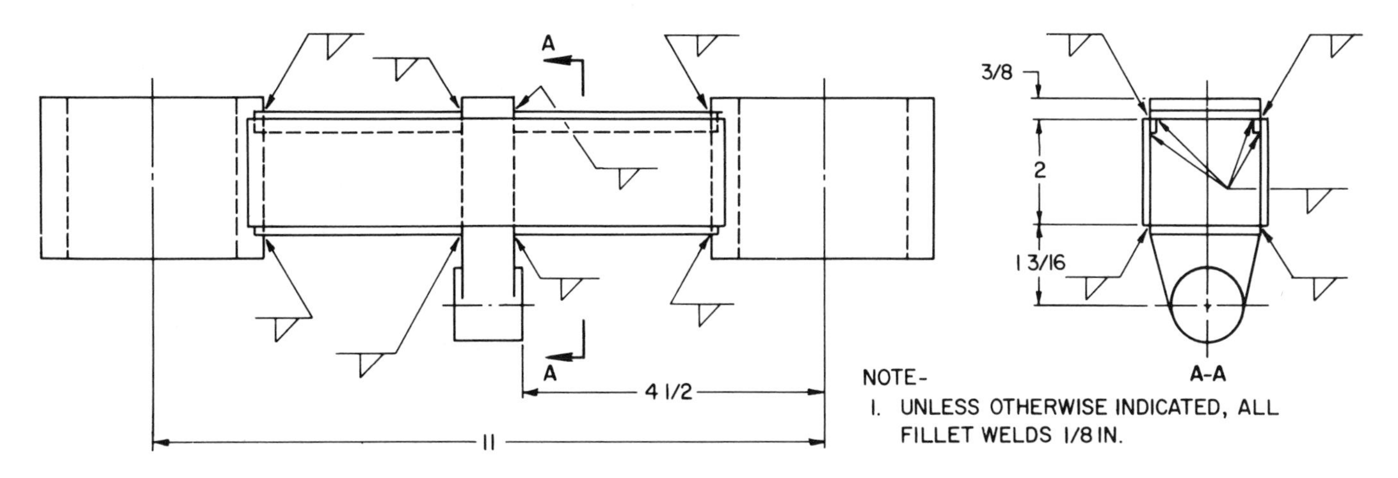

Fig. 13-3. Weld size will be somewhere else on print or drawing when not given on welding symbol.

lustrated in Fig. 13-4.

When used, *weld length* is included to the right of the weld symbol. The weld is made completely across the joint when no weld length dimension is given with the welding symbol. Refer to Fig. 13-5.

The extent and location of a fillet weld may be given graphically on a print view by *hatching*. Hatching is shown in Fig. 13-6.

Regardless of the geometric shape of the joint, a weld-all-around symbol means that the weld is to be made continuously around the joint. Refer to Fig. 13-7.

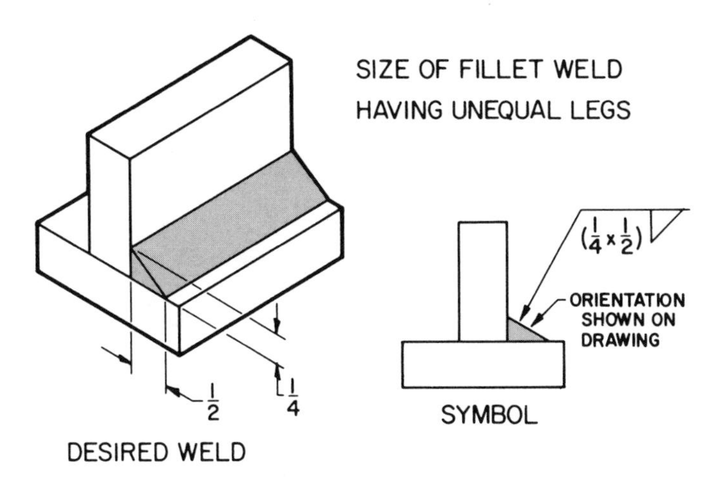

Fig. 13-4. Welding symbol for fillet weld having unequal legs.

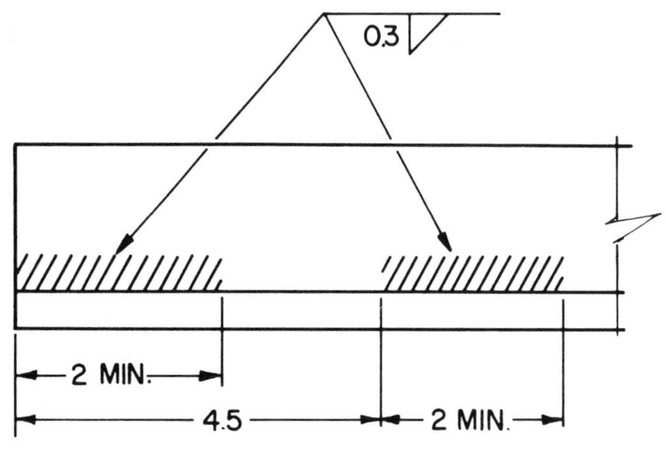

Fig. 13-6. Hatching can denote extent and location of a fillet weld.

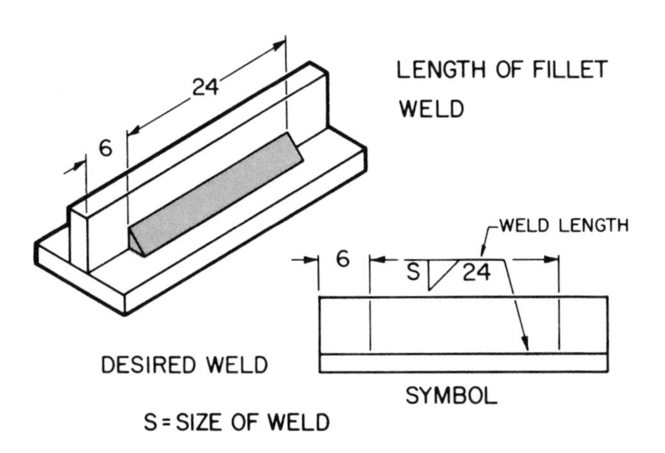

Fig. 13-5. Weld length is given by number to right of weld symbol.

Additional arrows pointing to each section of the joint will be used when you must make abrupt changes in weld direction. This is illustrated in Fig. 13-8.

The *pitch* (center-to-center spacing) of intermittent fillet welds is expressed as the distance between centers of increments on ONE side of the joint. See Fig. 13-9. The pitch in inches or millimeters is shown to the right of the length dimension.

Fig. 13-10 shows the weld symbol for staggered intermittent fillet welding.

A fillet weld symbol is also used for a fillet weld in a hole or slot, Fig. 13-11.

The FACE CONTOUR (flat, convex, or concave) of a fillet weld may be specified. The shape required

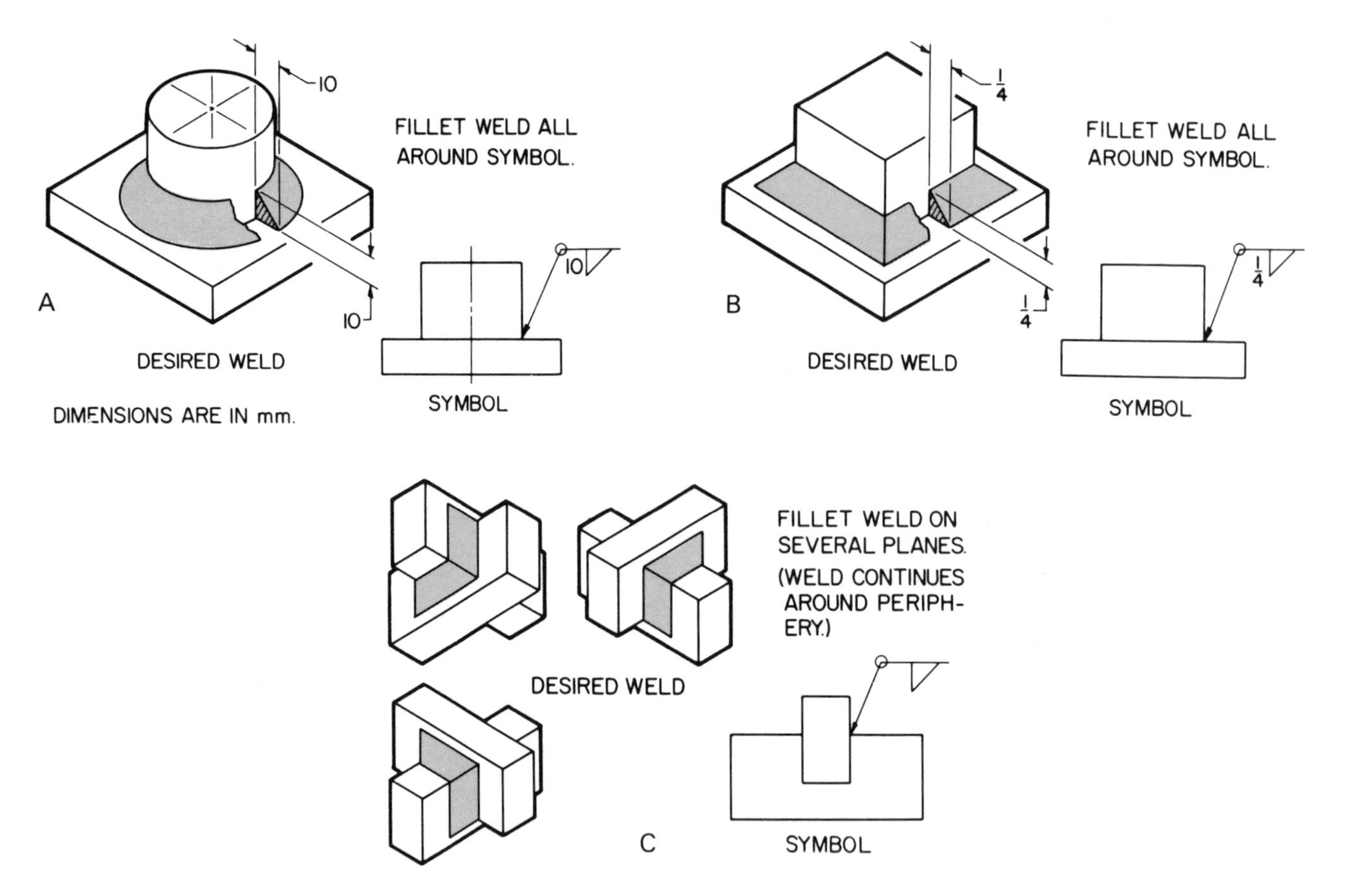

Fig. 13-7. A—Welding symbol for all-around fillet weld on round stock. B—Same welding symbol is used when stock is square or rectangular in shape. C—Note all-around fillet weld on several planes.

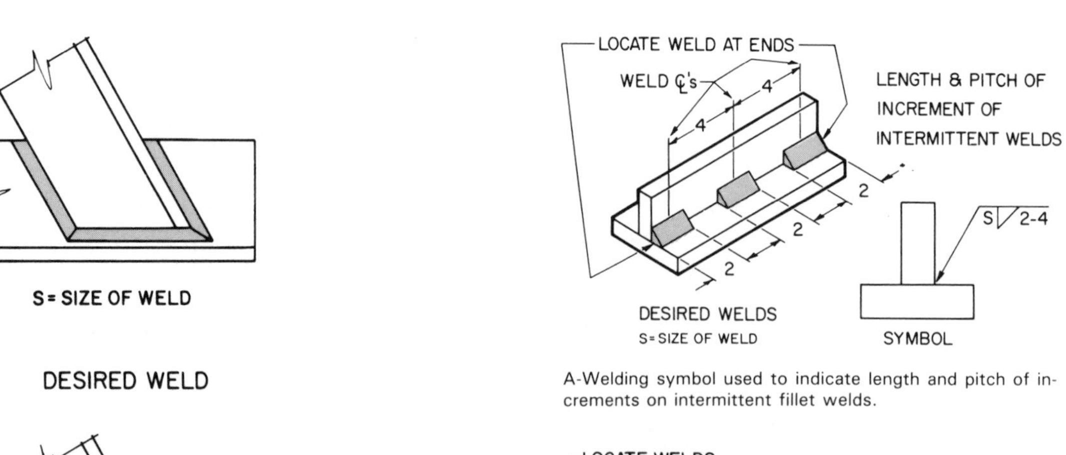

A-Welding symbol used to indicate length and pitch of increments on intermittent fillet welds.

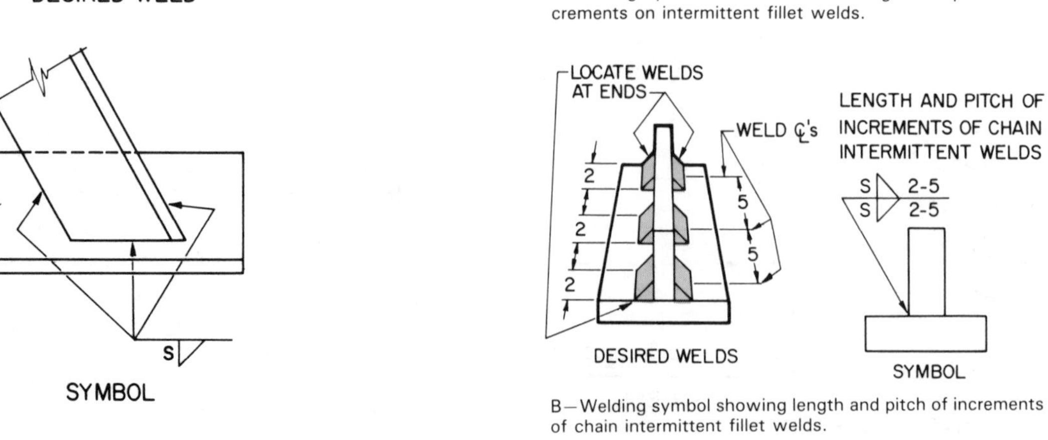

B—Welding symbol showing length and pitch of increments of chain intermittent fillet welds.

Fig. 13-8. When abrupt changes in weld direction are made, welding symbol arrows will point to each section of joint.

Fig. 13-9. Note intermittent fillet weld symbol use.

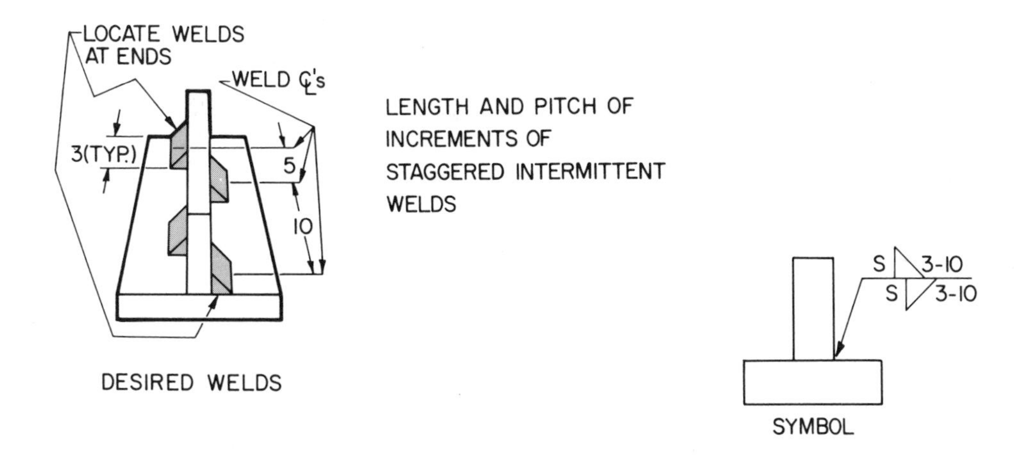

Fig. 13-10. Note welding symbol for staggered intermittent fillet welds.

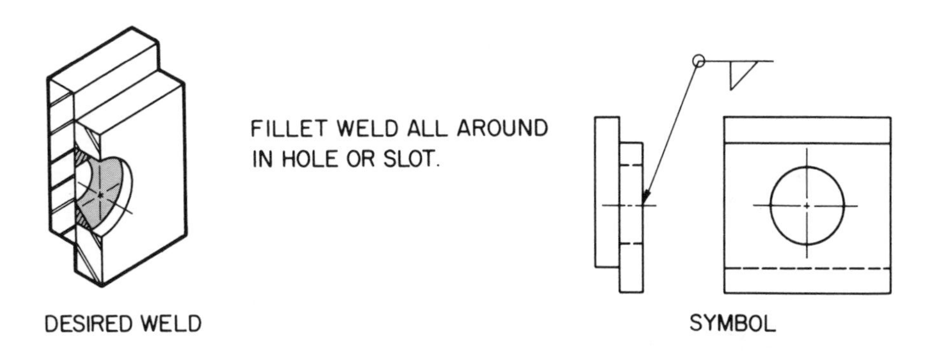

Fig. 13-11. Study welding symbol and desired weld for all-around fillet weld in a hole or slot.

TYPE	SYMBOL	EXAMPLE	AS WELDED
FLUSH			
CONCAVE			
CONVEX			

Fig. 13-12. Note weld symbols for specifying face contour of fillet weld. Required shape is shown by appropriate contour symbol included with fillet weld symbol.

METHOD	SYMBOL	EXAMPLE
CHIPPING	C	C
GRINDING	G	G
HAMMERING	H	H
MACHINING	M	M
ROLLING	R	R
PEENING	P	P

Fig. 13-13. When face contour of a weld is to be finished mechanically, a symbol (letter) of appropriate finish technique is included with fillet weld and face contour weld symbols.

is indicated by a *contour symbol* used with the fillet weld symbol. Fig. 13-12 shows these symbols.

If the face contour is to be FINISHED MECHANICALLY, a *finish symbol* for the method of finishing the weld is added to the contour symbol. Refer to Fig. 13-13.

While the *degree of finish* (smoothness) is NOT part of American Welding Society standards, some prints do indicate the desired finish. See Fig. 13-14.

When another type of weld is to be used with a fillet weld, weld symbols for BOTH TYPES are included on the welding symbol. Fig. 13-15 gives an example.

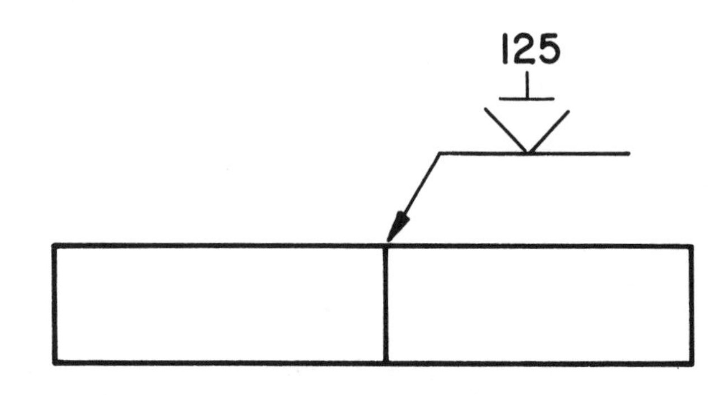

Fig. 13-14. Degree of finish or weld smoothness is indicated by a number that specifies (in microinches) roughest surface acceptacle for this particular application. The larger the number, the rougher the surface permitted. A 125 surface is smoother than a 250 finish. It is seldom used symbol.

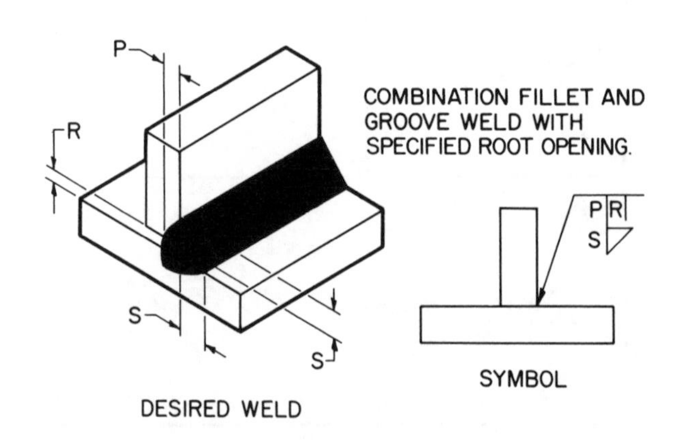

Fig. 13-15. More than one type of weld is specified for this joint. Note how both weld symbols are included on symbol.

UNIT 13—TEST YOUR KNOWLEDGE

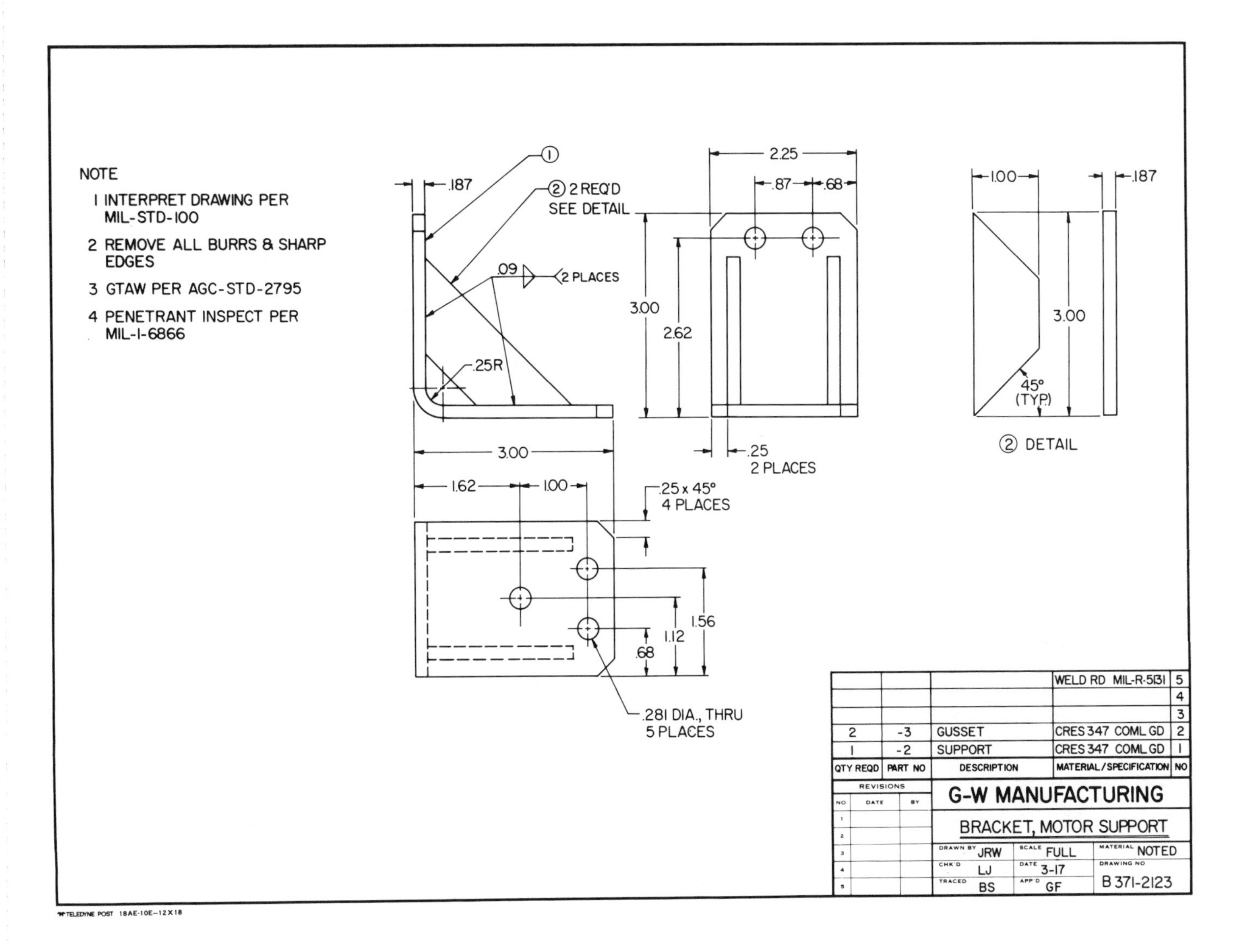

Part I

Refer to the drawing to answer the following questions.

1. What is the drawing title? ____________
2. Name the drawing number. ____________
3. How many parts make up the assembly? ______
4. What welding process is to be used? ______
5. The gussets are located ________ in from the edges of the support.
6. The gusset is made from _____x_____x_____ stock.

   There are ______ required in each assembly.
7. The support is made from _____x_____x_____ stock.
8. What type of weld is specified? ____________
9. Name the type of welding rod that is specified. ________
10. What is the size of the specified weldments. ________
11. Sketch the specified welds on the front and side views in the drawing.
12. Each corner of the bracket is chamfered _____x _____.
13. What are the number of holes to be drilled? ________
14. Name the hole size. _____ dia.
15. The complete bracket is to be finished by ____ ______________________________.

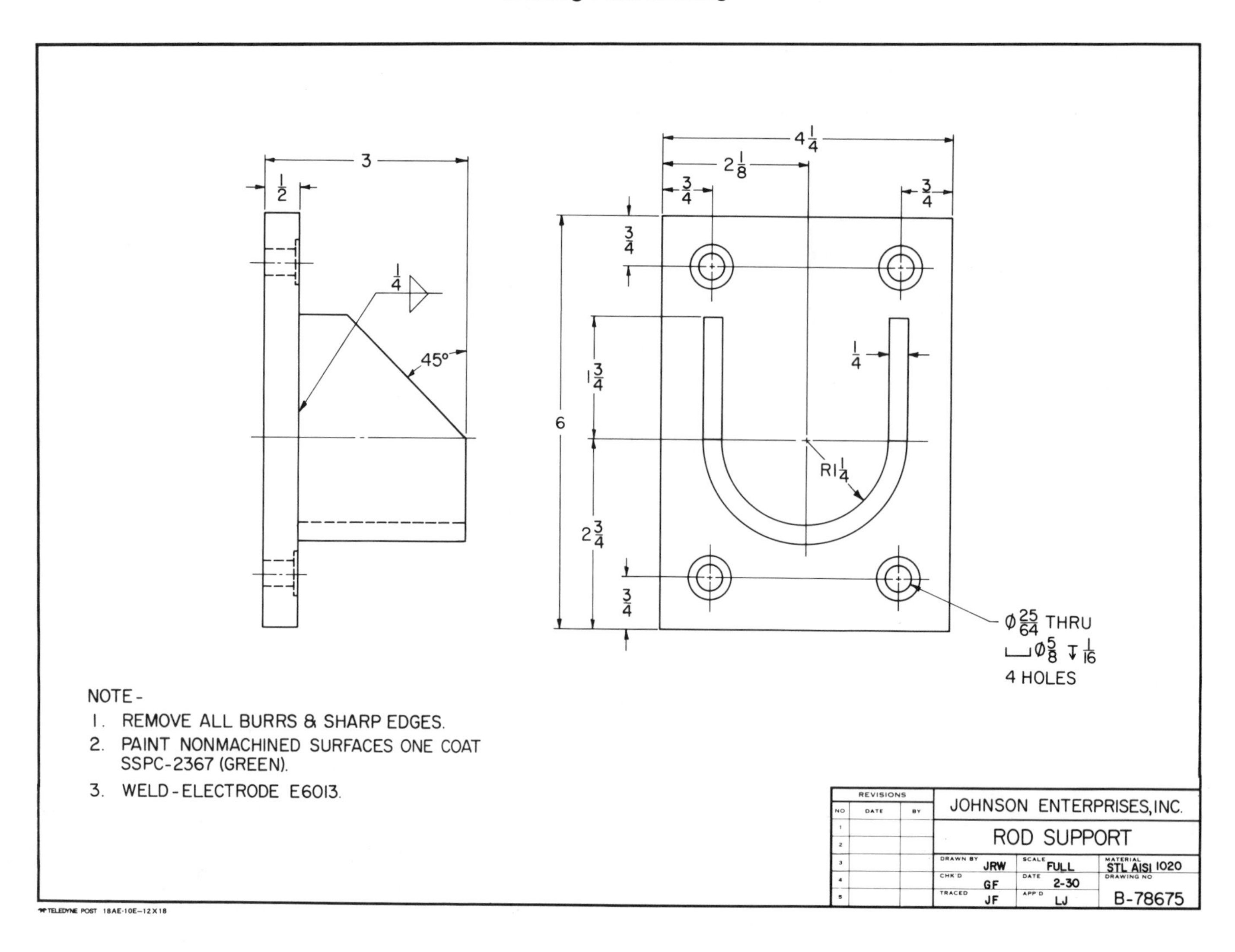

Part II

Refer to the drawing to answer the following questions.

1. What is the name of the product to be made?

   ____

2. Name the drawing or print number. __________
3. List the number of parts in the assembly. ____
4. What material is specified to make the product?

   ____

5. What type of weld rod is specified? __________
6. Name the size and type of the weld specified.

   ____

7. On the drawing shown above, make a sketch of the required weld.
8. What is the size of the mounting plate?

   ______x______x______.

9. List the size of the material required to make the curved section. ______x______x______.
10. The centerline of the curved section is located ______ in. up from bottom of mounting plate and ______ in. in from the side.
11. What are the number and size of holes to be made in mounting plate. a. _________ b. _________
12. What additional operation must be performed on drilled holes? ______________________
13. After welding and machining, what finishing operations are specified?

    a. ______________________________

    b. ______________________________

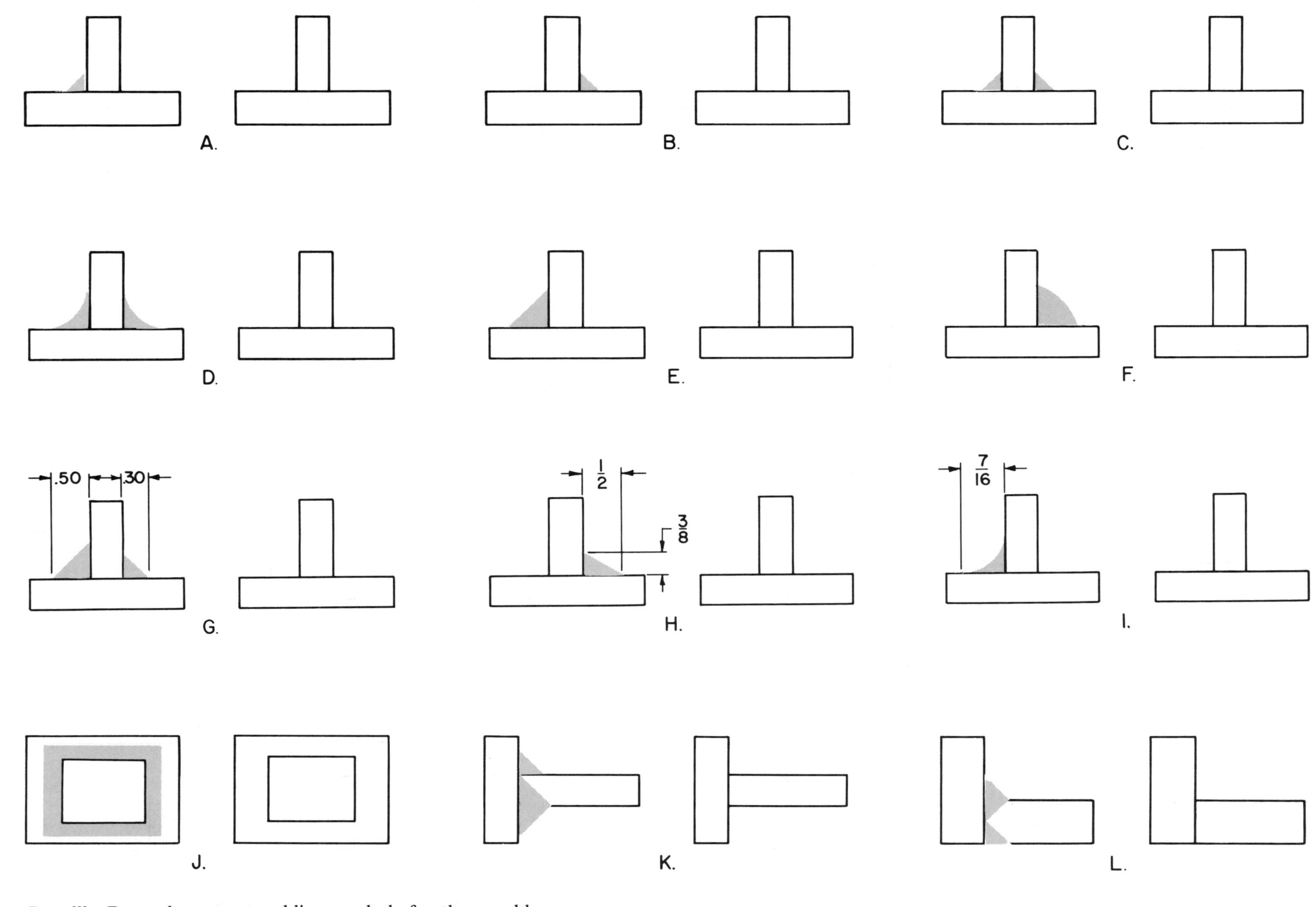

Part III Draw the correct welding symbols for these welds.

Part IV Draw the correct weld as indicated by the welding symbol.

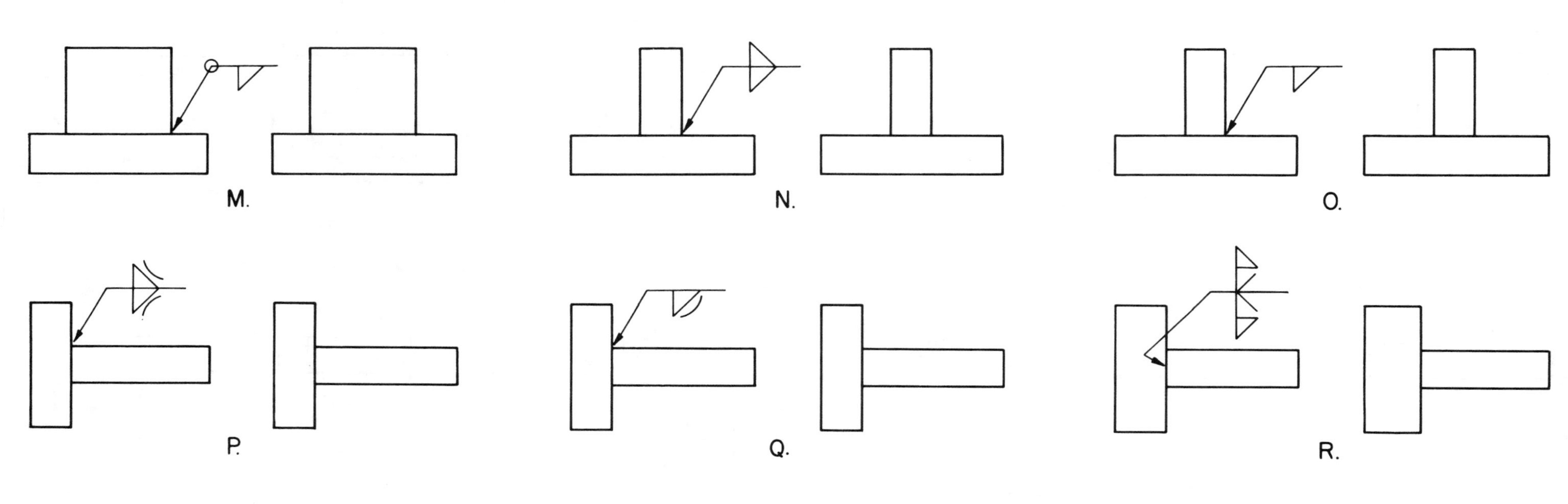

Part V Draw the correct welding symbol for the following welds.

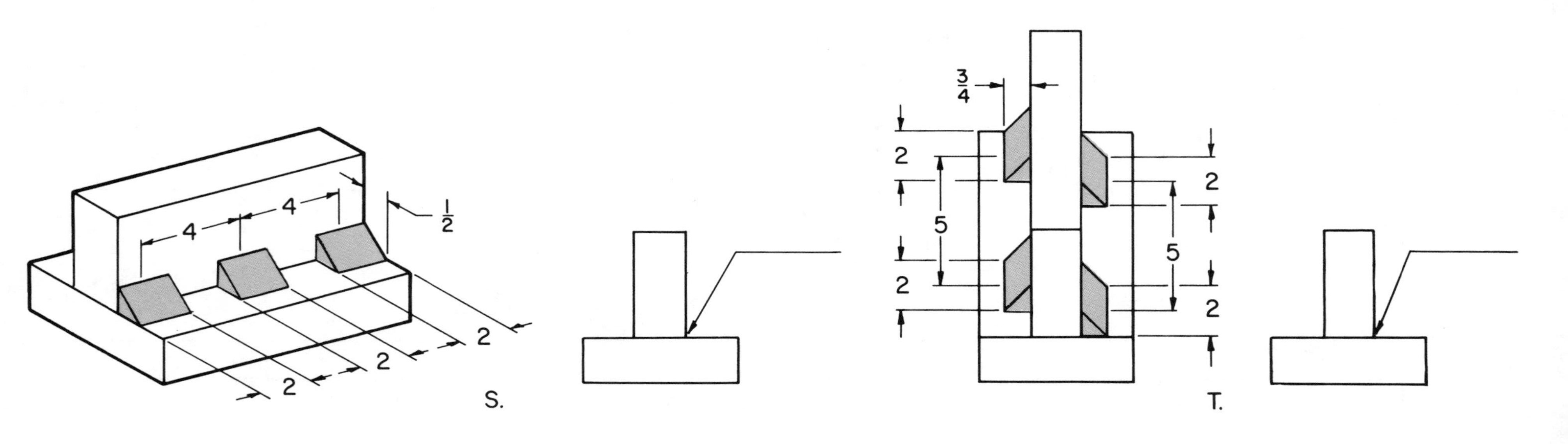

Fillet welds are frequently used in fabricating components for larger assemblies. (Nederman, Inc.)

Unit 14

PLUG WELDS

A *plug weld* is a circular weld made through a hole in one piece of metal, joining it to another piece. An example is given in Fig. 14-1.

The walls of the hole may be parallel or angular and the hole may be completely or only partially filled with weld metal.

Note! A plug weld is NOT to be confused with an all-around fillet weld made in a hole.

GENERAL USE OF PLUG WELD SYMBOL

A plug weld in the *arrow side member* of a joint is specified when the weld symbol is placed on the side of the reference line TOWARDS the reader. This is shown in Fig. 14-2.

A plug weld in the *other side member* of a joint is indicated when the weld symbol is placed on the side of the reference line AWAY from the reader, Fig. 14-3.

Plug weld dimensions are shown on the same side of the reference line as the weld symbol. Fig. 14-4 gives an example.

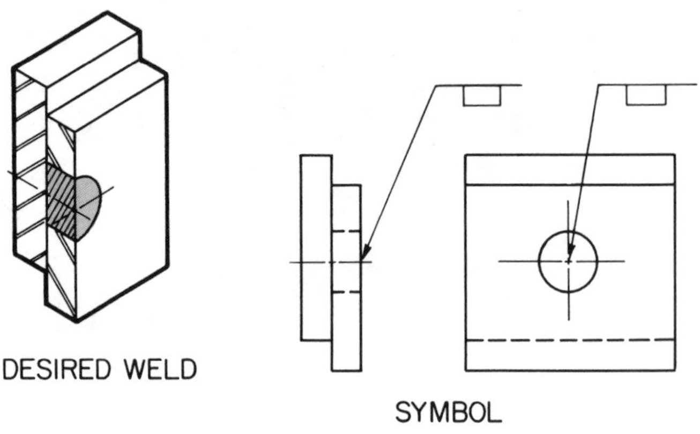

Fig. 14-2. Note significance of arrow side plug welding symbol.

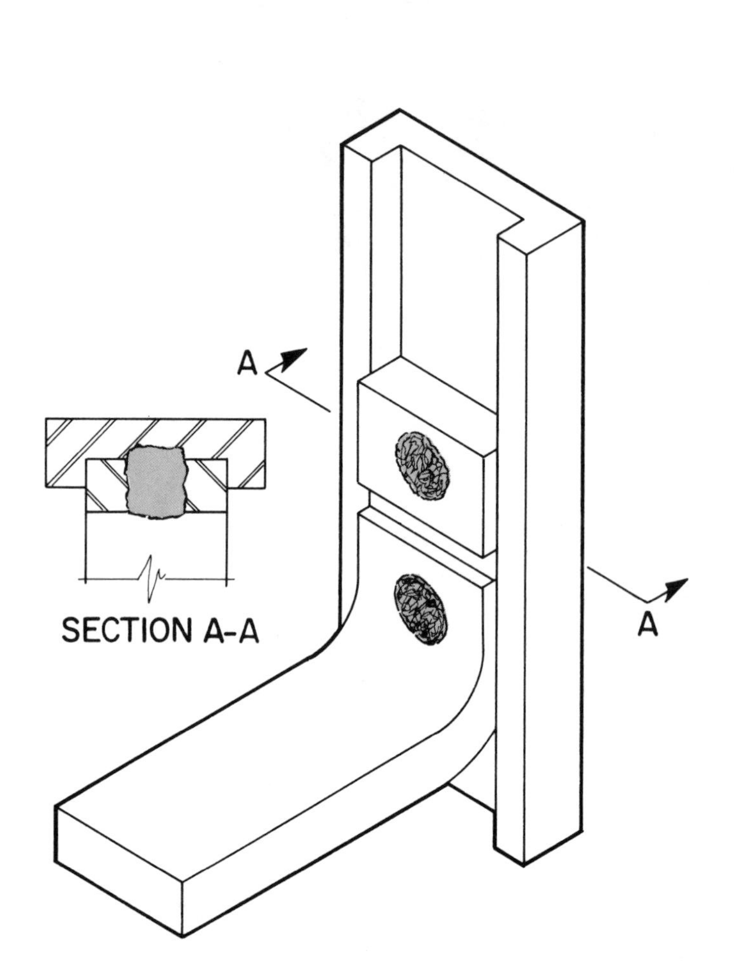

Fig. 14-1. Study typical plug weld.

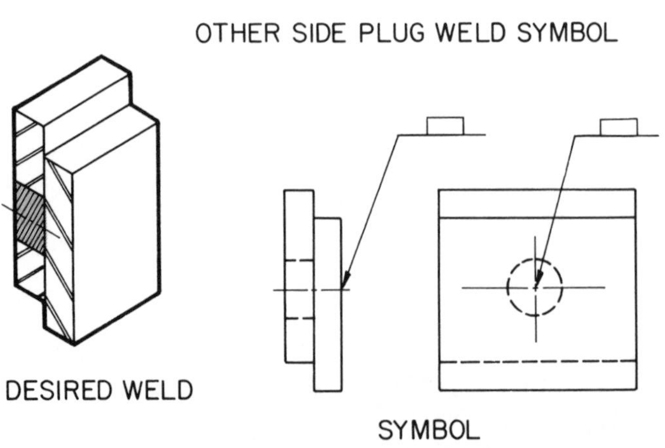

Fig. 14-3. Note significance of other side plug welding symbol.

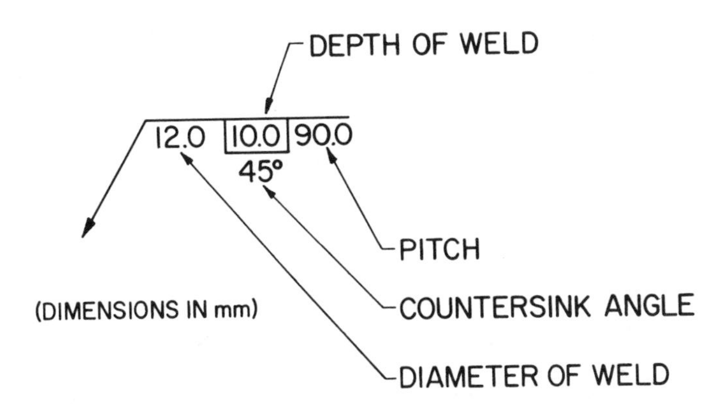

Fig. 14-4. Plug weld dimensions are on same side of reference line as weld symbol. Dimensions are located as shown. Study their positions around symbol.

SIZE OF PLUG WELD

Plug weld size refers to the weld diameter at the base of the weld. It is given to the LEFT of the weld symbol, as in Fig. 14-5.

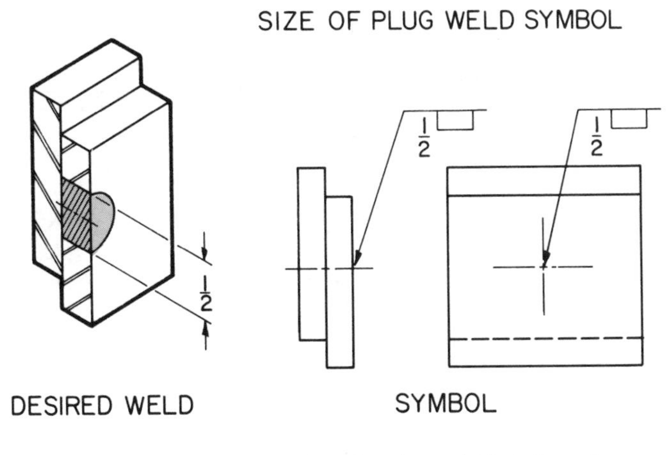

Fig. 14-5. Note how size of a plug weld is indicated.

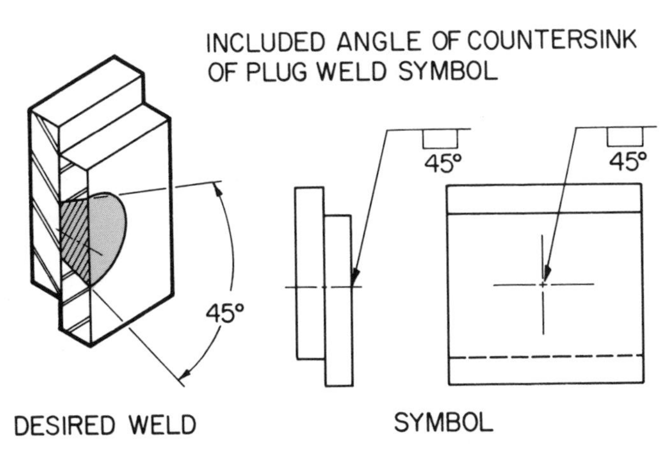

Fig. 14-6. When a plug weld is to be made in a countersunk hole included angle of countersink is specified in relation to weld symbol.

COUNTERSINK ANGLE OF PLUG WELD

When the side of a plug weld hole is tapered, the included angle of the taper is specified and shown on the welding symbol. Refer to Fig. 14-6. The *countersink angle* is given below the plug weld symbol.

DEPTH OF FILLING OF PLUG WELD

Plug weld holes are not always completely filled with weld material. When the area is NOT to be filled completely, *filing depth* is indicated within the plug weld symbol. Fig. 14-7 shows an example.

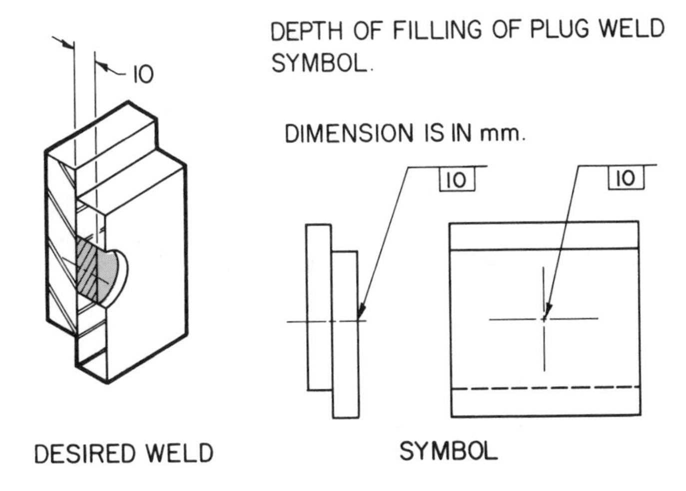

Fig. 14-7. When depth of plug weld is less than complete, specified weld depth is shown inside weld symbol.

SPACING OF PLUG WELDS

The *pitch* (center-to-center spacing) of plug welds is indicated by the dimension placed on the same side as, and to the right of, the weld symbol. See Fig. 14-8.

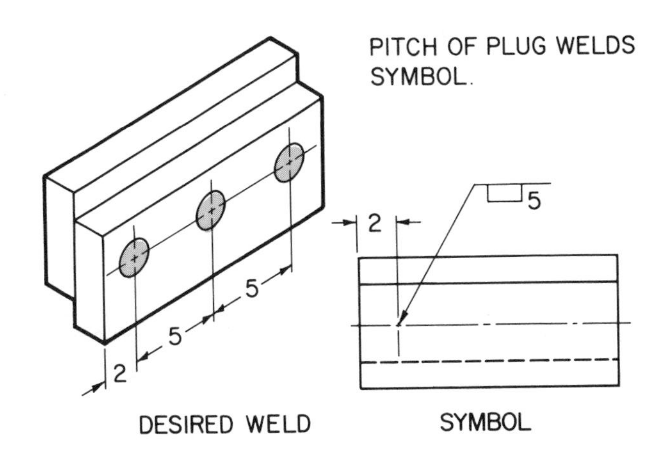

Fig. 14-8. The pitch or center-to-center spacing of multiple plug welds is shown to right of weld symbol.

Fig. 14-9 shows the use of combined dimensions for plug welds. Study this illustration carefully.

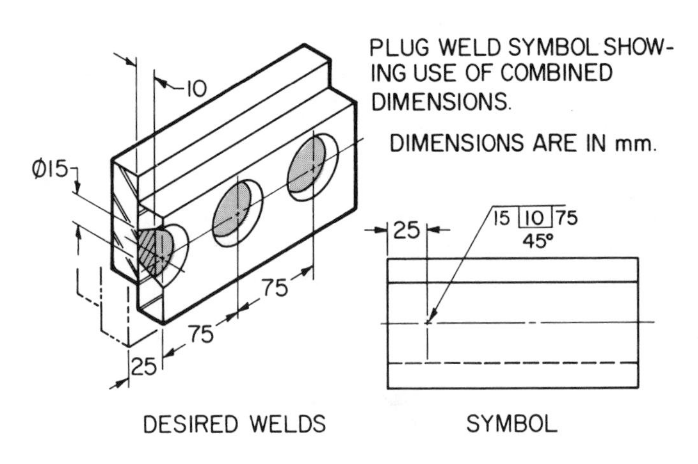

Fig. 14-9. Plug welding symbol showing combined dimensions. Note position of dimensions around plug weld symbol.

SURFACE CONTOUR OF PLUG WELDS

When plug welds are to be welded approximately *flush* (no mechanical finishing required), only the flush contour symbol is added to the weld symbol. However, if the welded surface is to be finished flush by MECHANICAL MEANS, the finishing technique must also be added to the weld symbol. See Fig. 14-10.

SYMBOL SHOWING SURFACE CONTOUR OF PLUG WELDS

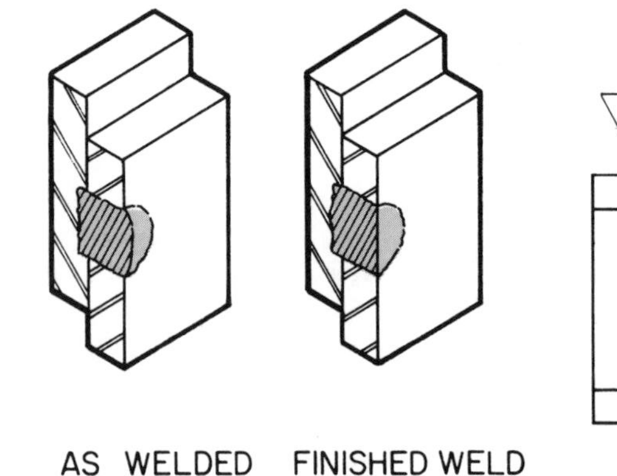

Fig. 14-10. Symbol showing how flush surface contour plug weld is specified. When weld surface is to be made flush with surrounding surfaces mechanically, a letter indicates finishing technique.

UNIT 14—TEST YOUR KNOWLEDGE

Part I

Refer to the drawing (167E102) and answer the following questions. (See page 139.)

1. What is the name of the part? ______________
2. Where is the drawing number located? ______________
3. What is the scale of drawing? ______________
4. How many individual parts make up the unit? __________
5. Except as noted, what general tolerances are allowed? a. Decimal______ b. Fractional______ c. Angular______
6. How many plug welds are required?__________
7. What specifications are given for the plug welds? ______________
8. What other type welds are indicated? ________
9. What type welding rod is specified? __________
10. List the sub-assemblies that make up the completed unit.

	Sub-Assembly	Part No.
a.	______________	________
b.	______________	________
c.	______________	________
d.	______________	________

11. The overall length of the unit is ____________.
12. The total width of the unit is ____________.
13. The length of the tongue is ____________.
14. The distance from the base of the end plate to the centerline of the bore is ____________.
15. Bore diameter is ____________.

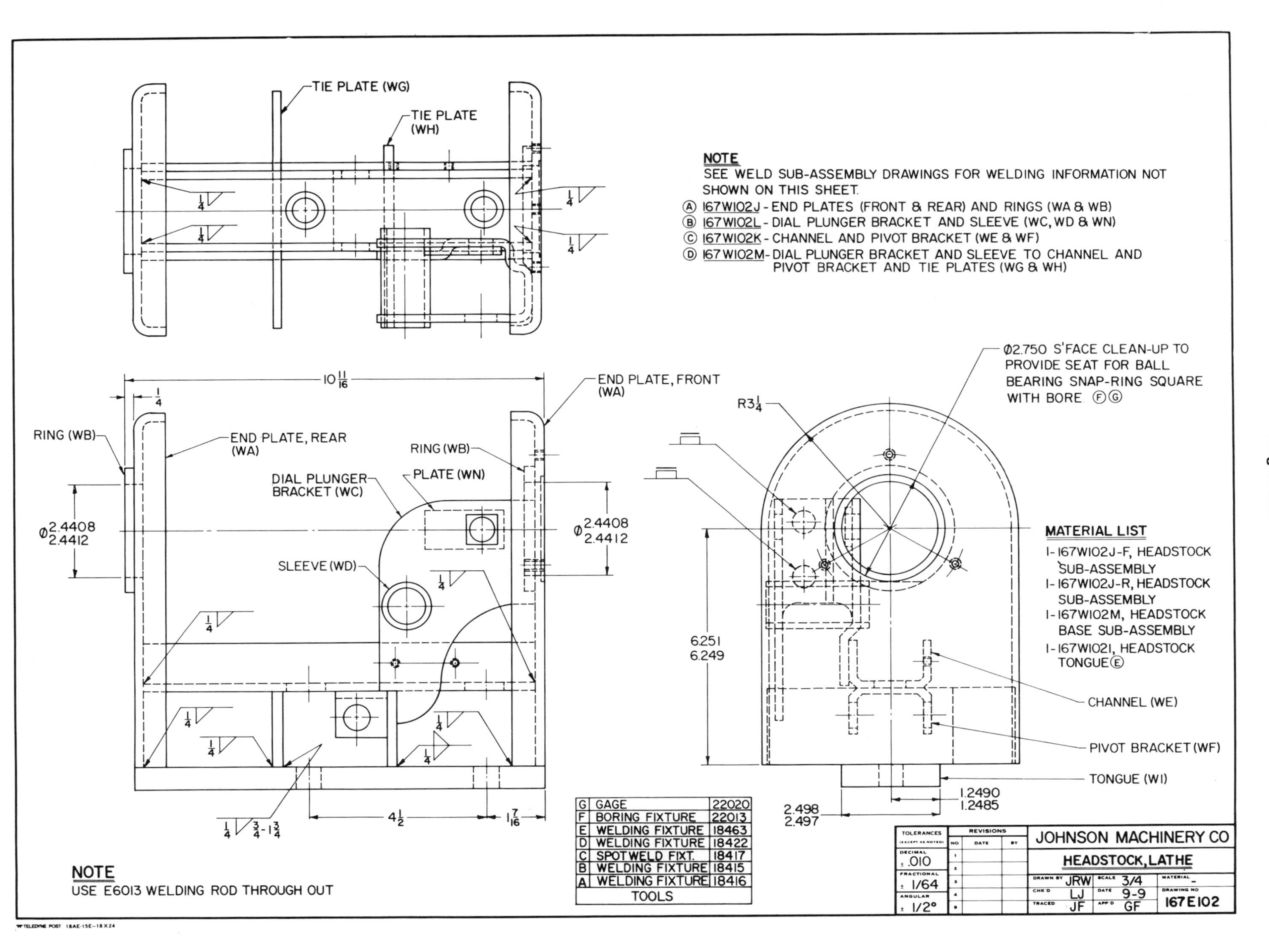
TIE PLATE (WG)
TIE PLATE (WH)
NOTE
SEE WELD SUB-ASSEMBLY DRAWINGS FOR WELDING INFORMATION NOT SHOWN ON THIS SHEET.
(A) 167W102J - END PLATES (FRONT & REAR) AND RINGS (WA & WB)
(B) 167W102L - DIAL PLUNGER BRACKET AND SLEEVE (WC, WD & WN)
(C) 167W102K - CHANNEL AND PIVOT BRACKET (WE & WF)
(D) 167W102M - DIAL PLUNGER BRACKET AND SLEEVE TO CHANNEL AND PIVOT BRACKET AND TIE PLATES (WG & WH)
10 11/16
1/4
END PLATE, FRONT (WA)
RING (WB)
END PLATE, REAR (WA)
RING (WB)
DIAL PLUNGER BRACKET (WC)
PLATE (WN)
Ø2.4408
2.4412
Ø2.4408
2.4412
SLEEVE (WD)
4 1/2
1 7/16
1/4 3/4-1 3/4
Ø2.750 S'FACE CLEAN-UP TO PROVIDE SEAT FOR BALL BEARING SNAP-RING SQUARE WITH BORE (F)(G)
R3 1/4
6.251
6.249
MATERIAL LIST
1-167W102J-F, HEADSTOCK SUB-ASSEMBLY
1-167W102J-R, HEADSTOCK SUB-ASSEMBLY
1-167W102M, HEADSTOCK BASE SUB-ASSEMBLY
1-167W102I, HEADSTOCK TONGUE (E)
CHANNEL (WE)
PIVOT BRACKET (WF)
TONGUE (WI)
1.2490
1.2485
2.498
2.497
G GAGE 22020
F BORING FIXTURE 22013
E WELDING FIXTURE 18463
D WELDING FIXTURE 18422
C SPOTWELD FIXT. 18417
B WELDING FIXTURE 18415
A WELDING FIXTURE 18416
TOOLS
TOLERANCES (EXCEPT AS NOTED)
DECIMAL ±.010
FRACTIONAL ±1/64
ANGULAR ±1/2°
REVISIONS NO DATE BY
JOHNSON MACHINERY CO
HEADSTOCK, LATHE
DRAWN BY JRW SCALE 3/4 MATERIAL -
CHK'D LJ DATE 9-9 DRAWING NO 167E102
TRACED JF APP'D GF
NOTE
USE E6013 WELDING ROD THROUGH OUT

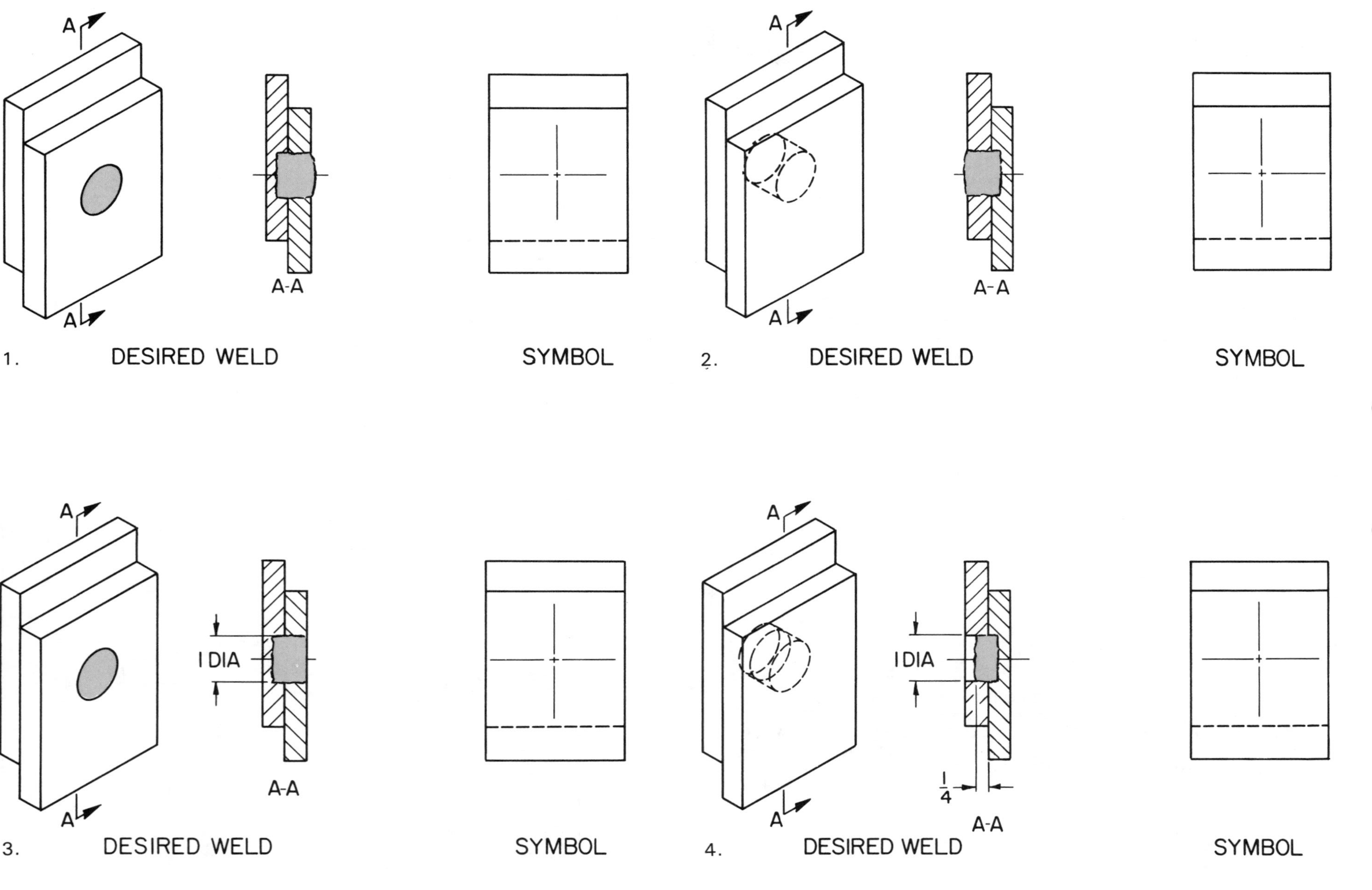

Part II Draw in the correct welding symbol for the above plug welds.

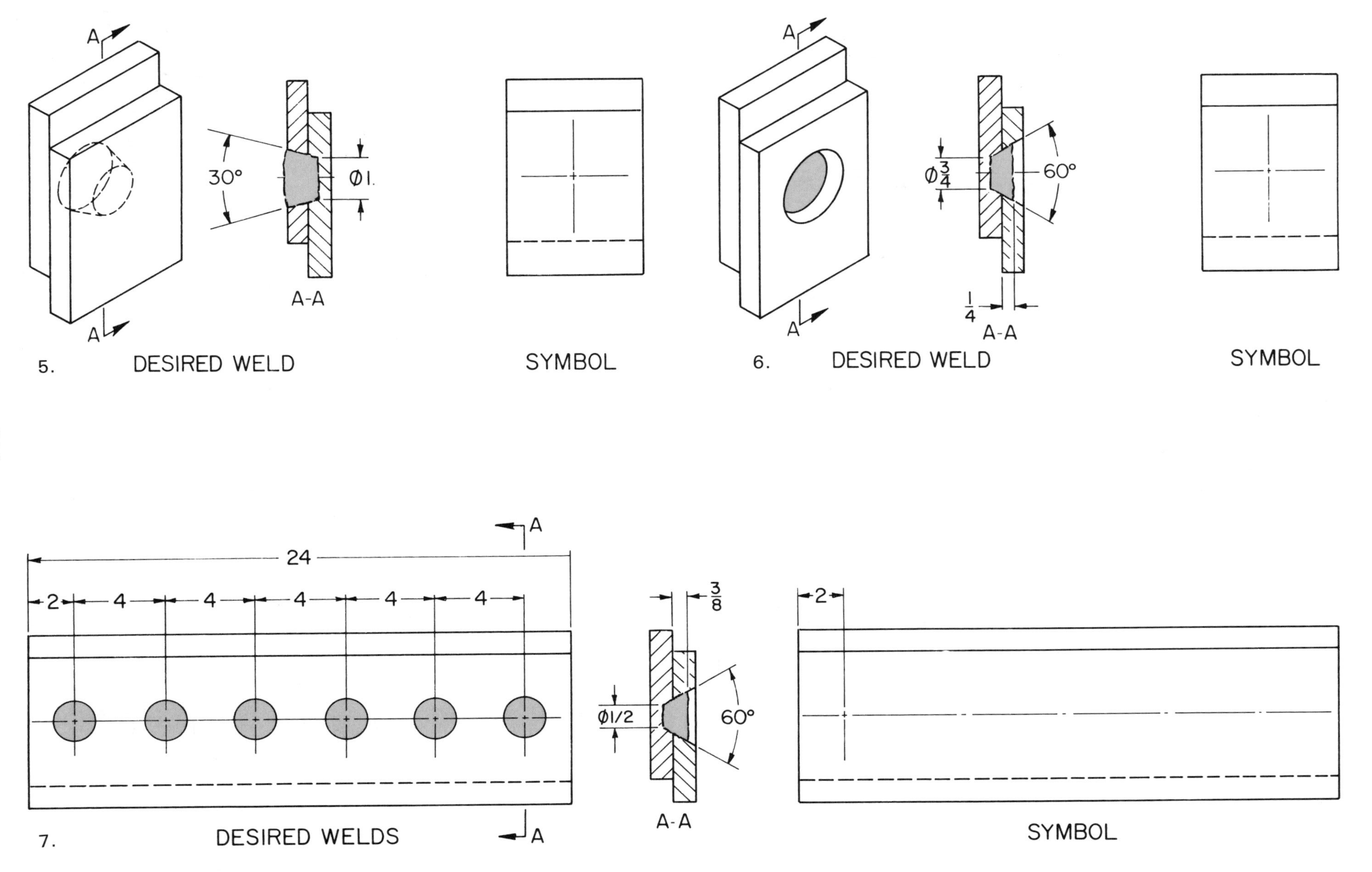

Part II (continued) Draw the correct symbol for these plug welds.

Unit 15

SLOT WELDS

A *slot weld* is made in a slot cut through one member of a joint to join that part of the metal to another joint member. Fig. 15-1 shows a typical example.

The slot may be partially or completely filled with weld.

The same weld symbol is used to indicate BOTH plug and slot welds.

Note! A slot weld should not be confused with a plug weld. A slot weld is made in an elongated hole. A plug weld is made in a round hole.

GENERAL USE OF SLOT WELD SYMBOL

A slot in the *arrow side* member of a joint is indicated when the slot weld symbol is located on the side of the reference line TOWARDS the reader. See Fig. 15-2.

A slot in the *other side* member is signified when the slot weld symbol is placed on the side of the reference line AWAY from the reader. Refer to Fig. 15-3.

Slot orientation on either arrow side or other side slot welds will be shown on the drawing.

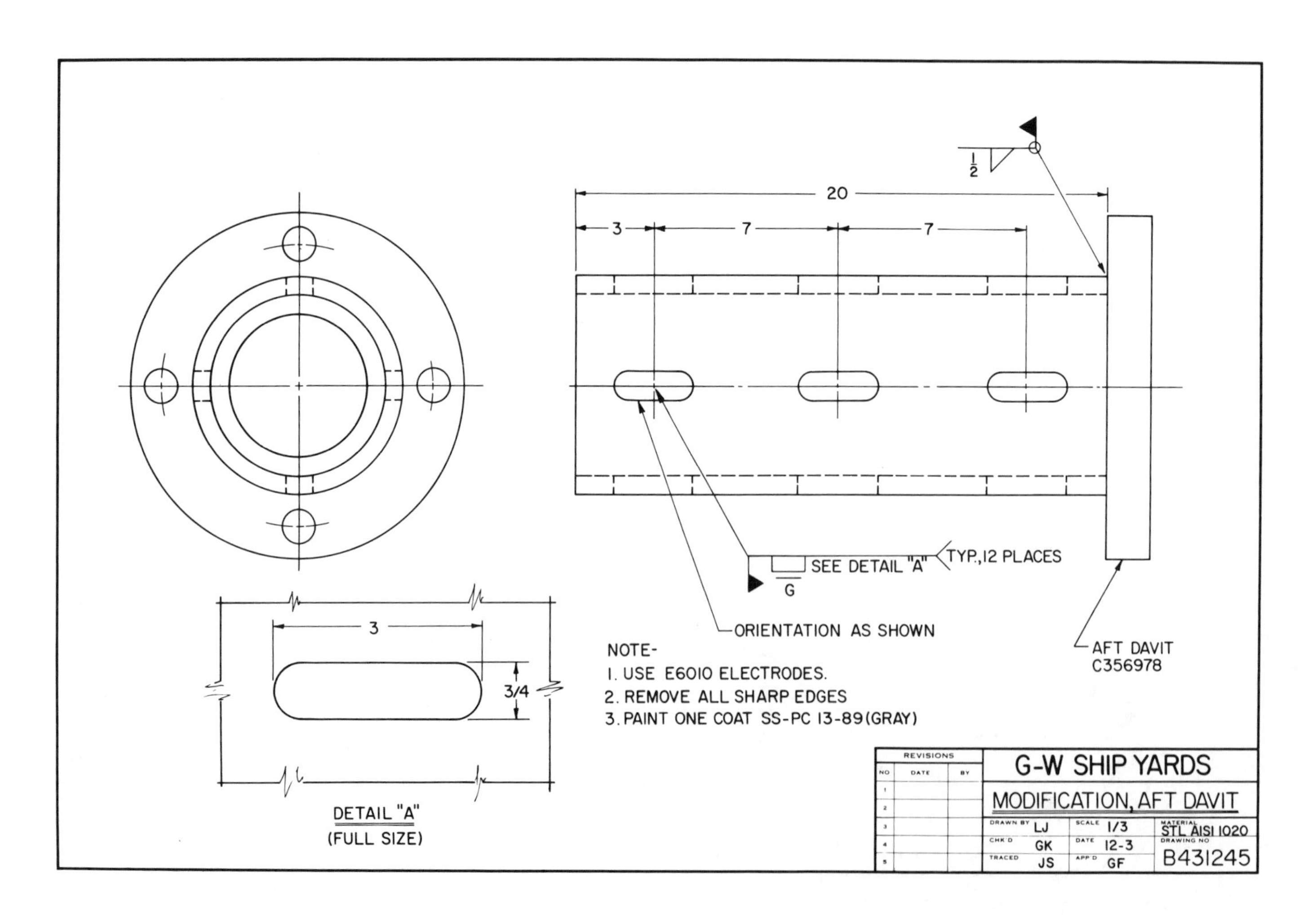

Fig. 15-1. This is a component modification drawing that requires slot welds. Where are welds to be made?

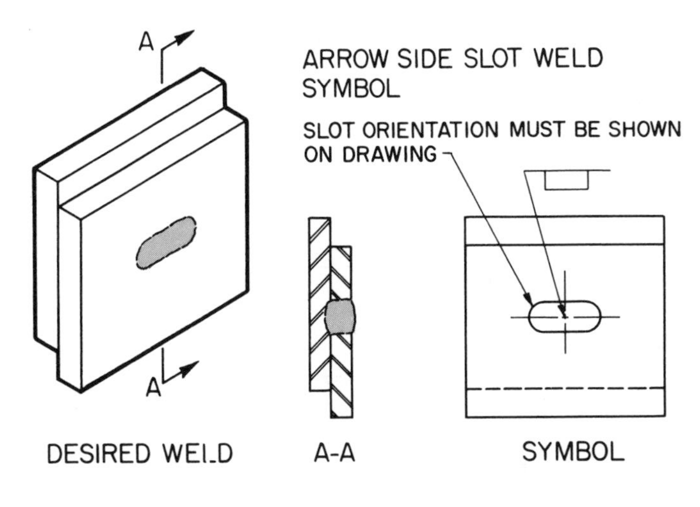

Fig. 15-2. Note significance of arrow side slot welding symbol.

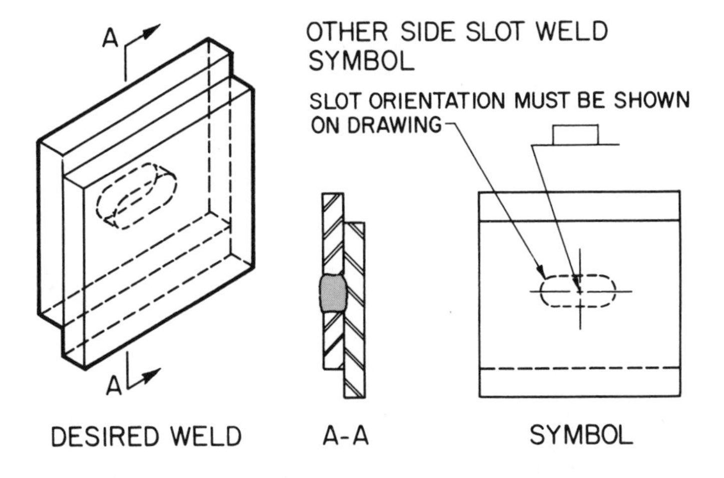

Fig. 15-3. Note other side slot welding symbol.

Slot weld dimensions will be shown on the same side of the reference line as the weld symbol.

DEPTH OF FILLING

The required *depth* of a partially filled slot weld will be shown inside the weld symbol, as in Fig. 15-4. If no dimension is shown in the weld symbol, UNLESS OTHERWISE SPECIFIED, the slot weld should completely fill the slot.

SLOT WELD DETAILS

There is usually not enough room on the welding symbol to furnish full details on length, width, spacing, countersink angle, orientation, and location of required slot welds. This information is normally shown elsewhere on the drawing. Reference to the location of weld details will be made on the welding symbol. See Fig. 15-5.

SURFACE CONTOUR AND FINISH OF SLOT WELDS

A *flush contour symbol* will be added to the slot weld symbol when the welded area is to be approximately flush with the surrounding metal (no mechanical finishing required).

If the slot weld is to be made *flush mechanically*, the flush contour symbol AND method of making the weld flush will be added to the slot weld symbol.

Fig. 15-6 shows an example of a slot weld made mechanically flush.

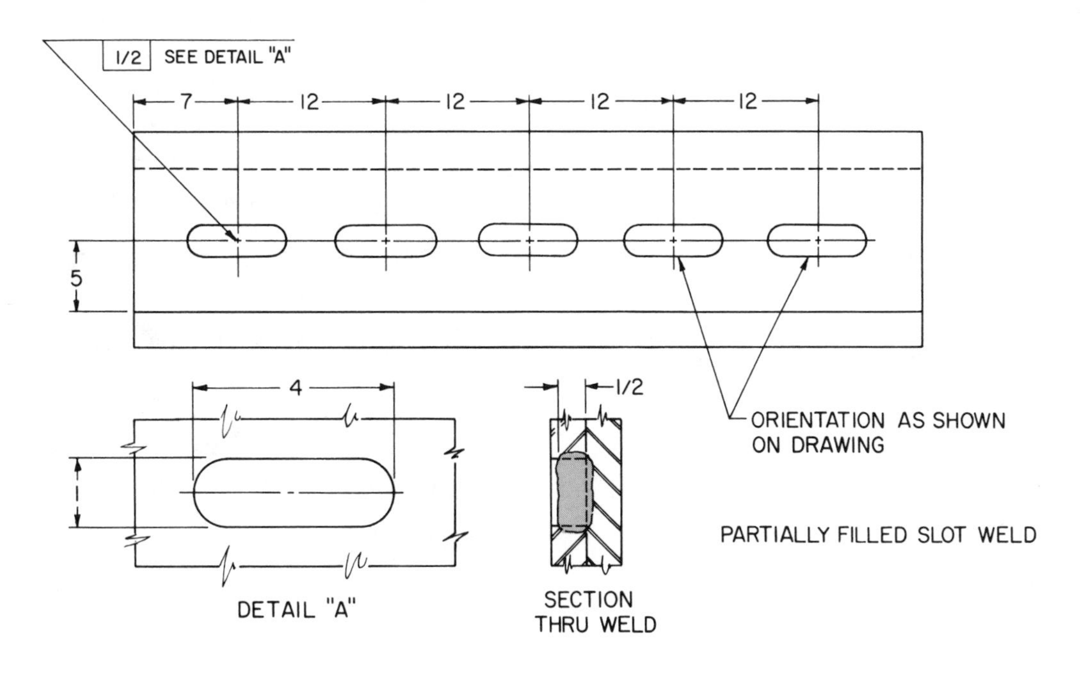

Fig. 15-4. Drawing shows partially filled slot weld and welding symbol to specify required welds.

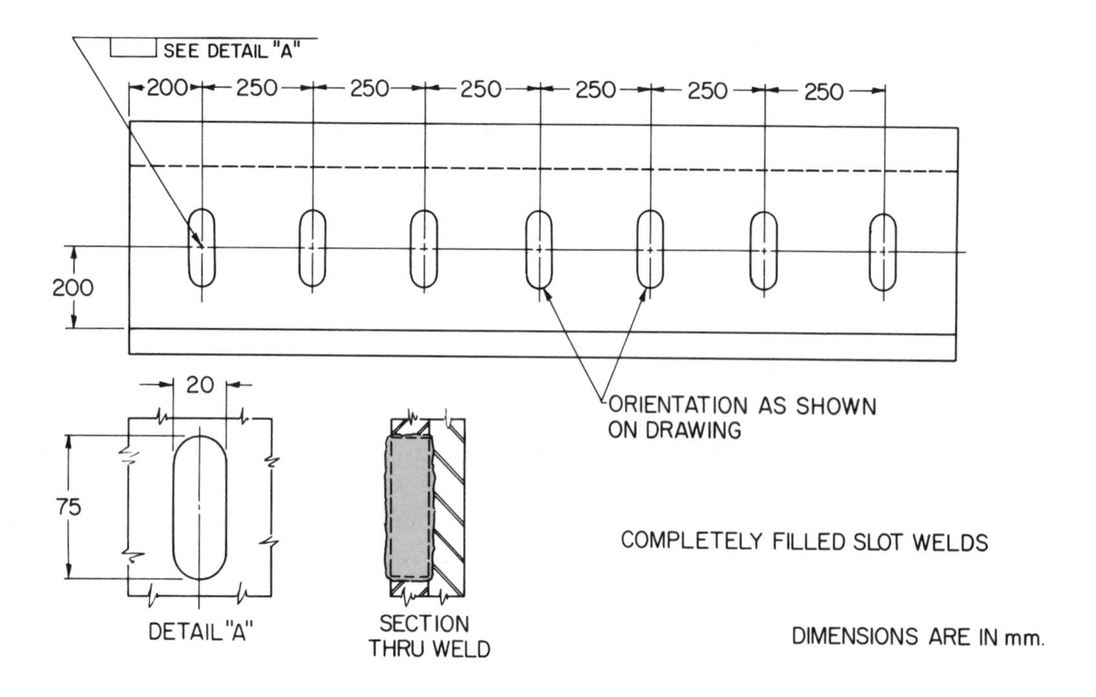

Fig. 15-5. Slot weld information is commonly shown elsewhere on drawing. Location of information is noted on welding symbol.

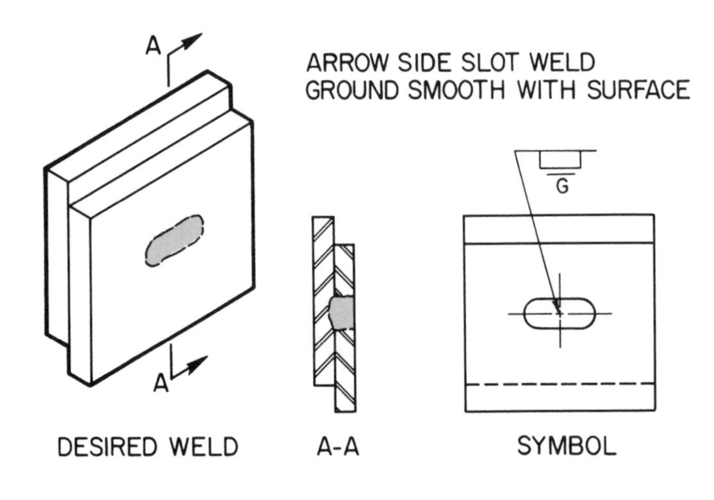

Fig. 15-6. Flush contour symbol and method of making weld flush are added to slot weld symbol when made flush by mechanical means.

UNIT 15—TEST YOUR KNOWLEDGE

Part I

Refer to the drawing (167C100) and answer the following questions.

1. Name the unit shown on the drawing. ______

______

2. Where is the drawing number located? ______

______

3. The scale of the drawing is ______.
4. What is the next assembly? ______
5. How many individual parts make up the completed unit? ______
6. Name the bed length of the unit shown. ______

______

7. What information is given on the slot welds?

______

8. What are the special instructions indicated in Note "A"? ______

______

9. The distance from the base of the unit to the top of the rails is ______.
10. What special instructions are to be followed when welding the rails into place?

    a. ______

    ______

    b. ______

    ______

11. How many different types of welds are indicated?

____

____

____

12. The total width of the unit is __________.
13. The width of the mounting plates is ______.
14. What instructions are given concerning the hole positions on the mounting plate? __________

____

____

____

15. Welding Note "B" is concerned with__________

____

______________________________.

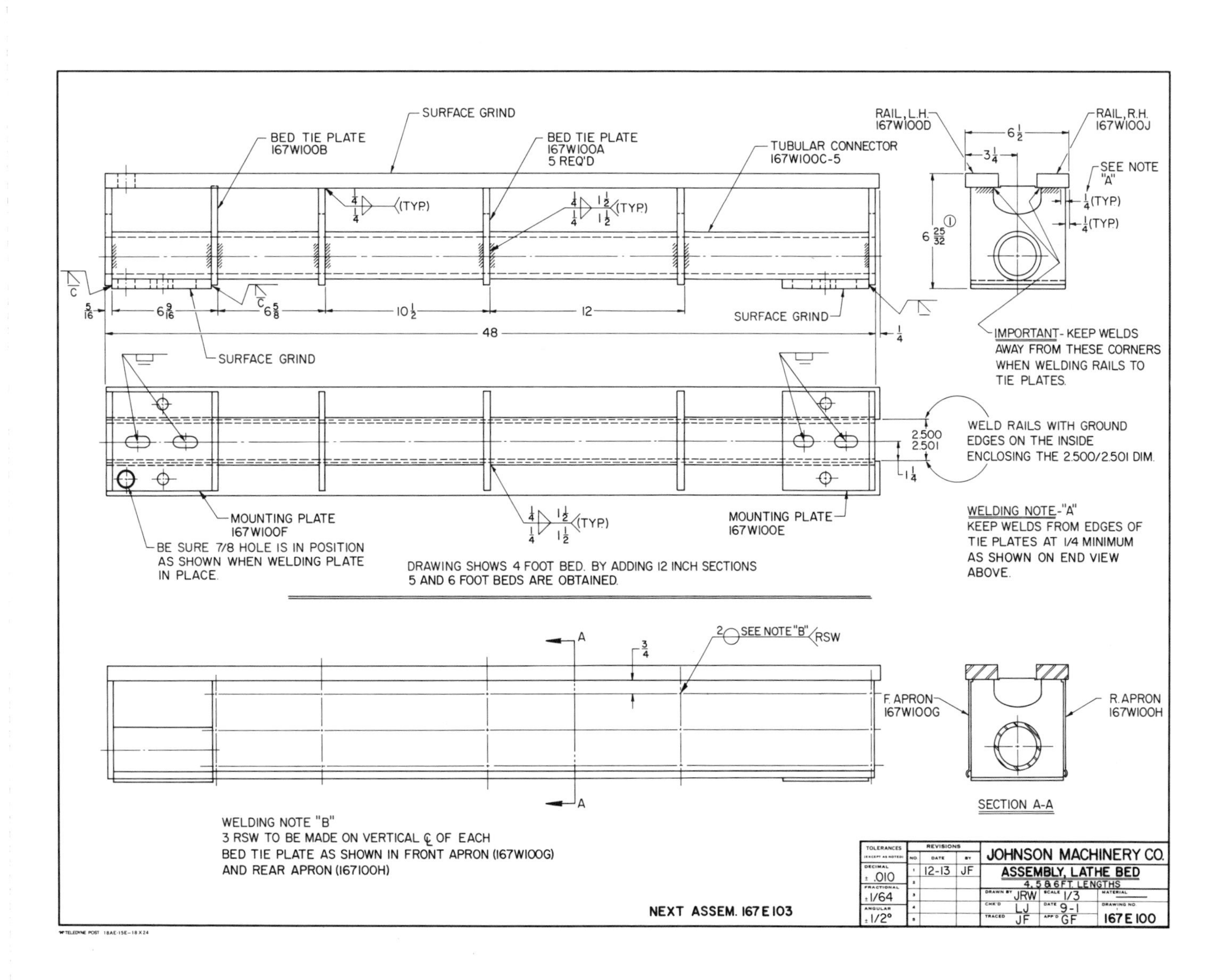

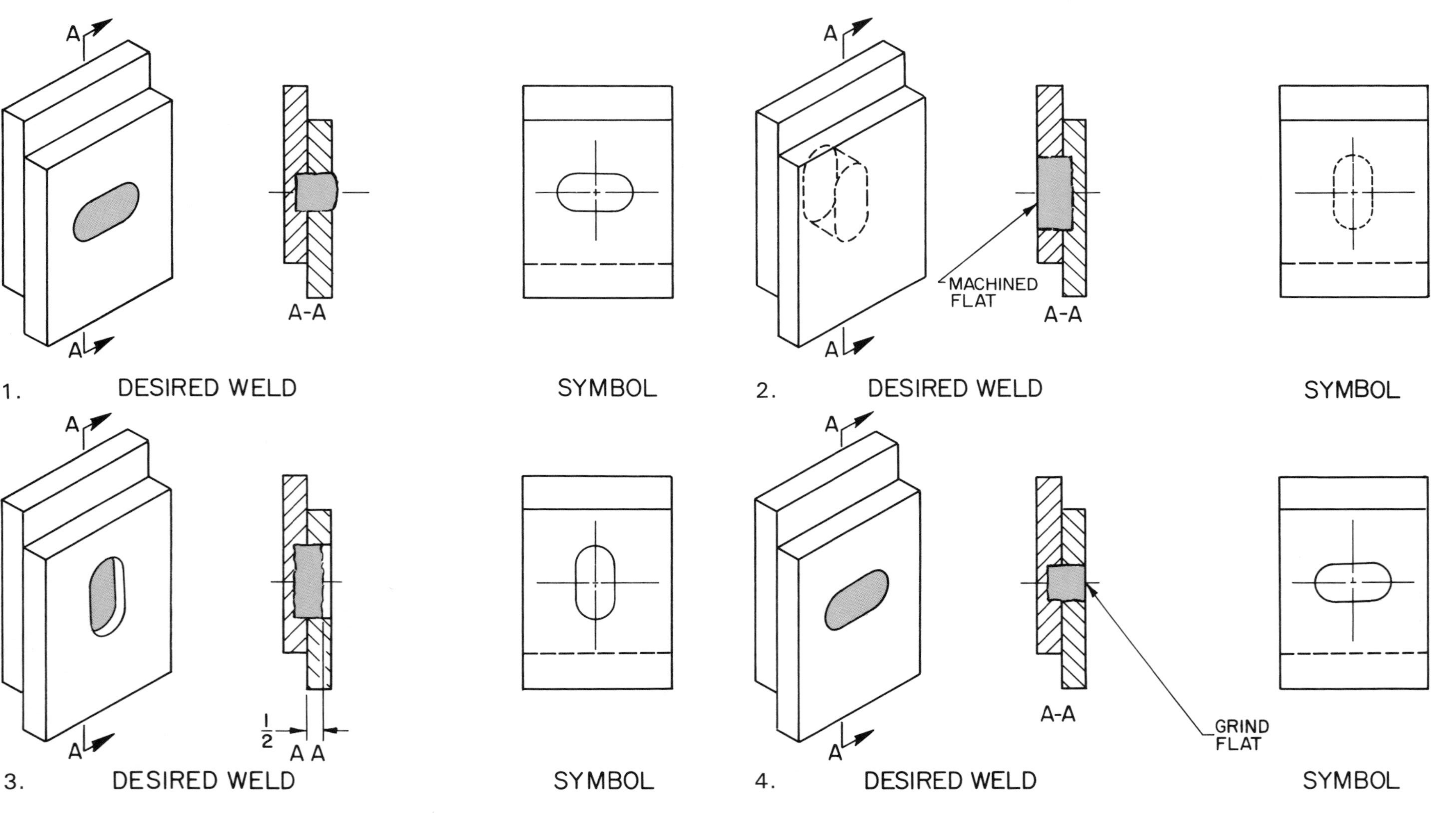

Part II Draw correct welding symbol for the slot welds on pages 146 and 147.

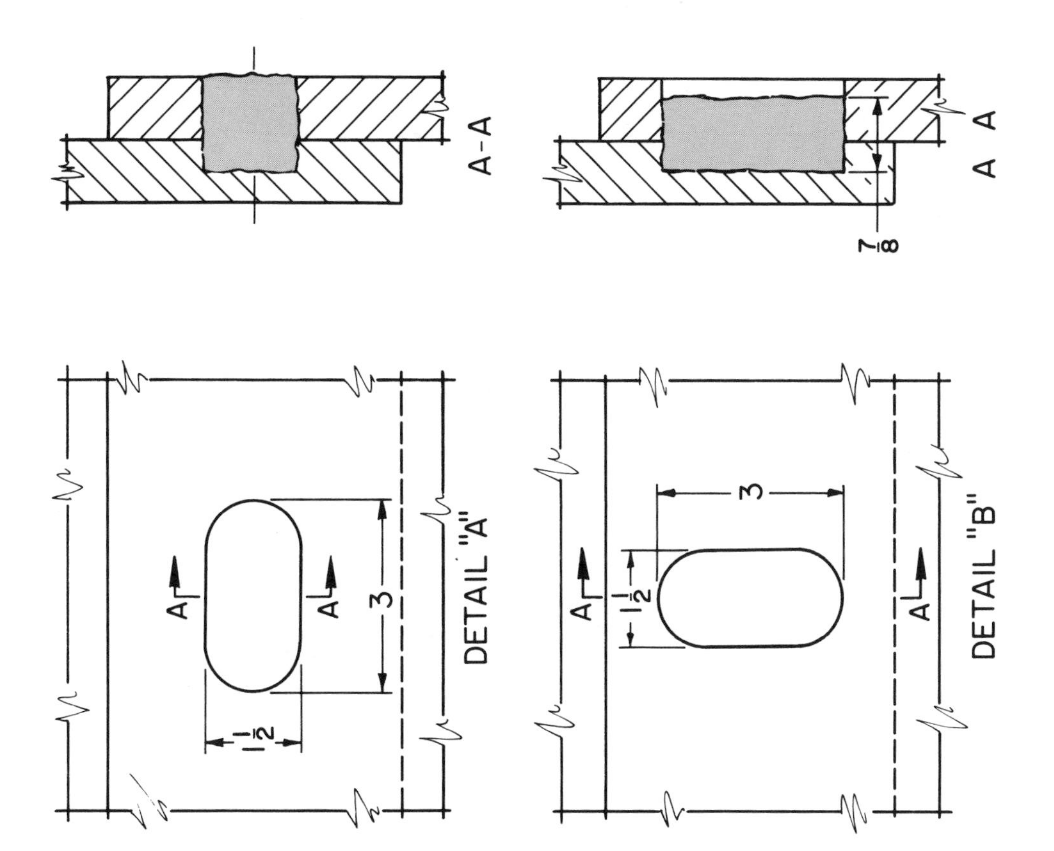
A-A
A A
DETAIL "A"
DETAIL "B"

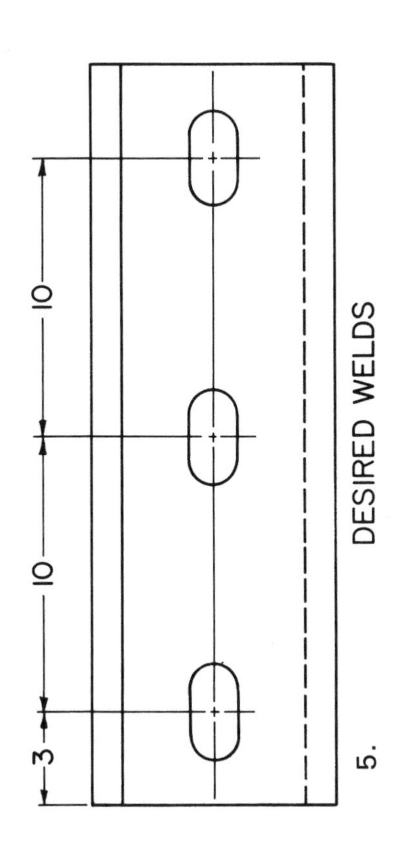
DESIRED WELDS
5.

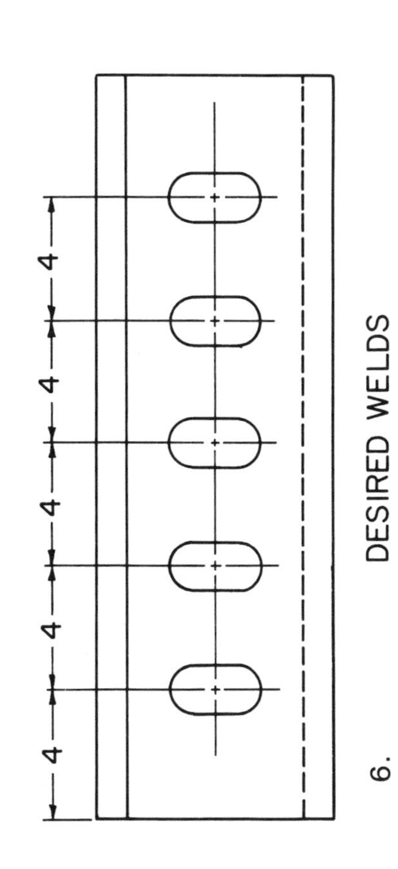
DESIRED WELDS
6.

Unit 16

SPOT WELDS

Spot welds may be made by any of several welding processes: resistance, arc, projection, electron beam, etc. With a spot weld, the welding process is indiciated in the TAIL of the welding symbol. See Fig. 16-1.

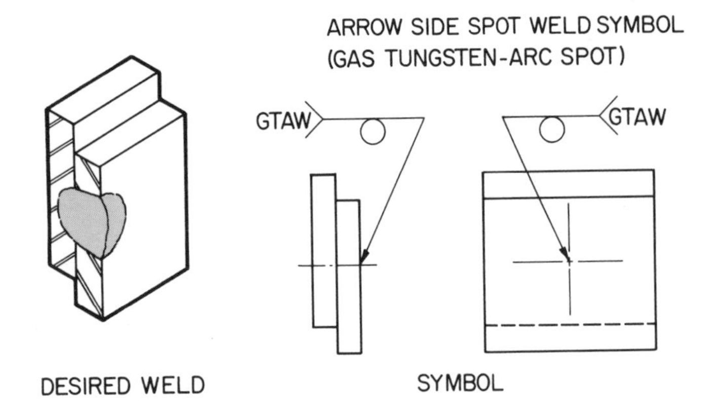

Fig. 16-1. When spot weld is specified, welding process is indicated in tail of welding symbol.

The location of the spot weld symbol in relation to the reference line of the welding symbol may or may NOT have arrow side or other side significance. Refer to Fig. 16-2.

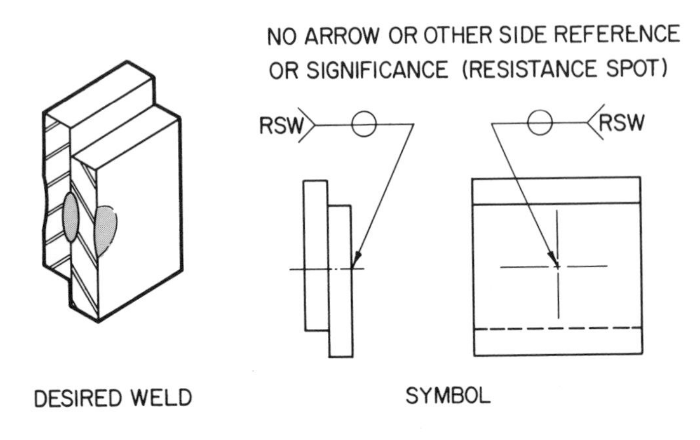

Fig. 16-2. Location of spot weld symbol in relation to reference line may or may not have arrow side or other side significance. (Also see Fig. 16-1.)

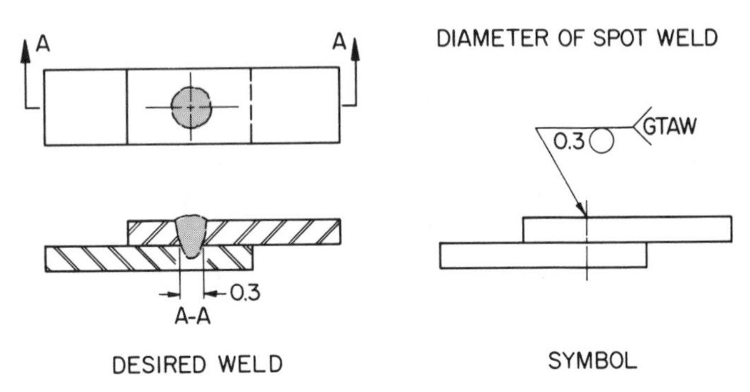

Fig. 16-3. Spot weld size is placed to the left of weld symbol.

SIZE AND STRENGTH OF SPOT WELDS

A spot weld is dimensioned by either size or strength.

Spot weld size is the diameter of the weld at the *interface zone* (point where members are joined). It is expressed in fractions, decimals, or millimeters. Spot weld size is shown to the LEFT of the weld symbol. This is illustrated in Fig. 16-3.

Spot weld strength is indicated in pounds or kilograms per spot in tension. It is also shown to the LEFT of the weld symbol, Fig. 16-4.

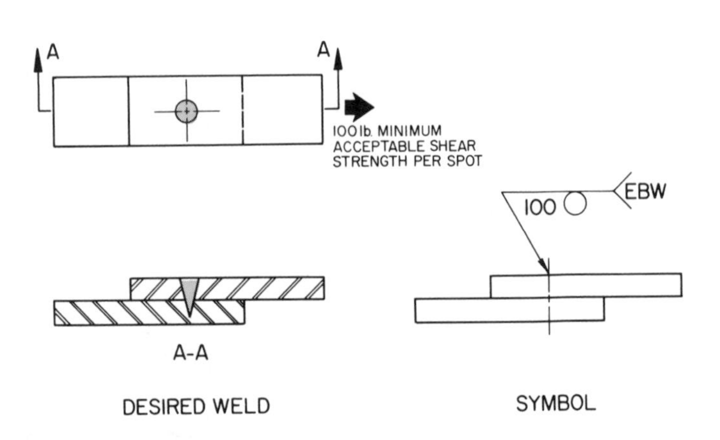

Fig. 16-4. Strength of spot weld is placed to the left of weld symbol.

SPACING OF SPOT WELDS

The *pitch* (center-to-center spacing) of spot welds is indicated to the RIGHT of the weld symbol. Fig. 16-5 shows an example.

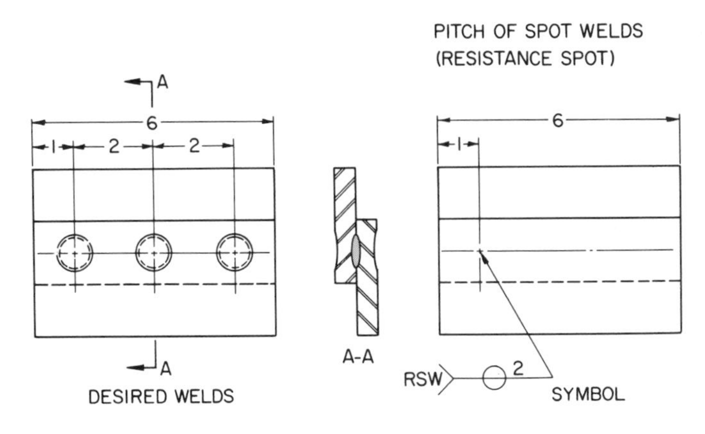

Fig. 16-5. Center-to-center spacing or pitch of spot welds is given to right of weld symbol.

EXTENT OF SPOT WELDING

When a series of spot welds are specified for less than the full length of a joint, the extent of the welds is dimensioned as in Fig. 16-6.

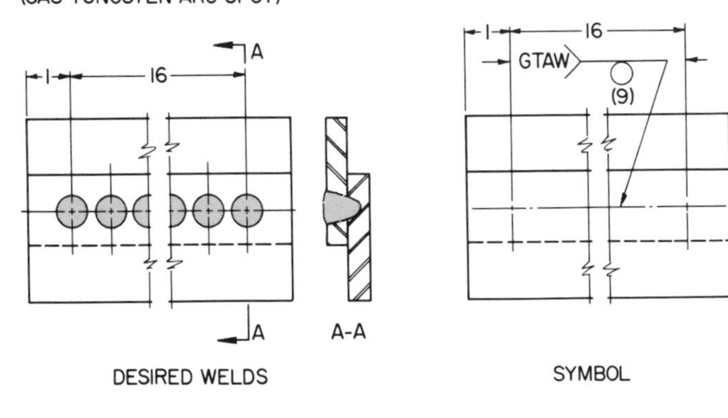

Fig. 16-6. Study weld symbol when a series of spot welds are specified and they are to take up less than full length of joint. Essential information for making welds is shown on plans as drawn.

NUMBER OF SPOT WELDS

If a specific number of spot welds is needed for a particular joint, the number of welds will be given in PARENTHESES either above or below the weld symbol. Refer to Fig. 16-7.

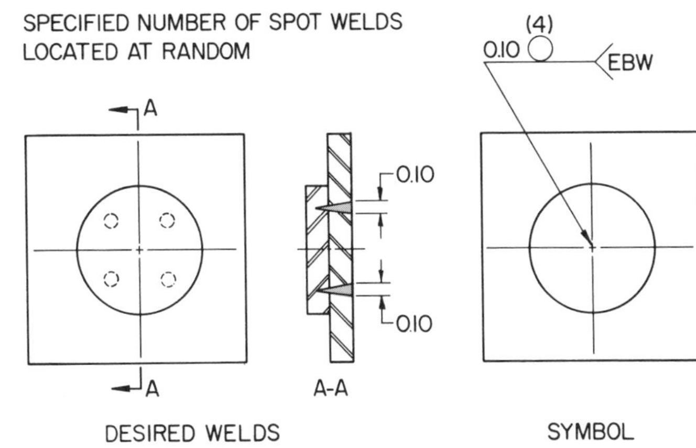

Fig. 16-7. Note how a specific number of spot welds is specified for this particular joint. The number of welds will be shown in parentheses either above or below weld symbol.

The location of a group of spot welds on intersecting centerlines will be indicated as shown in Fig. 16-8.

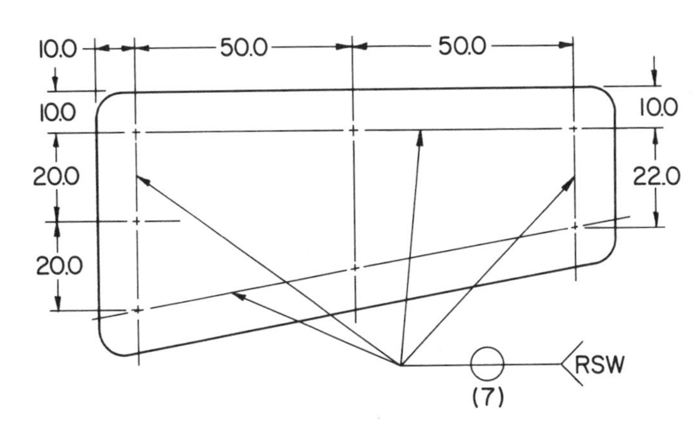

Fig. 16-8. Study how spot welds are specified when located on centerlines of component.

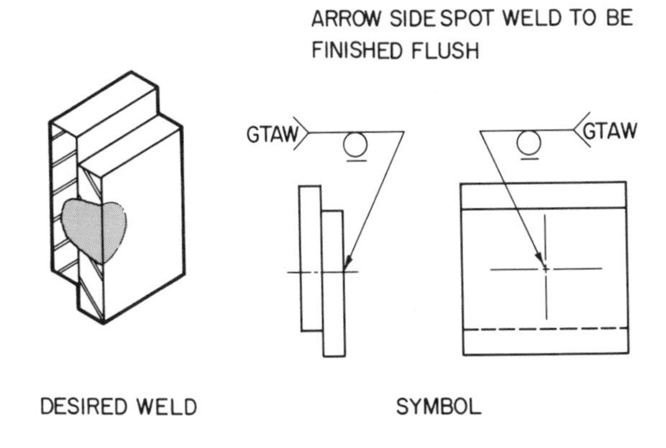

Fig. 16-9. When specifications call for exposed surface of spot welded joint to be flush, flush contour symbol is added to weld symbol.

FLUSH SPOT WELD JOINTS

The *flush contour symbol* is added to the weld symbol when specifications required the exposed surface of either member of a spot welded joint to be flush. Fig. 16-9 shows an example.

MULTIPLE MEMBER JOINT SPOT WELDS

With multiple member joint spot welds, the same spot weld symbol is used regardless of the number of pieces inserted between the two outer members of the required joint.

UNIT 16—TEST YOUR KNOWLEDGE

Part I

Refer to the drawing (167E104) to answer the following questions.

1. Name the drawing title. ________________
2. What is the drawing number? ____________
3. How many pieces make up the assembly? ____
   Name them.
   ________________ ________________
   ________________ ________________
   ________________ ________________
   ________________ ________________
4. What is the welding process to be used? ______
   ____
5. Is a welding sequence specified? __________
   If so, list it.
   ____
   ____
   ____
   ____
   ____
   ____
6. How many welds are specified to attach each side of the top, headstock pedestal (167W104C) to the back, headstock pedestal (167W104A) and front, headstock pedestal (167W104B)? ________ The weld pitch is __________.
7. What is the number of the next assembly? ____
8. How many welds are specified to attach each hinge (167W104E) to the front, headstock pedestal (167W104B)? __________ The weld pitch is __________.
9. What does the note concerning the placement of the hinge (167W104E) specify? ____________
   ____
   ____
   ____
10. This welding symbol is found on the print. What does it mean?

RSW = ________________________

(9) = ________________________

3 = ________________________

11. What is the full name and number of the base?
    ____
12. Section A-A is shown on the drawing. What does it show? ________________________
    ____
13. The following tolerances are allowed:
    Decimal: ________________________
    Fraction: ________________________
    Angular: ________________________
14. When positioning the top, headstock pedestal (167W104C) to be welded to the back, headstock pedestal (167W104A) and front, headstock pedestal (167W104B), what precaution must be observed about its placement? ________________
    ____

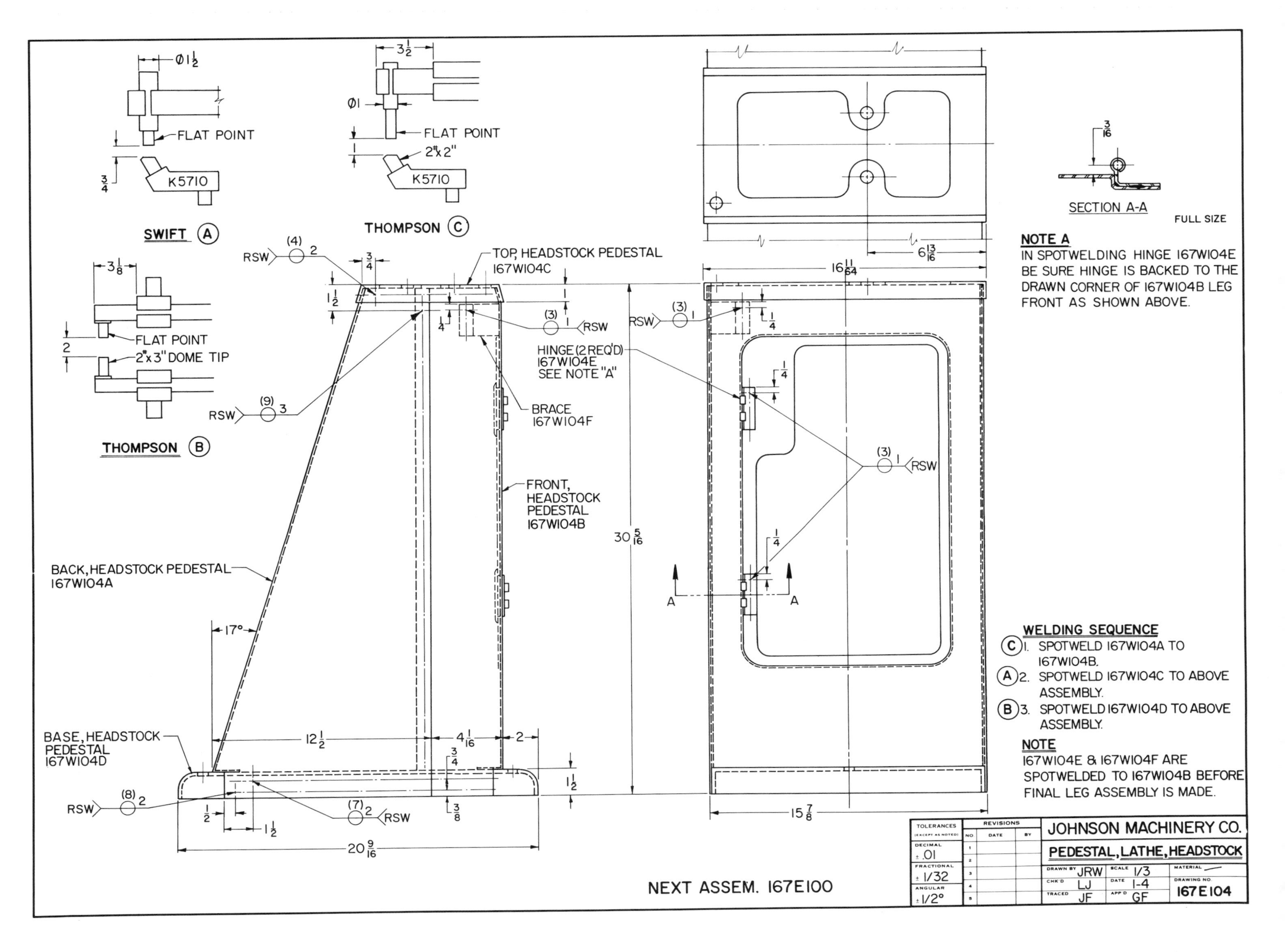

SWIFT (A)
FLAT POINT
K5710
THOMPSON (C)
FLAT POINT
2#x2"
THOMPSON (B)
FLAT POINT
2#x3" DOME TIP
TOP, HEADSTOCK PEDESTAL 167W104C
HINGE (2 REQ'D) 167W104E SEE NOTE "A"
BRACE 167W104F
FRONT, HEADSTOCK PEDESTAL 167W104B
BACK, HEADSTOCK PEDESTAL 167W104A
BASE, HEADSTOCK PEDESTAL 167W104D
RSW
SECTION A-A
FULL SIZE
NOTE A
IN SPOTWELDING HINGE 167W104E BE SURE HINGE IS BACKED TO THE DRAWN CORNER OF 167W104B LEG FRONT AS SHOWN ABOVE.
WELDING SEQUENCE
(C) 1. SPOTWELD 167W104A TO 167W104B.
(A) 2. SPOTWELD 167W104C TO ABOVE ASSEMBLY.
(B) 3. SPOTWELD 167W104D TO ABOVE ASSEMBLY.
NOTE
167W104E & 167W104F ARE SPOTWELDED TO 167W104B BEFORE FINAL LEG ASSEMBLY IS MADE.
NEXT ASSEM. 167E100
TOLERANCES (EXCEPT AS NOTED)
DECIMAL ±.01
FRACTIONAL ±1/32
ANGULAR ±1/2°
REVISIONS
JOHNSON MACHINERY CO.
PEDESTAL, LATHE, HEADSTOCK
DRAWN BY JRW
SCALE 1/3
CHK'D LJ
DATE 1-4
TRACED JF
APP'D GF
DRAWING NO. 167E104

Part II

1. What does the following welding symbol indicate?

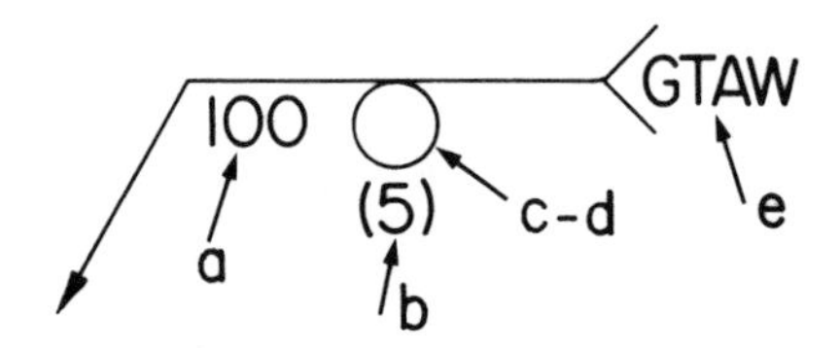

a. ______________________
b. ______________________
c. ______________________
d. ______________________
e. ______________________

2. Describe the weld(s) indicated by the following welding symbol.

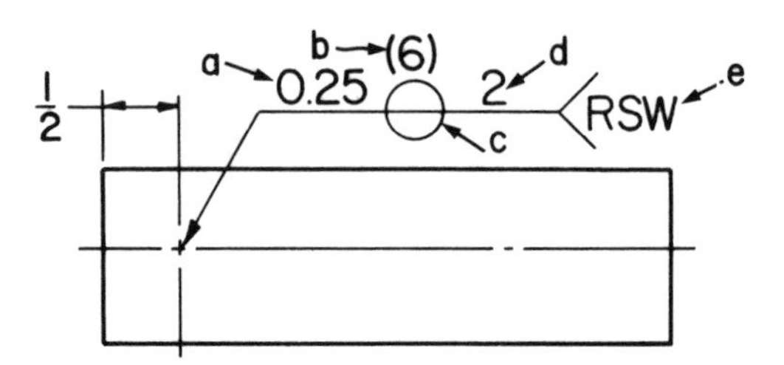

a. ______________________
b. ______________________
c. ______________________
d. ______________________
e. ______________________

3. Sketch in the weld(s) indicated by the welding symbol.

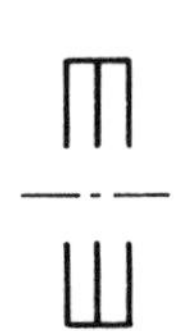

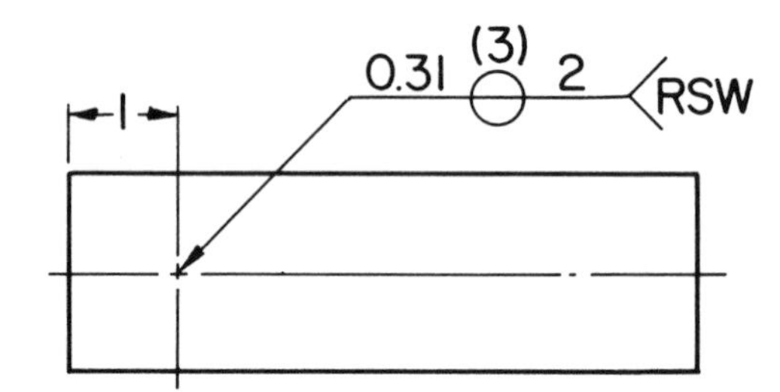

4. Describe the following weld(s).

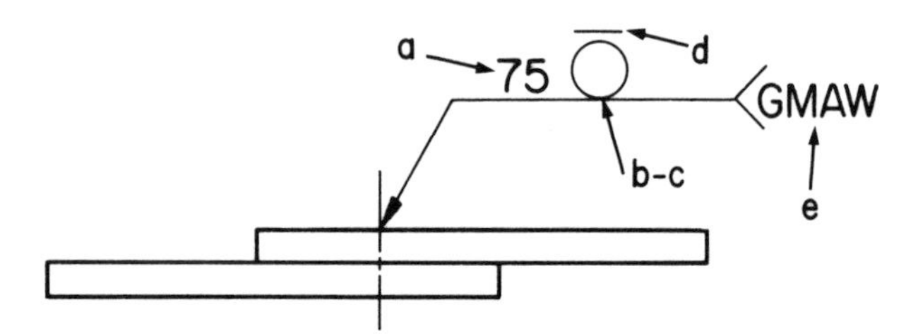

a. ______________________
b. ______________________
c. ______________________
d. ______________________
e. ______________________

5. Sketch the welding symbol that will describe the following spot weld(s).

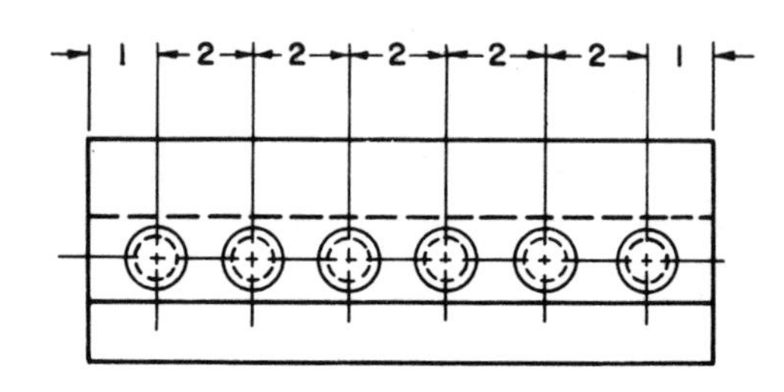

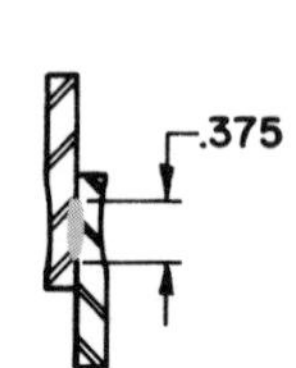

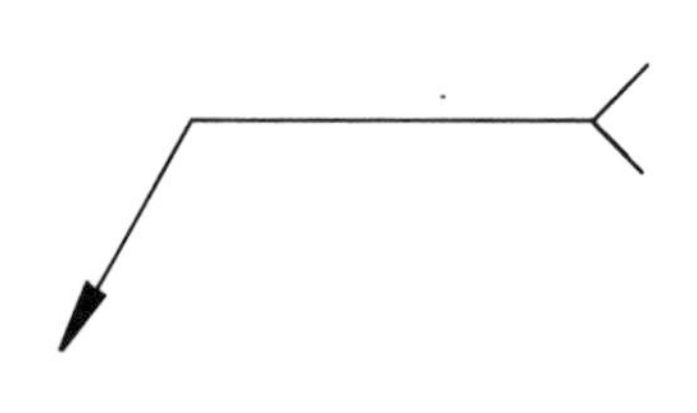

Unit 17

SEAM WELDS

A *seam weld,* like the spot weld, may be made by any of several welding processes. The process to be employed is indicated in the tail of the welding symbol, as shown in Fig. 17-1.

The location of the seam weld symbol, in relation to the reference line, may or may NOT have arrow side or other side significance. See Fig. 17-2.

SIZE AND STRENGTH OF SEAM WELDS

Seam welds are dimensioned by either size or strength. *Seam weld size* is designated as the width of the weld. It is expressed in fractions of an inch, decimals, or in millimeters. Seam weld size is shown to the LEFT of the weld symbol. Refer to Fig. 17-3.

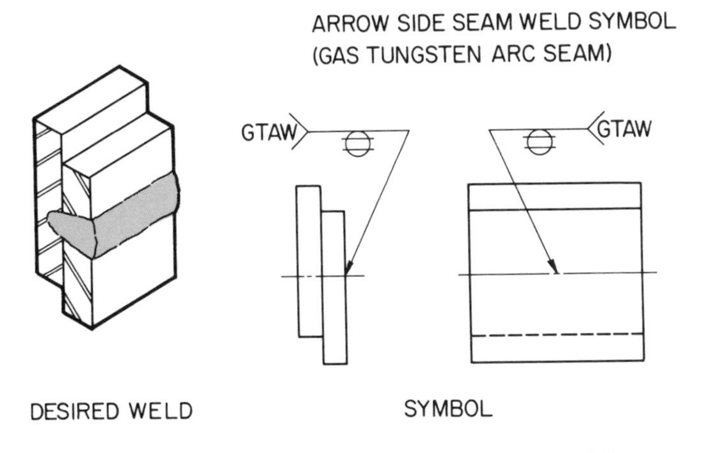

Fig. 17-1. Tail of the welding symbol contains type of welding process for seam weld.

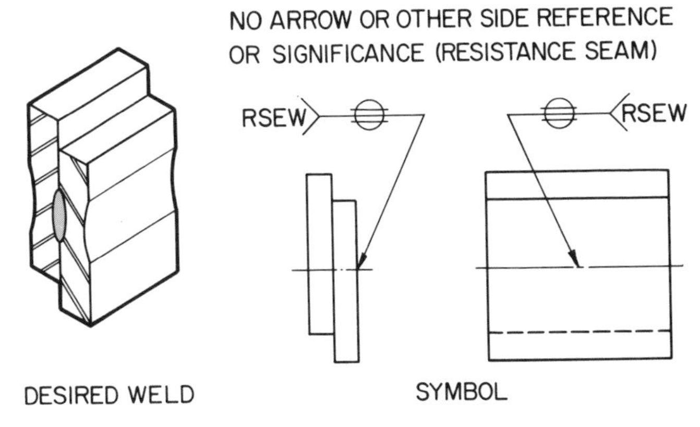

Fig. 17-2. Seam weld symbol location, in relation to reference line, may or may not have arrow side or other side significance. (Also see Fig. 17-1.)

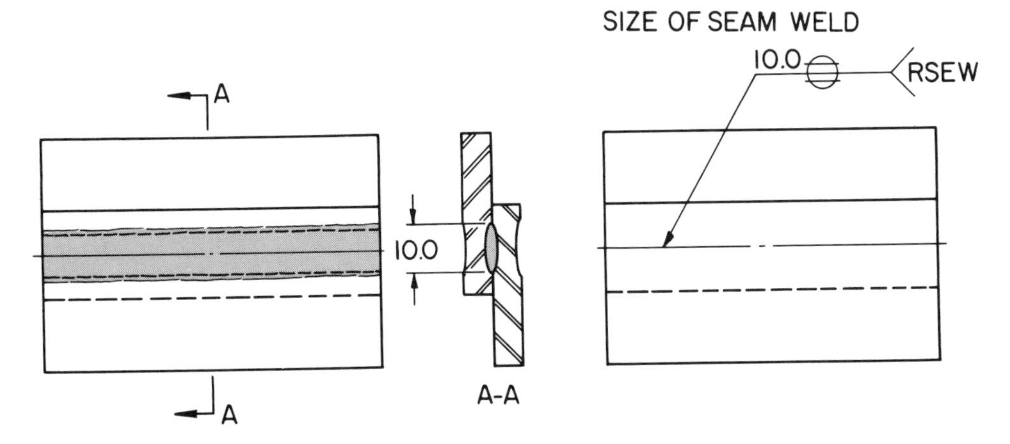

Fig. 17-3. Size of a seam weld is given as width of weld and is expressed in fractions of an inch, decimals, or millimeters. Note that it shown to left of weld symbol.

When used, *seam weld strength* is designated in pounds per inch of weld or in metric units. This information is also shown to the LEFT of the weld symbol. See Fig. 17-4.

LENGTH OF SEAM WELDS

Seam weld length, when indicated on the welding symbol, is shown to the RIGHT of the weld symbol. This is illustrated in Fig. 17-5.

A seam weld that extends less than the full length of the joint is dimensioned as shown in Fig. 17-6.

DIMENSIONING OF INTERMITTENT SEAM WELDS

The *pitch* (center-to-center spacing) of intermittent seam welds is given as the distance between centers of the weld increments. Fig. 17-7 gives an example. The

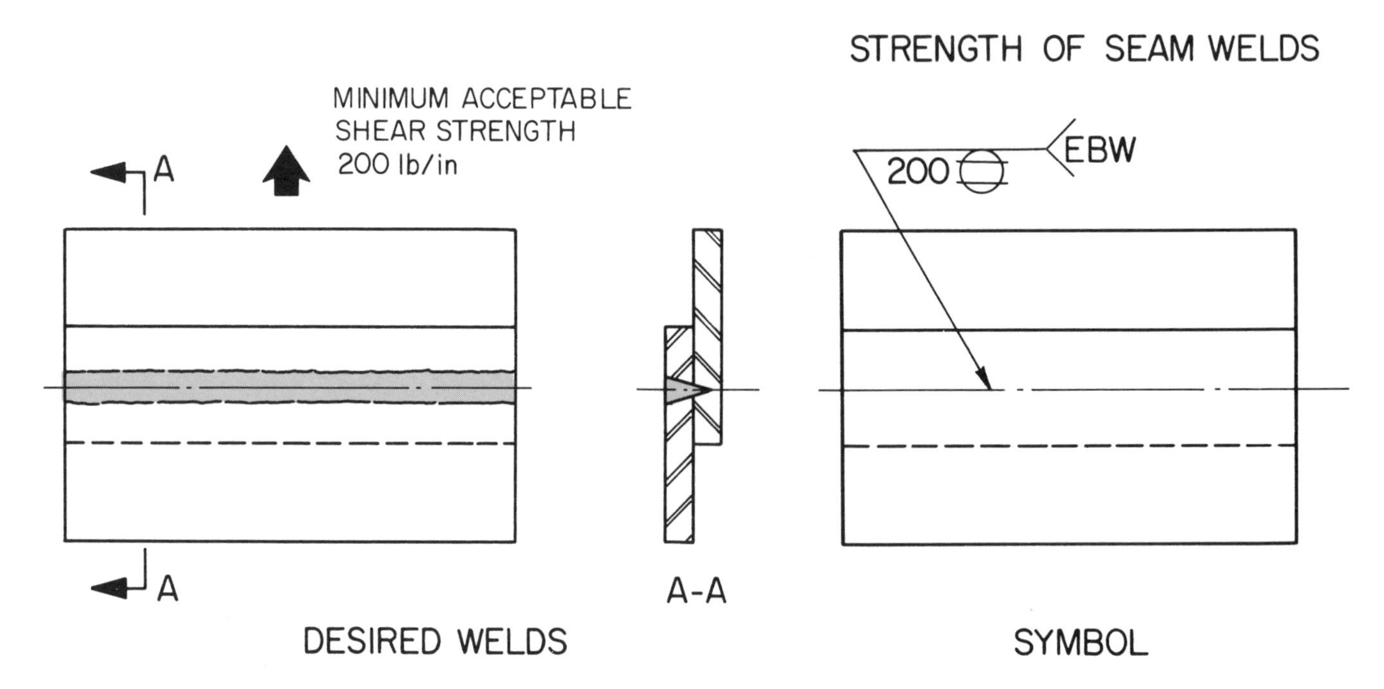

Fig. 17-4. Seam weld strength, when used, is given in pounds per linear inch or in metric units and is to left of weld symbol.

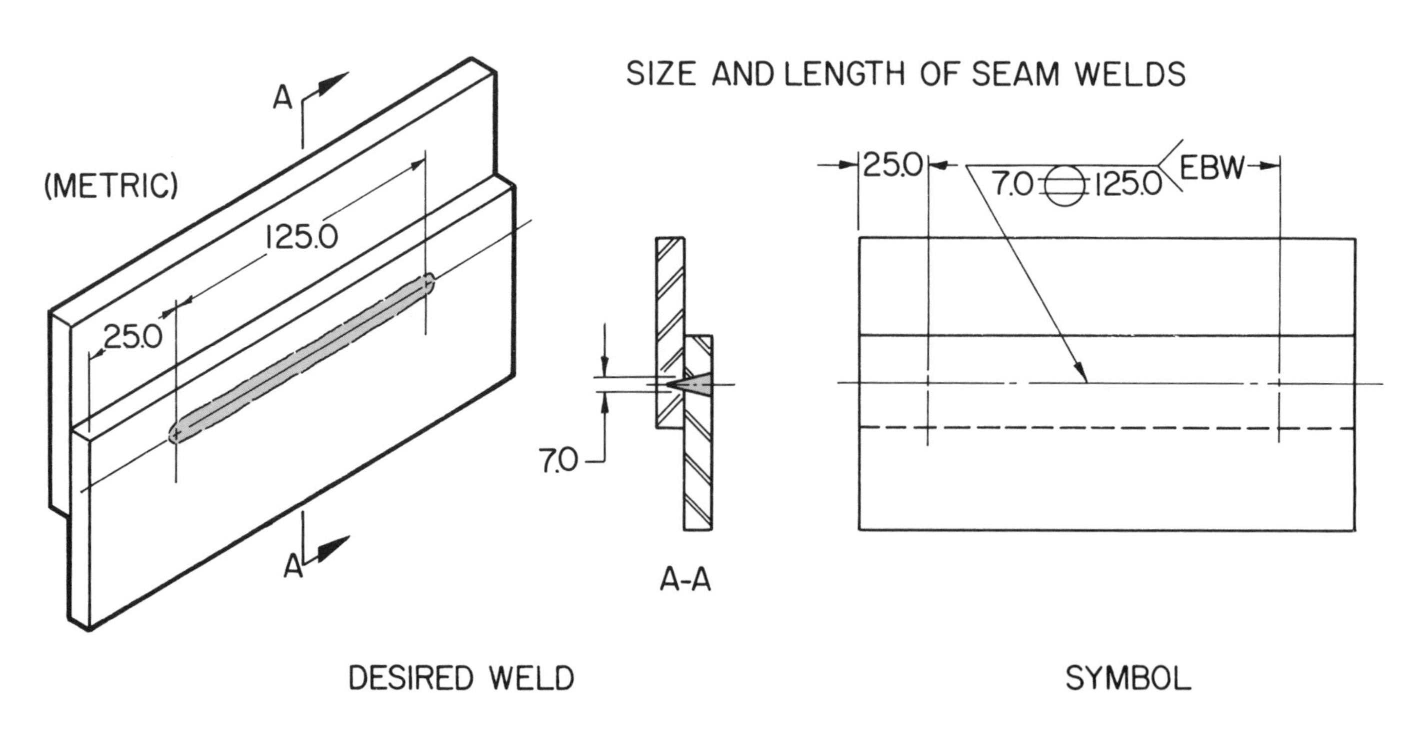

Fig. 17-5. When on welding symbol, seam weld length is on right of weld symbol.

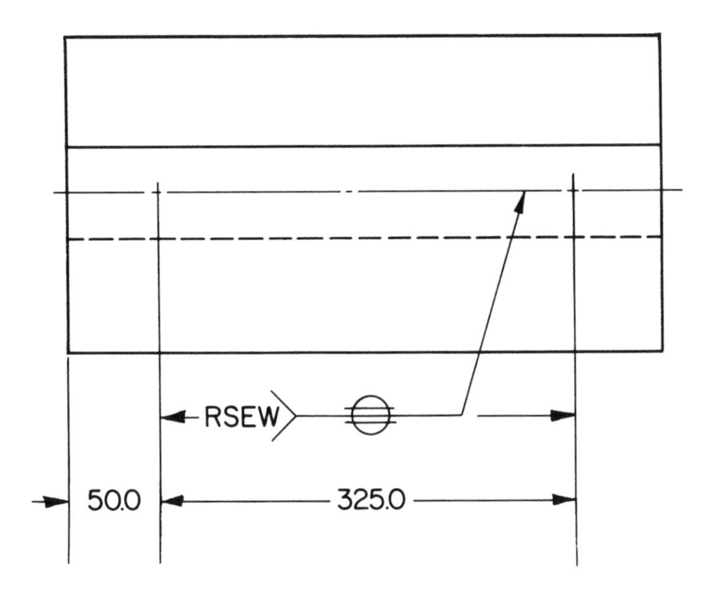

Fig. 17-6. Study dimensions for seam weld that extends less than the full length of the joint.

pitch is shown to the RIGHT of the weld length dimension.

ORIENTATION OF SEAM WELDS

Unless otherwise indicated, intermittent seam welds are understood to have length and width PARALLEL to the axis of the weld. Refer to Fig. 17-8.

FLUSH SEAM WELD JOINTS

The flush contour symbol is used when the exposed surface of either (or both) member is to be finished flush. See Fig. 17-9.

MULTIPLE JOINT SEAM WELDS

No special addition to the welding symbol is required when additional pieces are inserted between the two outer members.

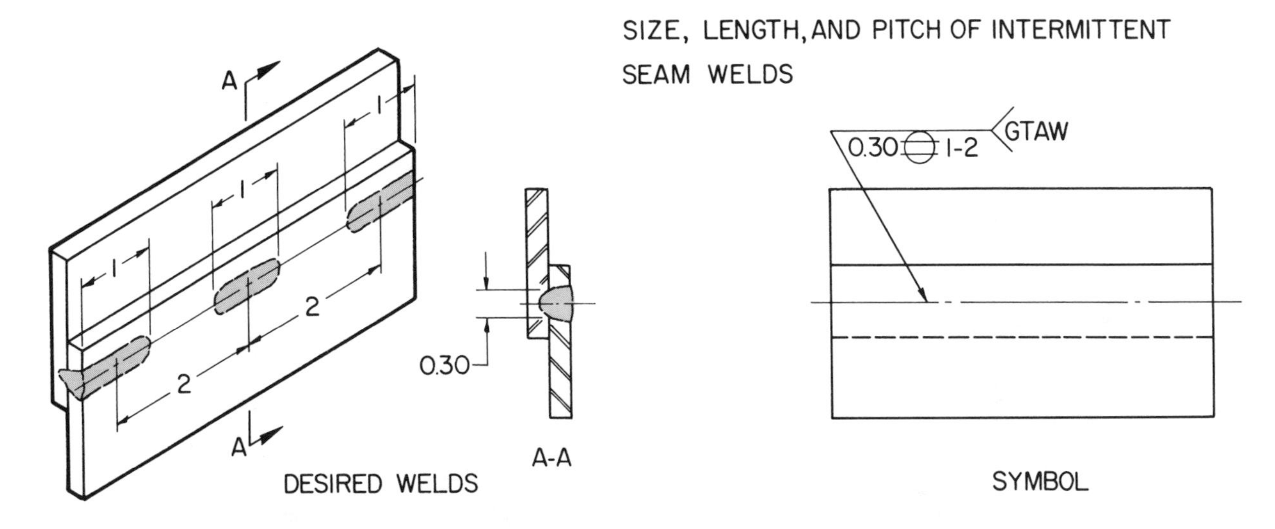

Fig. 17-7. Pitch or center-to-center spacing of intermittent seam welds is to right of weld length dimension.

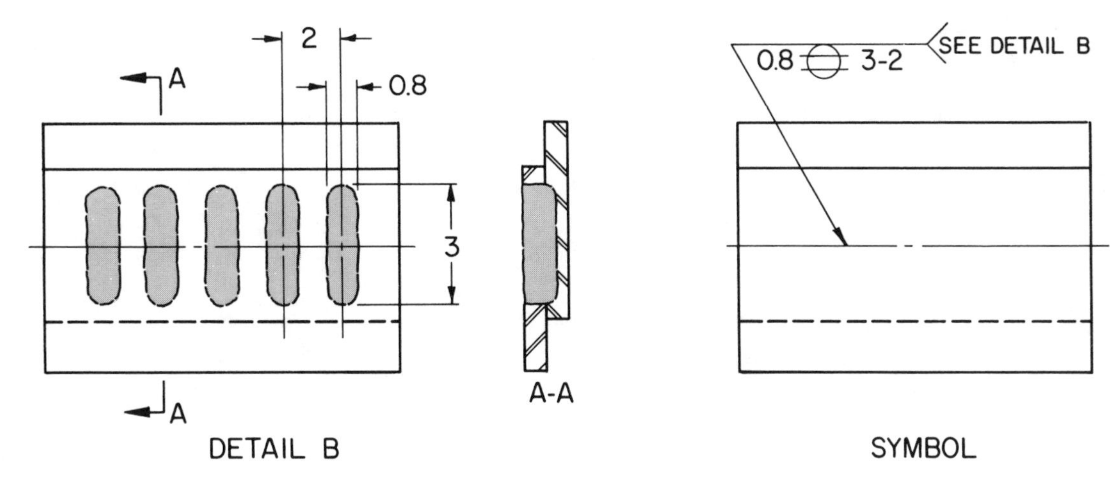

Fig. 17-8. Intermittent seam welds have length and width parallel to axis of weld, unless indicated otherwise.

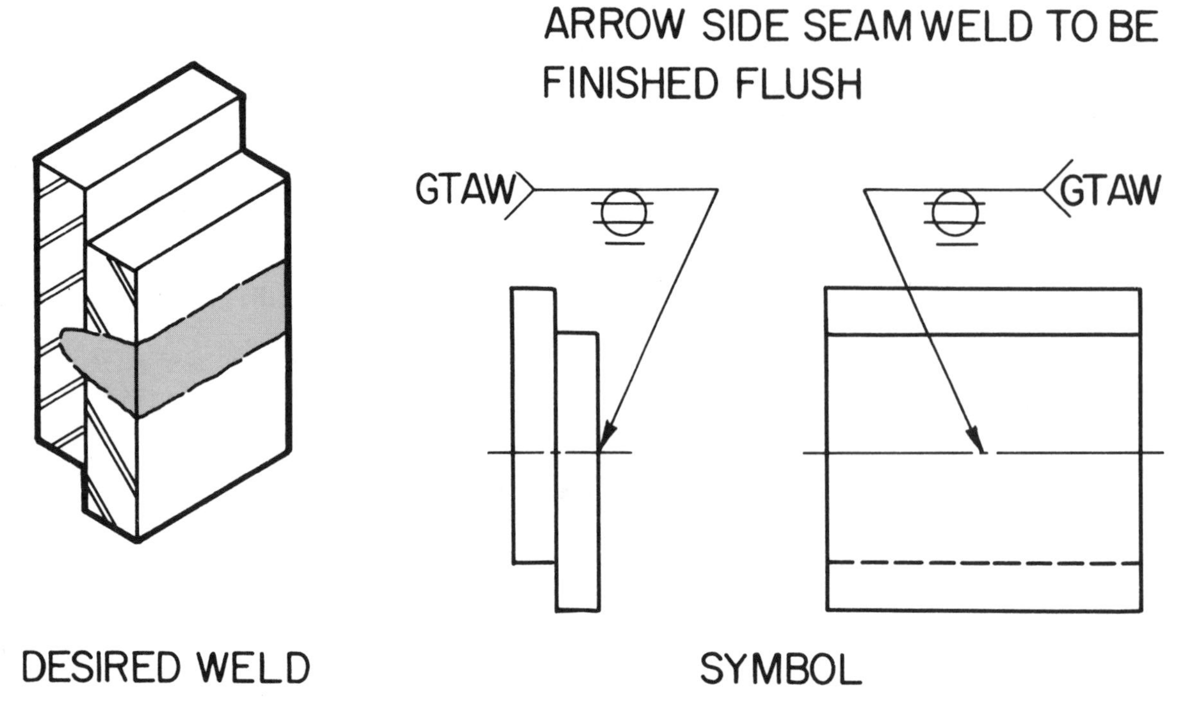

Fig. 17-9. Flush contour symbol denotes that exposed surface(s) is to be finished flush.

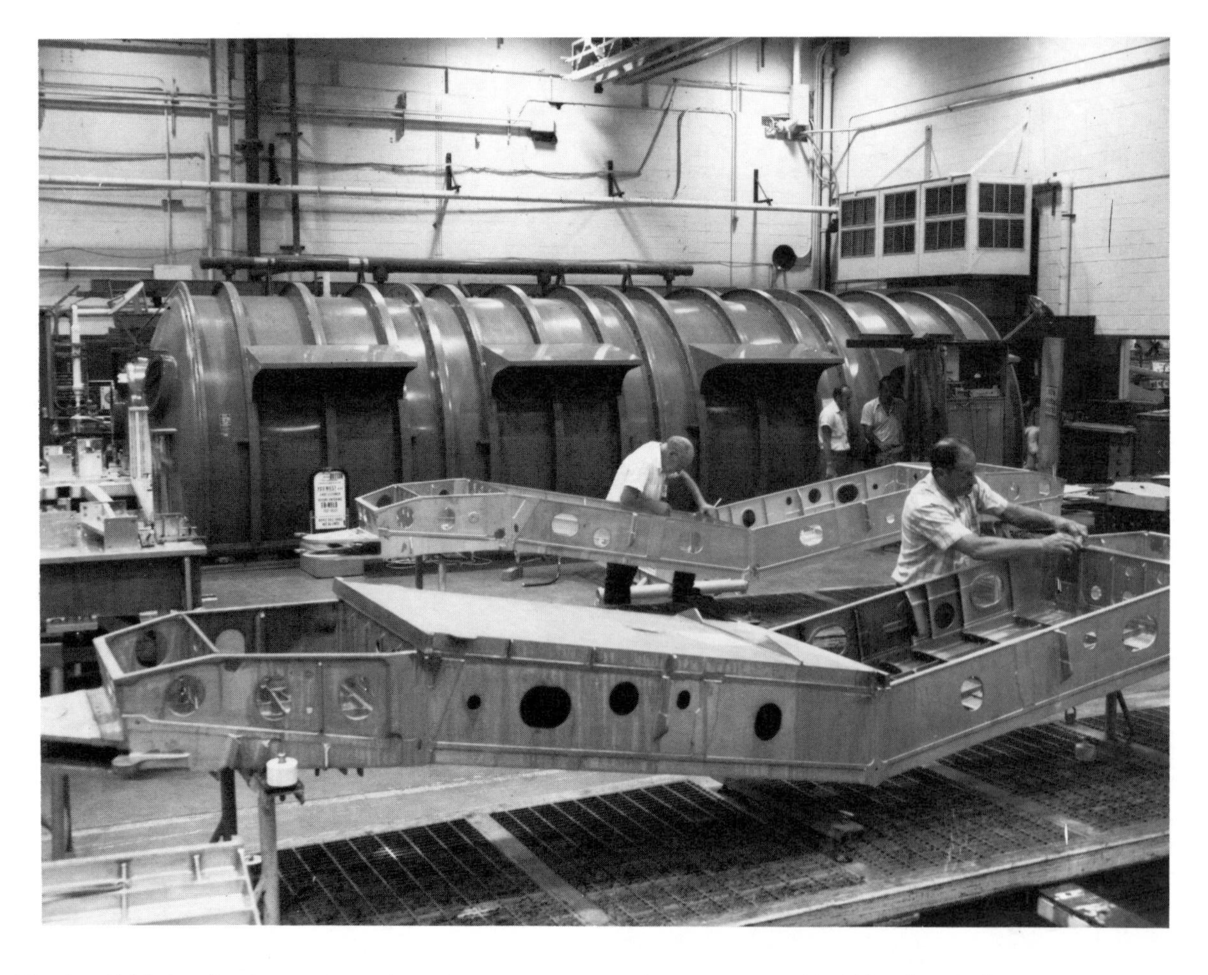

Titanium F-14 aircraft wing center sections after they have been electron beam welded. The clam shell vacuum chamber (it can accommodate work up to 12 x 26 ft. [3.66 x 7.93 ml]), can be seen in the background. (Grumman Aeorspace Corp.)

UNIT 17—TEST YOUR KNOWLEDGE

Part I

1. Sketch the welding symbol for the following seam weld on the adjacent incomplete symbol.
   a. Gas tungsten arc weld.
   b. Arrow side.
   c. Flush seam.
   d. 0.50 seam width.

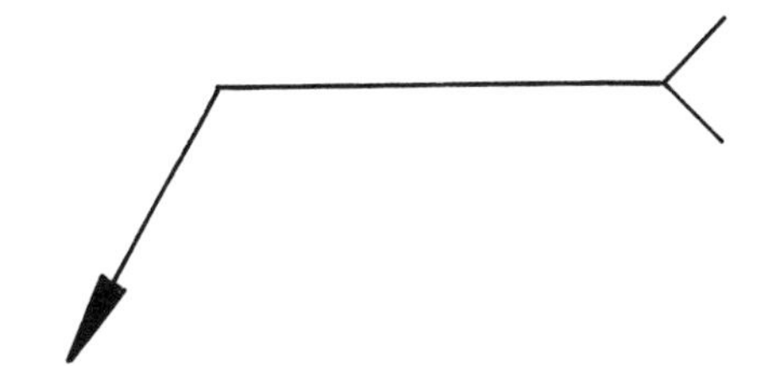

2. What does the following welding symbol indicate?

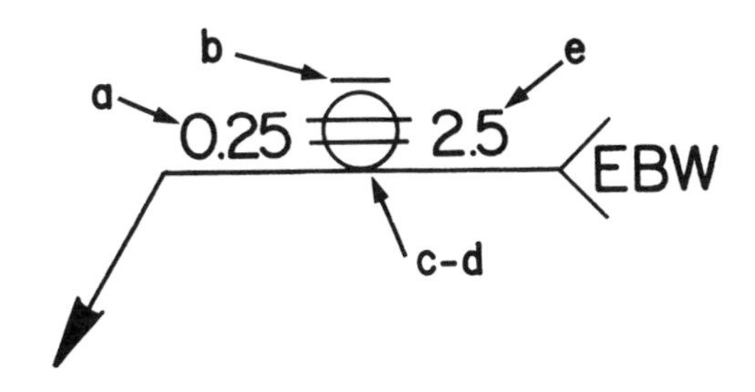

a. ______________________
b. ______________________
c. ______________________
d. ______________________
e. ______________________

3. Sketch in and dimension the weld(s) indicated.

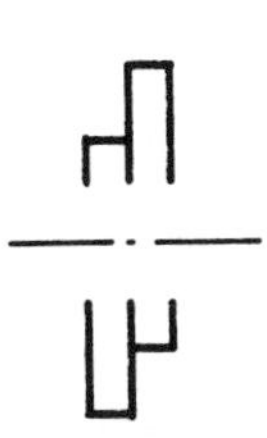

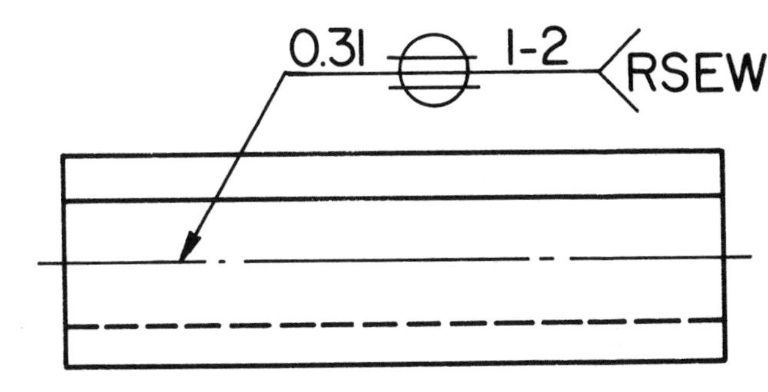

4. What does the figure (see arrow) indicate when used with a welding symbol?

____

____

____

____

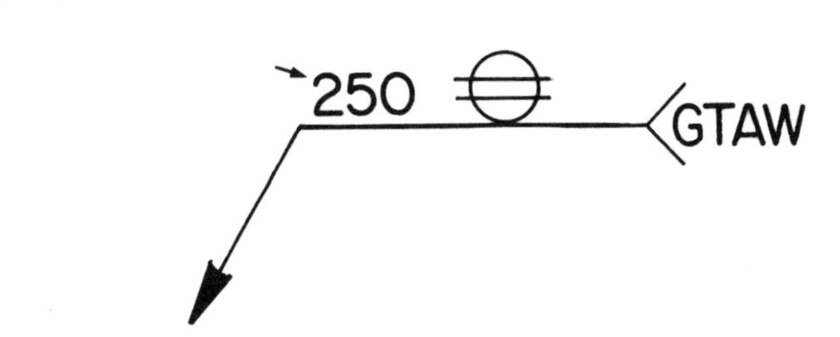

5. How does a welder determine seam weld orientation (direction weld is to run)?

____

____

____

____

Part II

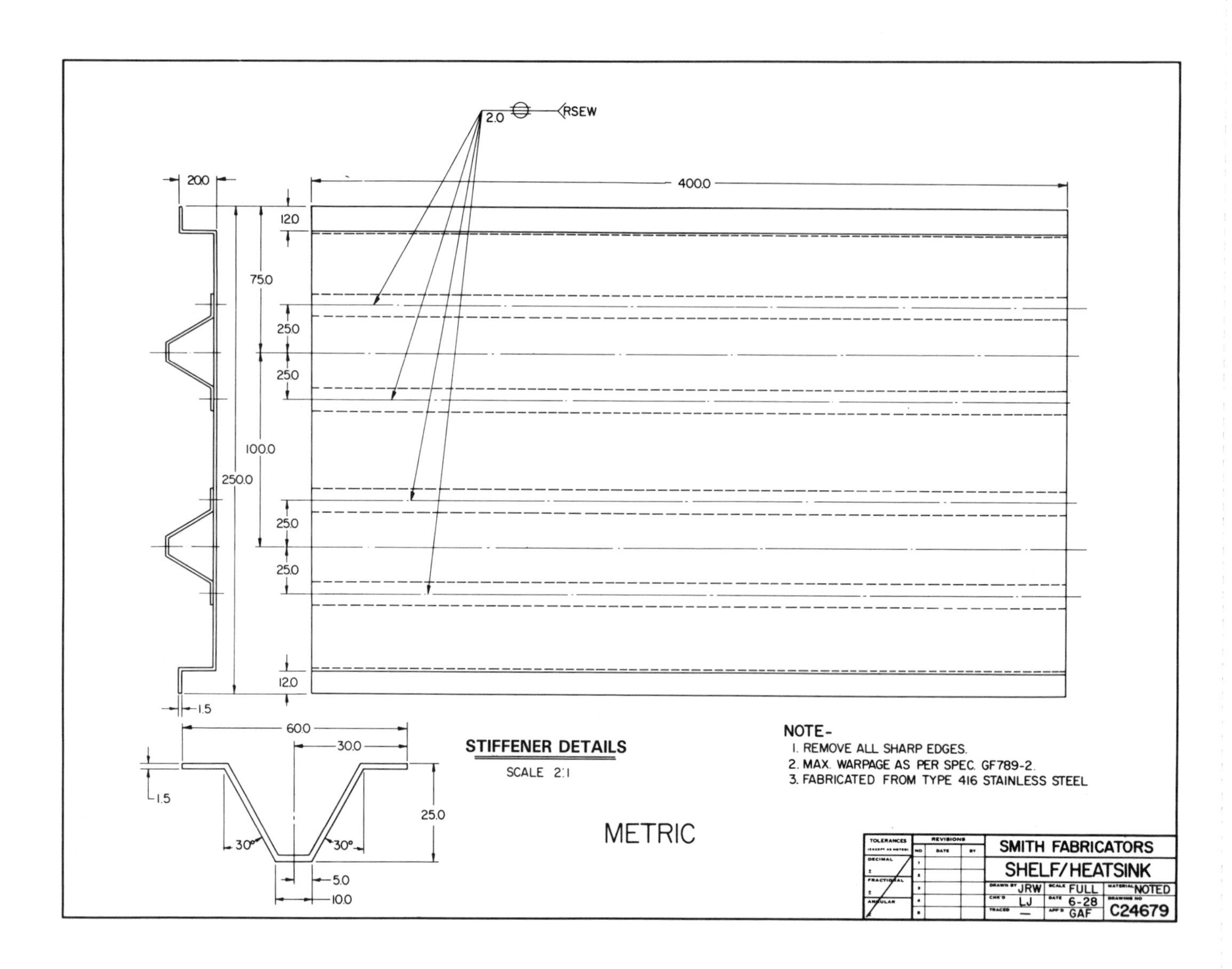

Refer to the drawing (C24679) to answer the following questions.

1. Name the product. ____________________
2. List the drawing number. ____________________
3. Dimensions are in ____________________.
4. What is the scale of the drawing? ____________
5. What is the scale of the stiffener details? ____
6. Name the drawing size? ____________________
7. What welding process is specified? __________
8. The product is fabricated from ________ ________ with a thickness of _________.
9. Name the size of the weld. ________________
10. Dimensions of the product are: __________ long, __________ wide, and __________ high.
11. Dimensions of the shelf before being formed are: __________ long and __________ wide.
12. The height of the stiffener is __________ and the width formed is __________.
13. What are the number of stiffeners specified? ______________________________
14. The centerline of the first stiffener is located __________ in from the edge of the shelf. The centerline of the second stiffener is __________ in from the edge of the shelf.

15. The note on the drawing specifies:
    a. ____________________
    b. ____________________
16. Study the drawing (C24679) carefully and prepare one question to be answered by members of the class. ____________________

This master computer control station continually monitors 58 robots on the robogate welding line at this Chrysler assembly plant. Plants like this have little need for uneducated and unskilled workers. (Chrysler Corporation)

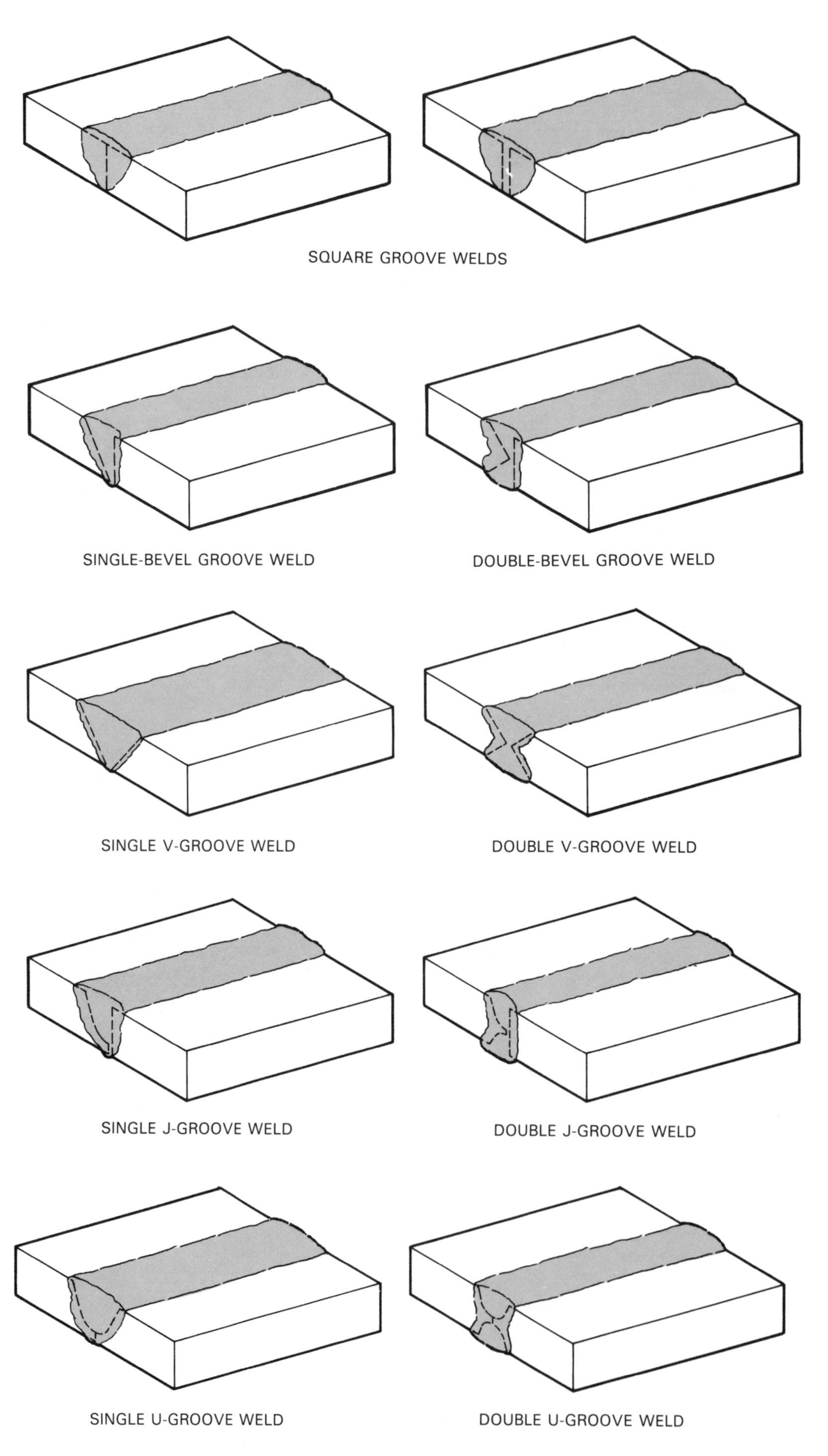

Fig. 18-1. Study groove weld types.

Unit 18

GROOVE WELDS

Groove welds are usually specified when thick metal sections are to be joined, Fig. 18-1. With the exception of the square groove and flare groove joints, one or more of the members being joined have metal removed to form a V, J, or U-shaped trough. The metal is removed by burning, grinding, chiseling, or machining.

GENERAL USE OF GROOVE WELD SYMBOL

Dimensions for the preparation of groove welds are shown on the SAME SIDE of the reference line as the weld symbol. This is illustrated in Fig. 18-2.

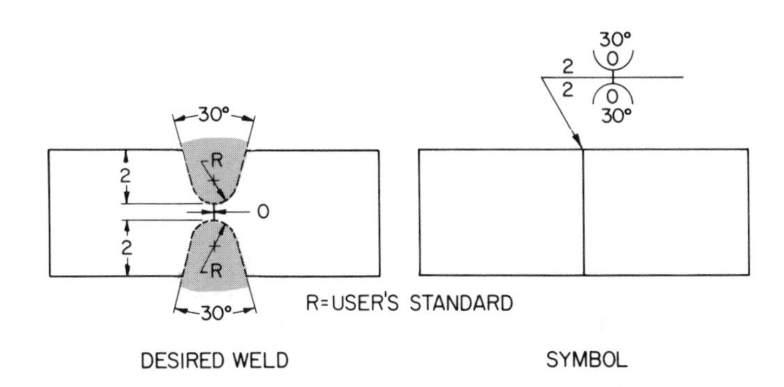

Fig. 18-3. Without general note on print, double-groove welds are dimensioned on both sides of the reference line.

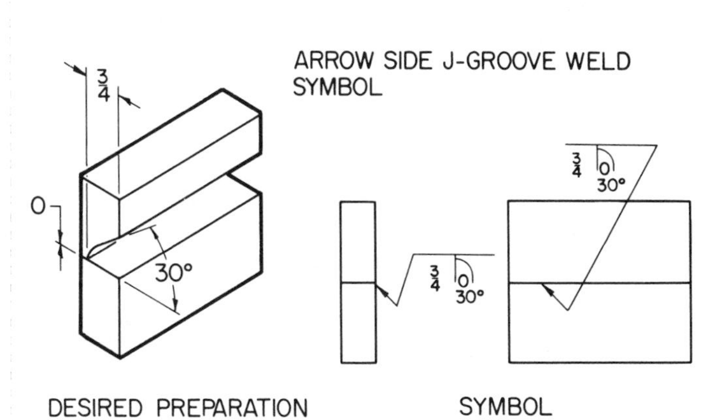

Fig. 18-2. Dimensions showing preparation of groove welds are on same side of reference line as weld symbol.

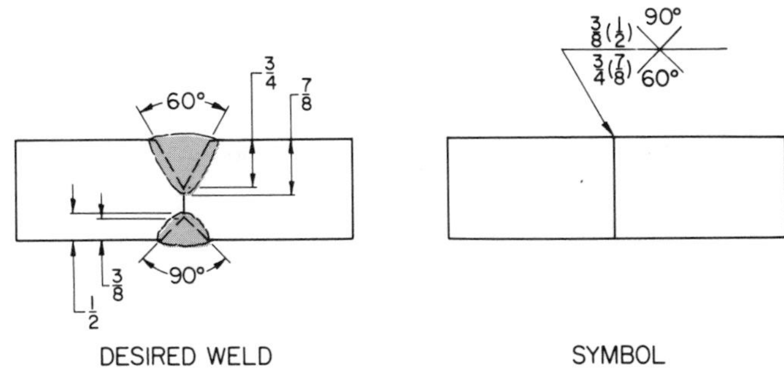

Fig. 18-4. Note how groove welds differing in size are dimensioned.

Double-groove welds, if no general note appears on the print, are dimensioned on both sides of the reference line. See Fig. 18-3. If the welds differ in size, they will be dimensioned as in Fig. 18-4.

Groove welding symbols will NOT include dimensions when a general note governing groove weld size appears on the print. Fig. 18-5 gives an example.

When a break in the arrow is used with bevel and J-groove welds, the arrow will point TOWARDS the member to be beveled. Refer to Fig. 18-6.

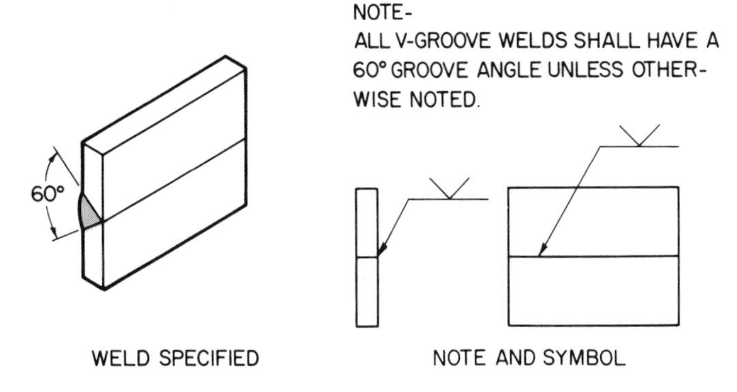

Fig. 18-5. No groove weld dimensions are given with welding symbol when a general note governing weld size is on print.

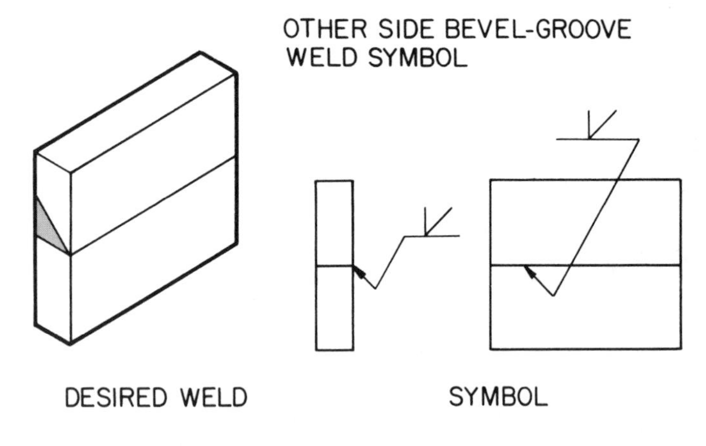

Fig. 18-6. Break in arrow always points towards member of single bevel and J-groove to be beveled.

SIZE AND EFFECTIVE THROAT OF GROOVE WELDS

The *effective throat* (weld penetration) of a groove weld is shown in parentheses when the weld extends only partially through the members being joined. It is to the LEFT of the weld symbol, as in Fig. 18-7.

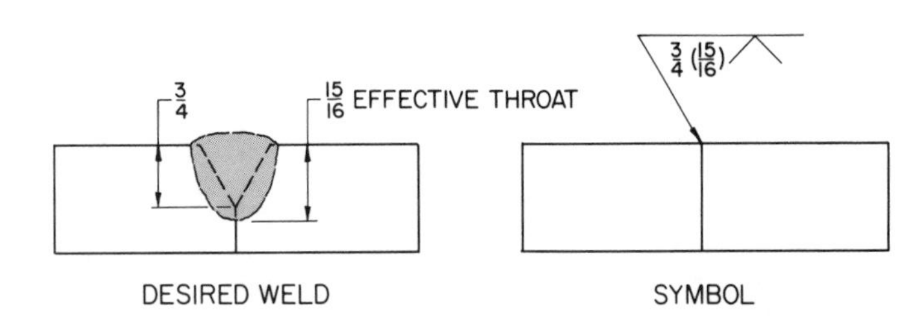

Fig. 18-7. Parentheses, to left of the weld symbol, gives effective throat (weld penetration) of the groove weld when weld extends partly through members being joined.

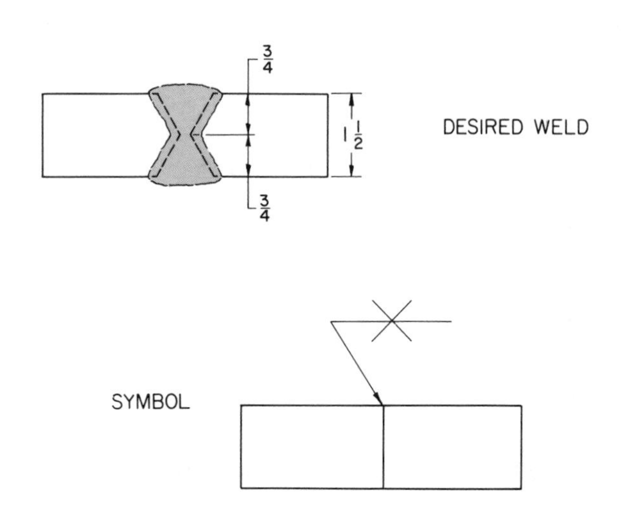

Fig. 18-8. When no dimension is given on welding symbol, weld should completely penetrate joint.

Complete joint penetration is indicated when no dimension is given on the welding symbol for a single groove or a symmetrical double-groove weld.

Fig. 18-8 illustrates complete weld penetration for a double-groove joint.

A dimension NOT in parentheses on the left of a bevel V, J, or U-groove weld symbol (effective throat not specified, or specified elsewhere on print) indicates the size of the weld preparation only. NO such dimension is needed with a square-groove weld. Refer to Fig. 18-9.

Optional groove preparation with complete penetration is indicated when the letters CP are shown in the tail of the reference line. No weld symbol is used, as in Fig. 18-10.

Weld size of a flare-groove weld is considered only to the tangent points. This is shown in Fig. 18-11.

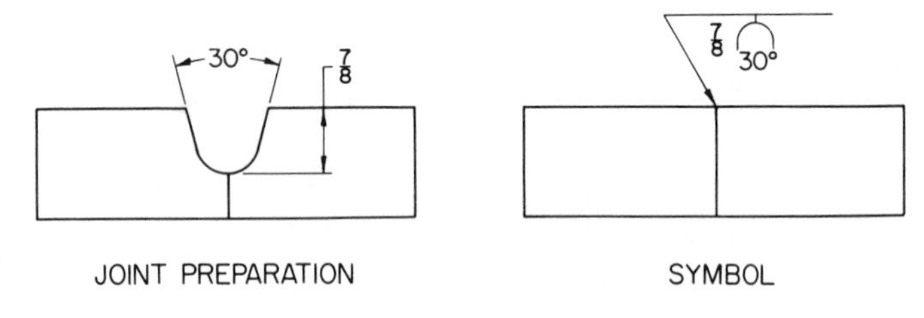

Fig. 18-9. Weld preparation only is indicated when a dimension is not in parentheses to left of bevel, V, J, or U-groove weld symbol.

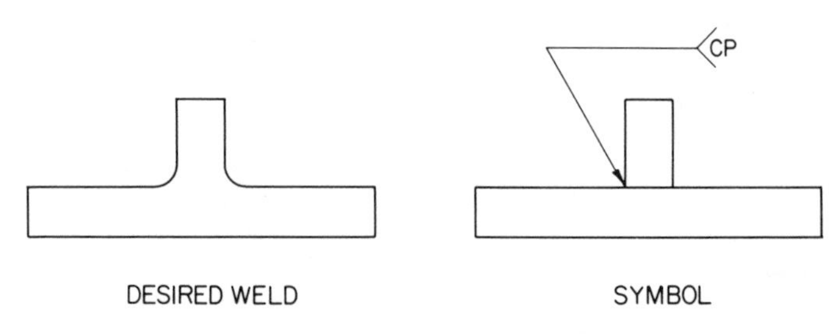

Fig. 18-10. Letters CP in tail of reference line require optional groove preparation with complete weld penetration.

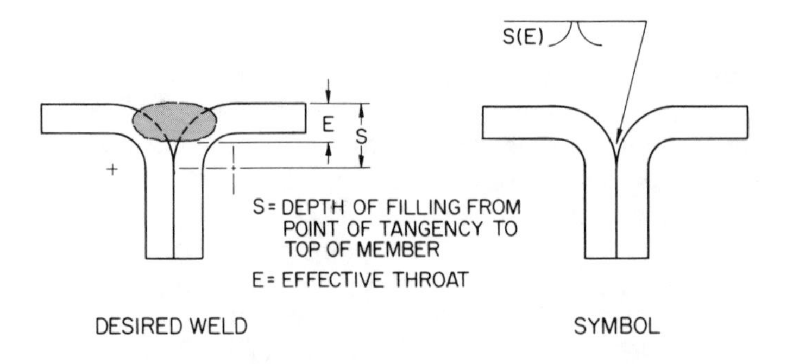

Fig. 18-11. Flare-groove weld size is only to tangent points of joint members.

GROOVE DIMENSIONS

Many users have established their own standards for groove weld dimensions. Unless otherwise noted on the print, these standards (the user's) are observed.

When user's standards for groove welds are NOT indicated, the following will apply:

1. Root opening is indicated inside the weld symbol, as illustrated in Fig. 18-12.
2. Groove angle is specified as shown in Fig. 18-13.
3. Groove radii and root faces of U and J-groove welds are shown by cross section, detail, or other means with a reference on the welding symbol, Fig. 18-14.

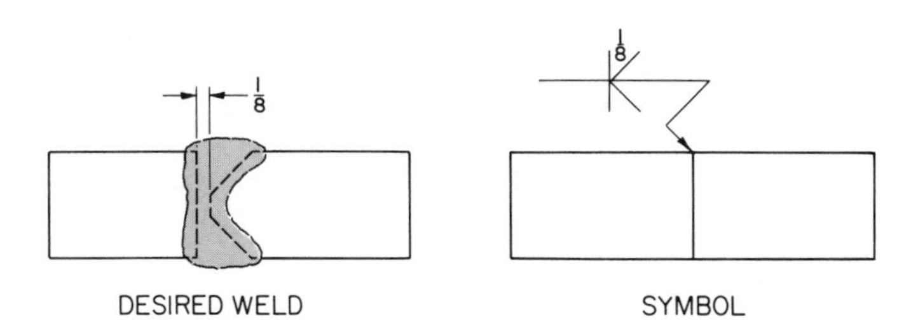

Fig. 18-12. Root opening of a groove weld is specified inside weld symbol.

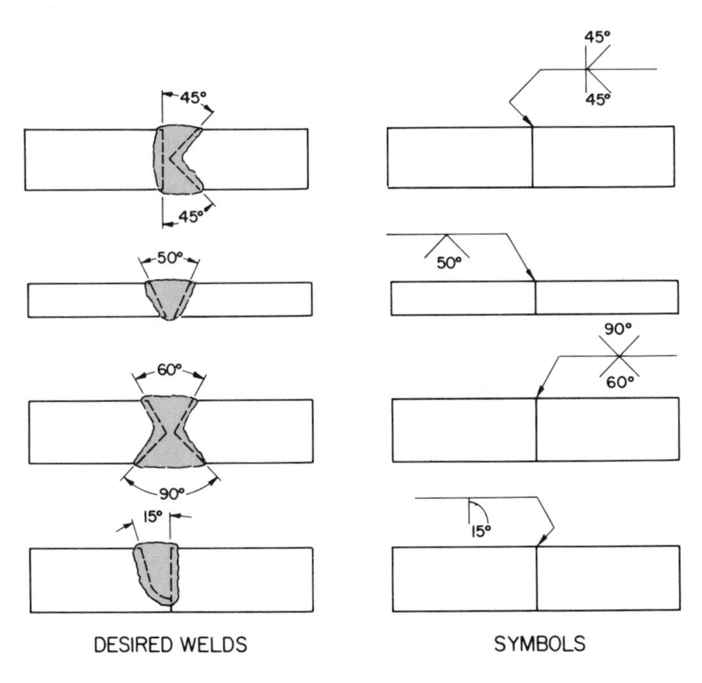

Fig. 18-13. Study how groove angles of groove welds are specified.

SURFACE FINISH AND CONTOUR OF GROOVE WELDS

Groove welds that are to be approximately flush, but not finished flush mechanically, are specified by a flush contour symbol. This symbol is placed above the weld symbol, as in Fig. 18-15.

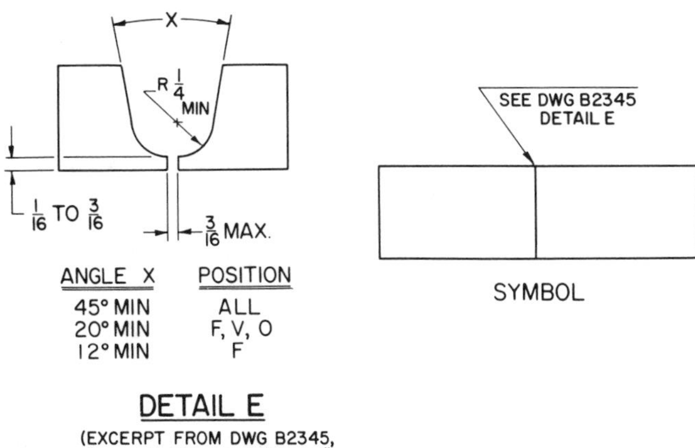

Fig. 18-14. Note that groove radii and root faces of U and J-groove welds are shown by cross section, detail, or other means with a welding symbol reference.

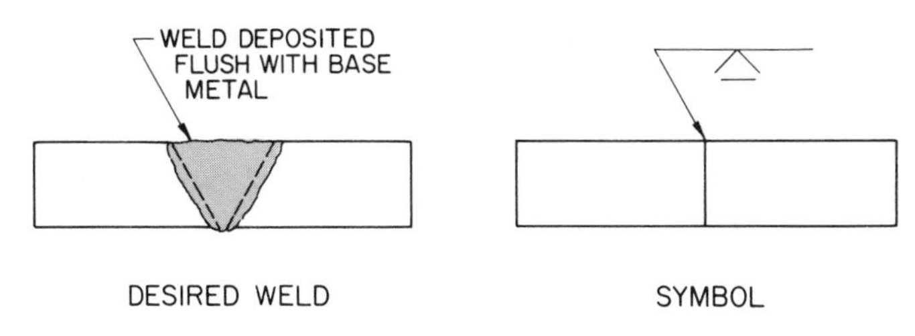

Fig. 18-15. A flush contour symbol is placed above weld symbol when groove weld is to be made approximately flush, without use of grinding, chipping, hammering, or machining.

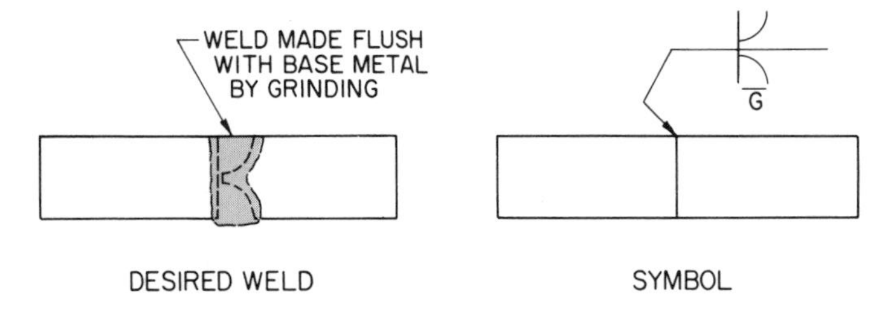

Fig. 18-16. Groove welds that must be made flush mechanically are specified by a flush contour symbol and by method to make weld flush.

Groove welds that are to be made flush by mechanical means are specified with a flush contour symbol and with the method of making the weld flush. Refer to Fig. 18-16.

Groove welds to be finished mechanically with a convex contour are specified by a convex contour symbol. The method of finishing the weld to a convex contour is also given. See Fig. 18-17.

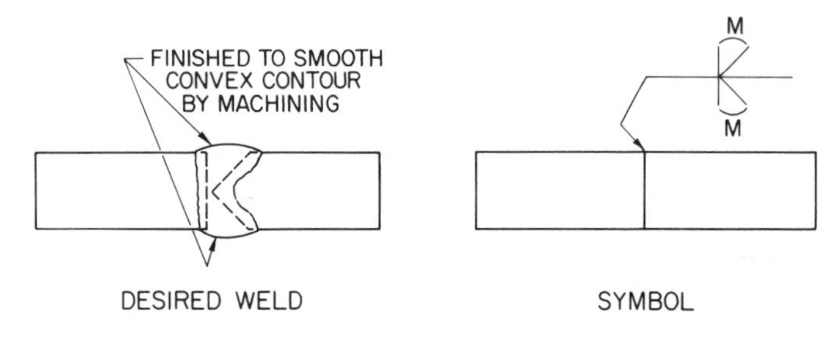

Fig. 18-17. When to be finished mechanically to a convex contour, groove welds are specified by a convex contour symbol and method of finishing.

BACK OR BACKING WELDS

The back or backing weld symbol shows that a bead type back or backing weld is needed with a single groove weld. You must remember that:

1. A *back weld* is made AFTER the groove weld.
2. A *backing weld* is made BEFORE the groove weld.

A note will state whether a back or backing weld is to be made. This note will be in the tail of the welding symbol, Fig. 18-18.

Either type weld is specified by a back or backing weld symbol located on the side of the reference line, opposite the groove weld symbol. Refer to Fig. 18-18.

NOTE-

1. GROOVE WELD MADE BEFORE WELDING OTHER SIDE.

2.

(ABOVE FOUND ELSEWHERE ON PRINT)

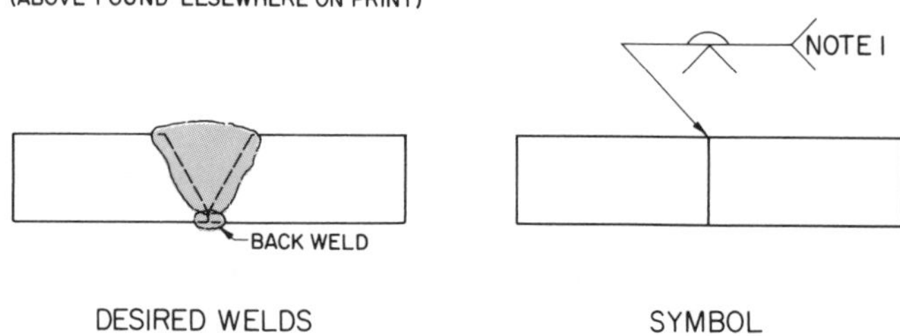

Fig. 18-18. Study this back or backing weld symbol. A note in tail of welding symbol tells you whether to make a back or backing weld.

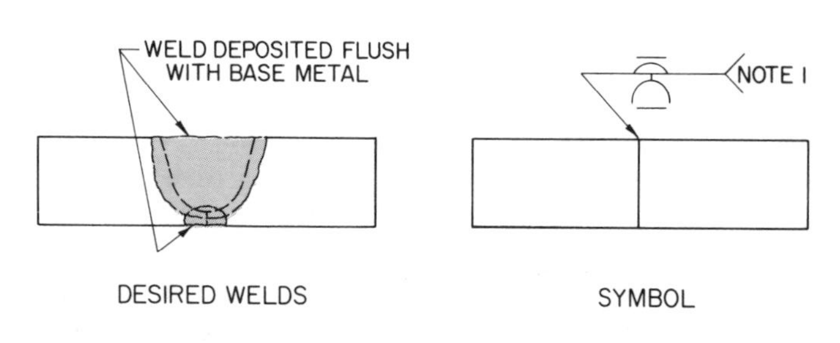

Fig. 18-19. A flush contour symbol denotes that weld is to be finished approximately flush as welded.

A flush contour symbol, added to the back or backing weld symbol, indicates the weld should be approximately flush with the base metal. See Fig. 18-19.

If the back or backing weld is to be made flush by mechanical means, the method of making the weld flush is added to the flush contour symbol. Fig. 18-20 gives an example.

When a back or backing weld is to be finished to a convex contour by mechanical means, a convex contour symbol and finish symbol are added to the weld symbol.

With the exception of HEIGHT (this is optional), no other back or backing weld dimensions will be shown with the weld symbol, Fig. 18-21. If other dimensions are required, they will be shown on the drawing.

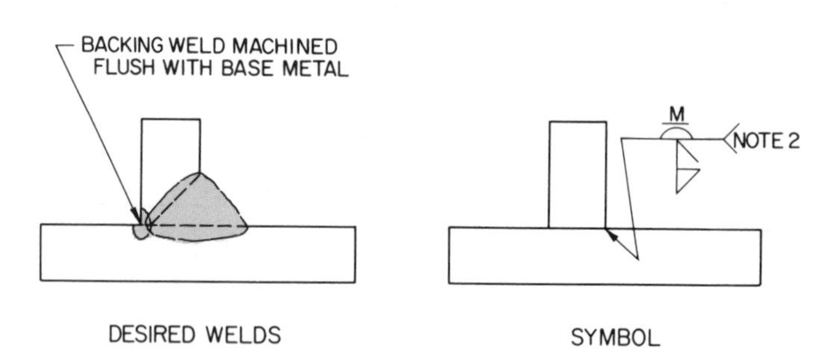

Fig. 18-20. Note how flush contour symbol has method of mechanically finishing flush above it.

BACKING BARS, SPACER BARS, AND EXTENSION BARS

Backing bars, Fig. 18-22, are employed when full penetration groove welds are required and welding can only be done from one side. They are thoroughly penetrated by the weld and usually left in place.

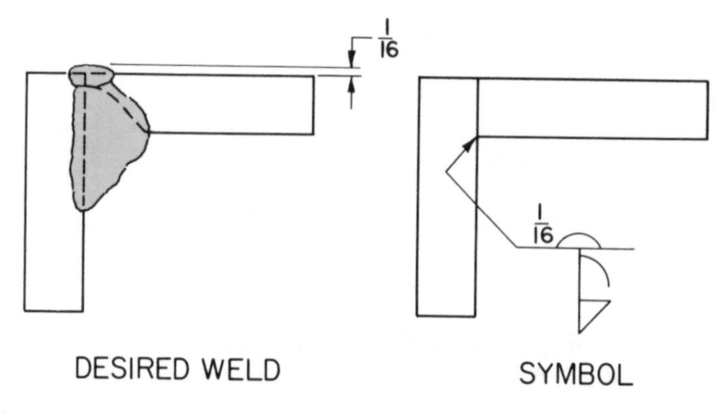

Fig. 18-21. Only back or backing weld height dimension is shown on welding symbol. If other dimensions are needed, they are given elsewhere on print.

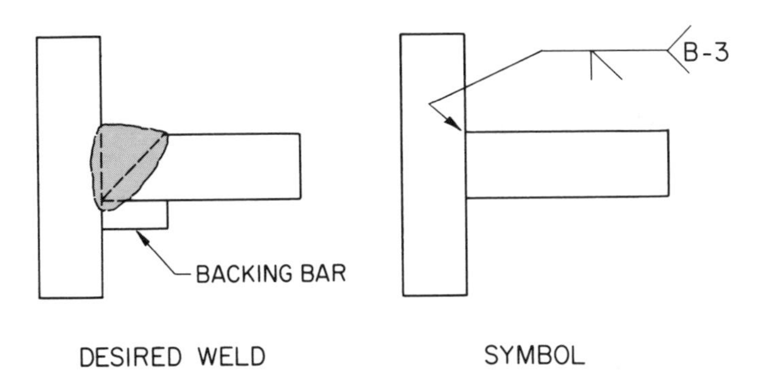

Fig. 18-22. Note utilization of backing bar. Specification indicating its use is shown in welding symbol tail.

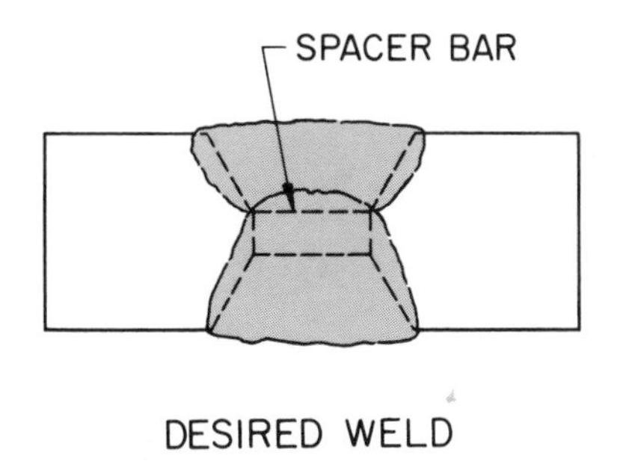

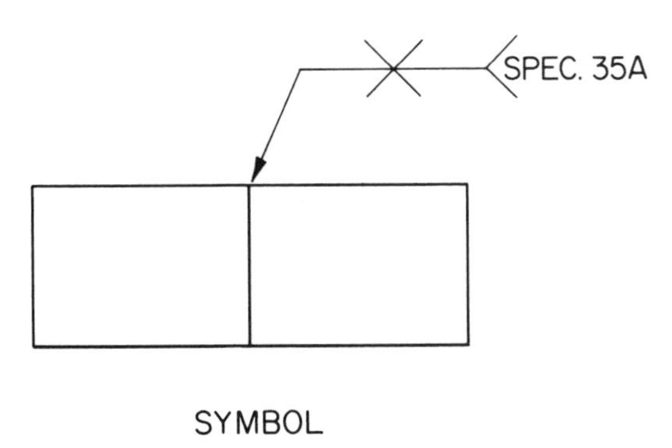

Fig. 18-23. Spacer bars may be specified when thick sections are welded. In such welds, the root, and spacer bar are gouged out before second side of groove is welded. Specifications are shown in welding symbol tail.

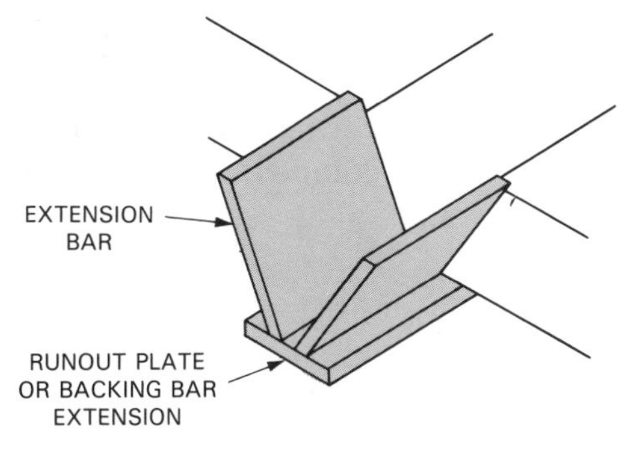

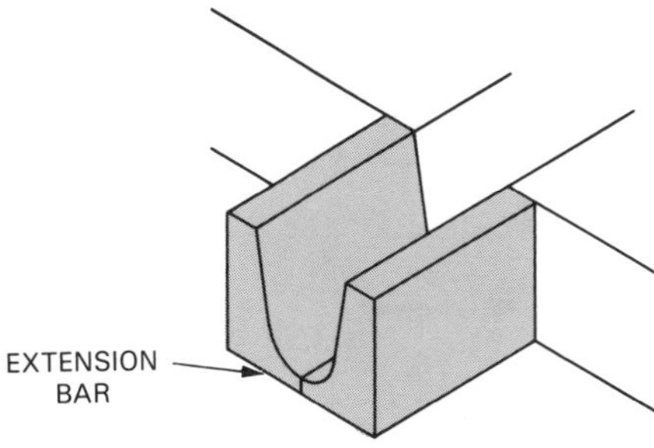

Fig. 18-24. An extension bar is used when full length groove weld is specified. It may be indicated by specifications in welding symbol tail or by sketches on drawing.

Spacer bars, Fig. 18-23, are sometimes used, especially if the weld is in thick material and the minimum possible vee angle is specified. In such welds, the root must be gouged out completed, including the spacer bar, before the second side of the groove is welded.

Extension bars, Fig. 18-24, provide an extension of the groove beyond the pieces being joined, when a full length groove weld is specified. The angle or contour of the extension bar must be identical with that of the groove.

Since welding symbols give no indication of the backing, spacer, or extension bar requirements, note that unless covered by reference to AWS prequalified joints or fabricators' standards, SPECIAL SKETCHES of the weld profile will be provided.

UNIT 18—TEST YOUR KNOWLEDGE

Carefully study the drawings of the groove weld joints shown. Then complete the information requested.

Part I

Identify the groove weld joints shown below.

1. ____________ 6. ____________
2. ____________ 7. ____________
3. ____________ 8. ____________
4. ____________ 9. ____________
5. ____________

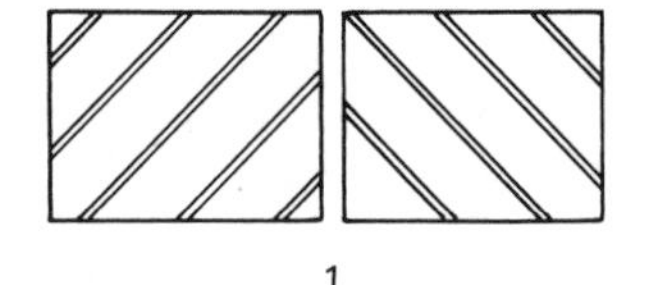

1

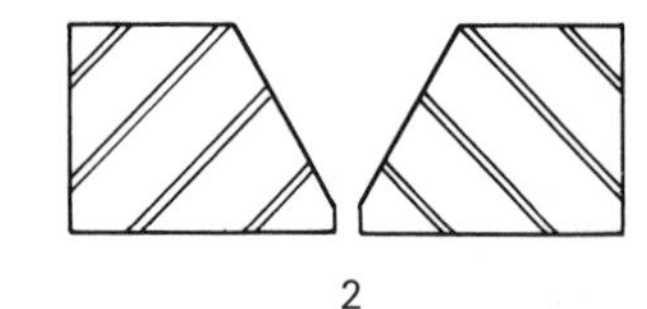

2

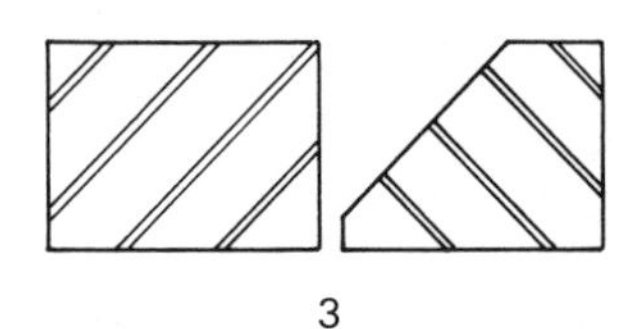

3

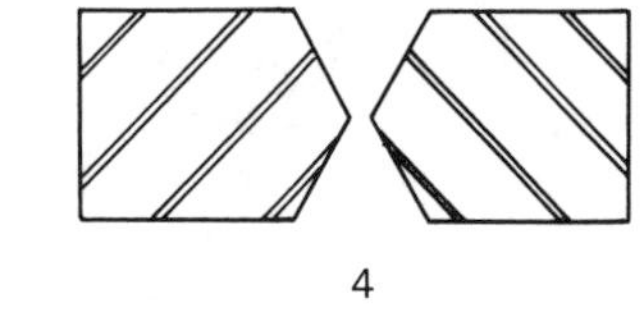

4

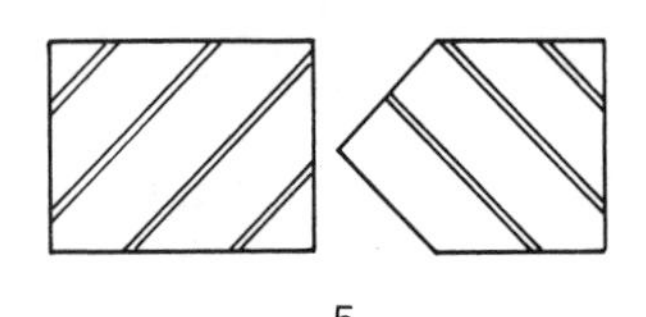

5

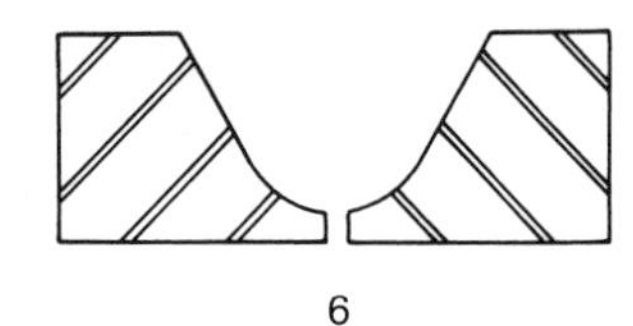

6

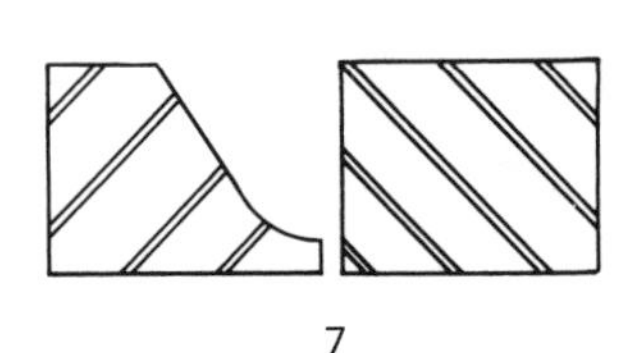

7

8

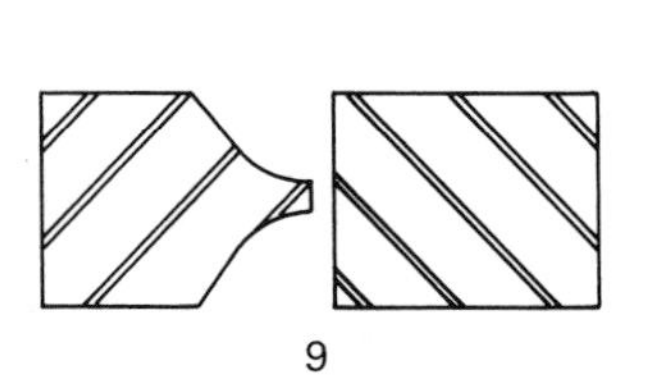

9

Part II

Sketch in the correct welding symbol on the groove welds shown in the left column.

Part III

Carefully study the drawings shown and sketch in the welding symbol(s) that will describe each joint.

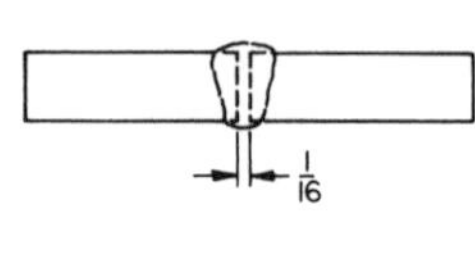

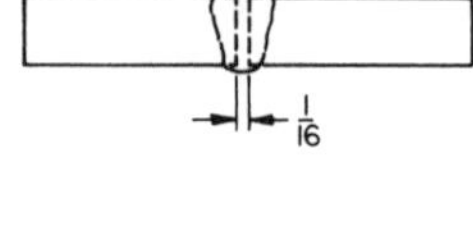

1

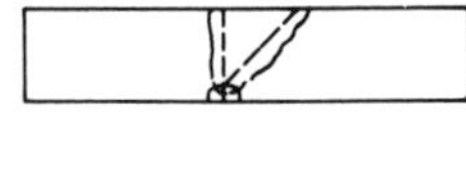

BEVEL ANGLE 45°
WELDS GROUND FLUSH

2

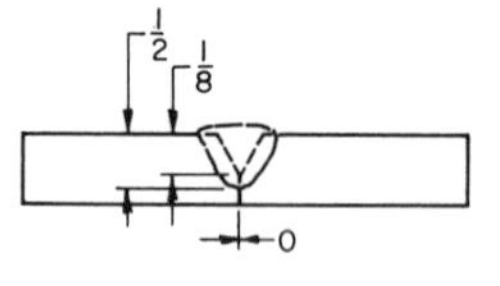

3

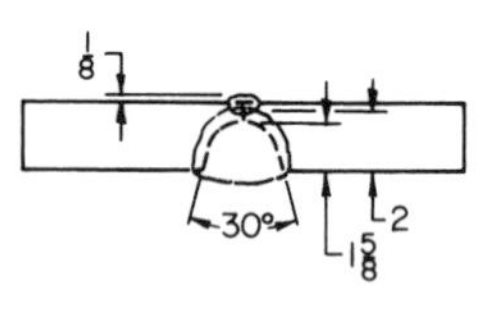

4

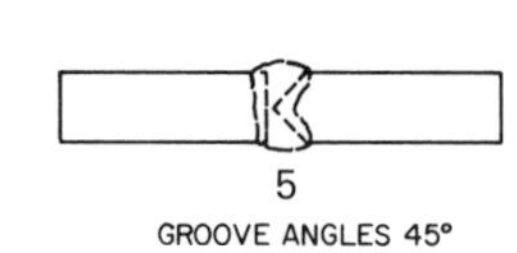

5
GROOVE ANGLES 45°

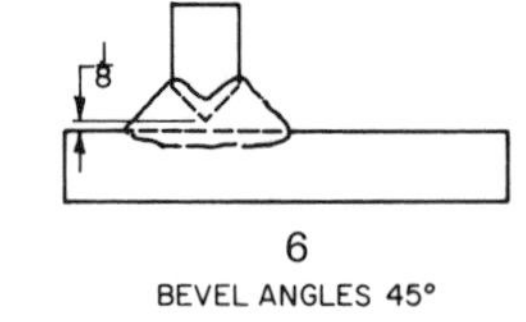

6
BEVEL ANGLES 45°

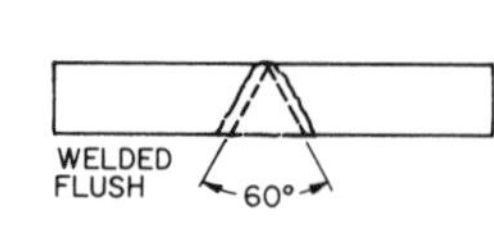

7

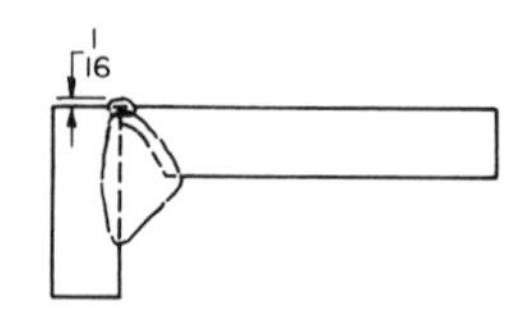

FIELD WELD

8

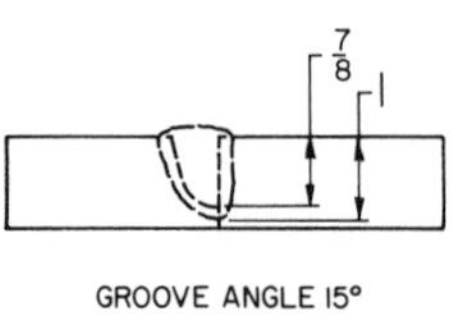

GROOVE ANGLE 15°

9

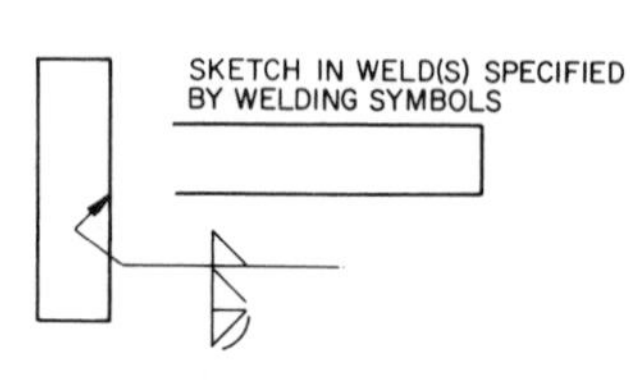

10

Part IV

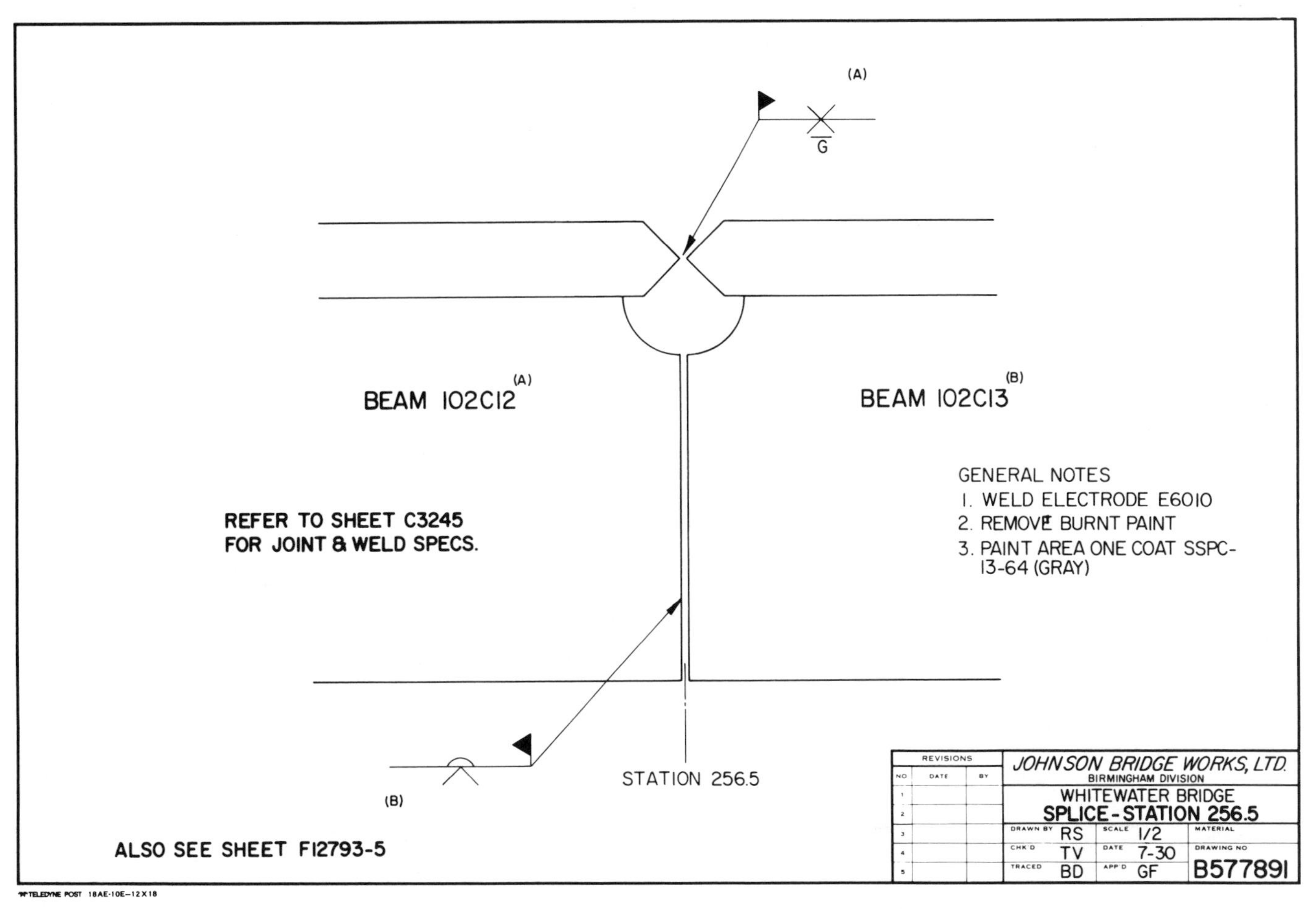

Use this print (B577891) to do Part IV of Unit 18—Test Your Knowledge.

Carefully study the drawing (B577891) and answer the following questions.

1. List the name and drawing number.
   a. ______________________
   b. ______________________
2. What parts are to be joined by welding?
   a. ______________________
   b. ______________________
3. Interpret the types of welds required to make the weldment (joint two sections).
   a. ______________________
   ____
   ____
   ____
   ____
   b. ______________________
   ____
   ____
   ____
   ____
4. Joint and weld specifications can be found ______ ______________________.
5. What type of welding rod is to be used? ______
6. What special requirements must be observed after the weldments are made?
   ____
   ____
7. Have any changes been issued against the drawing? ________ If there have been, list the number made. ________

Part V

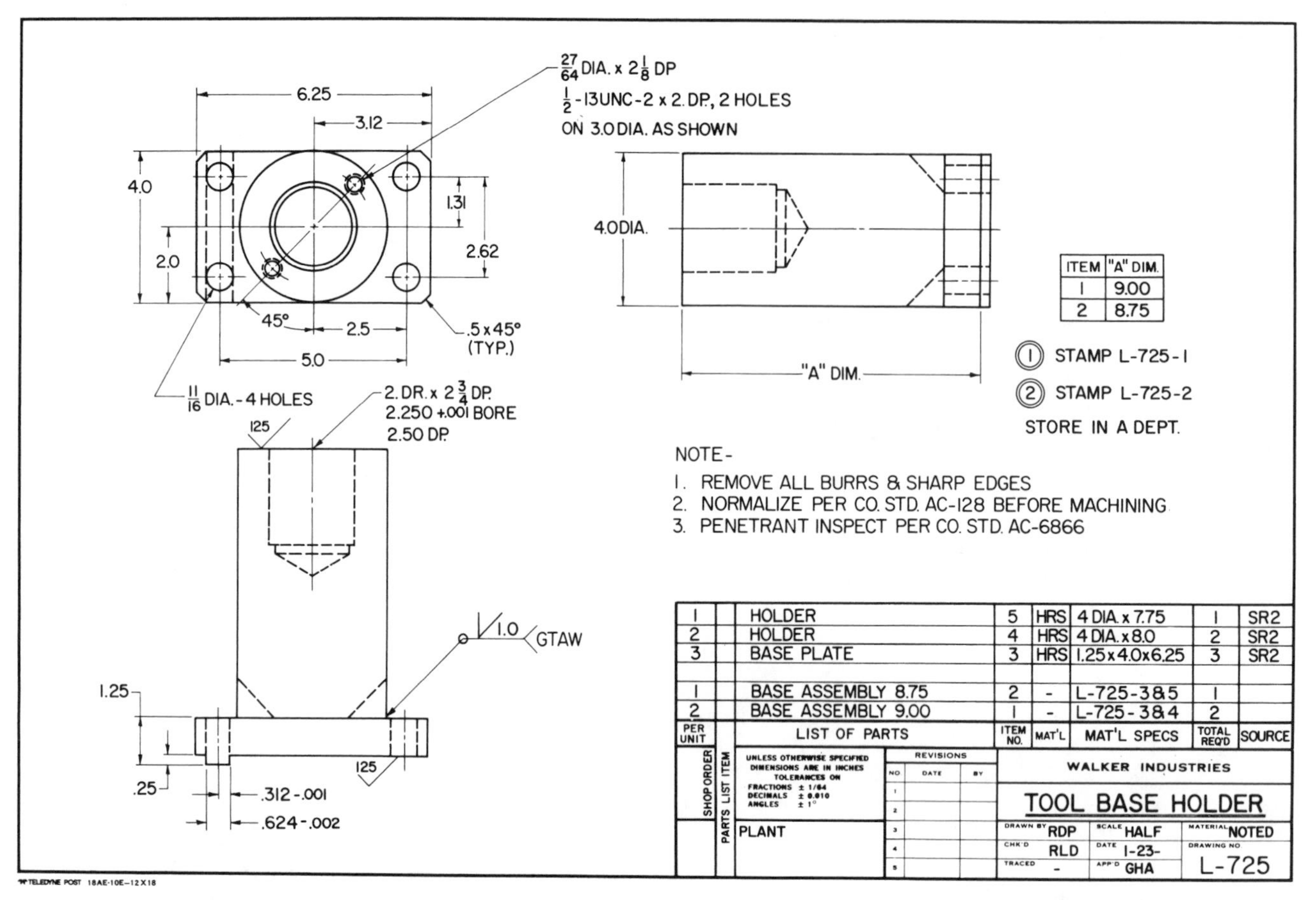

Use this print (L-725) to do Part V of Unit 18—Test Your Knowledge.

Carefully study the drawing (L-725) and answer the following questions.

1. List the name and drawing number of the print.
   a. ______________________
   b. ______________________
2. How many parts make up the assembly? ______________________
3. What are the names of the parts that make up the assembly? ______________________
4. Is more than one size unit indicated on the print? ______________________
5. If more than one size unit is indicated, how many are there and how is each unit identified?
   a. __________ b. __________
6. List the stock size required to make each part of the assembly. ______________________
7. Interpret the type weld(s) required to make the weldment(s). ______________________
8. What heat treatment is required after welding? ______________________
9. How is each weld to be inspected? ______________________
10. How many holes are drilled in the base? ______

11. The diameter of these holes is ____________.
12. How many threaded holes are indicated in the holder? ____________________
13. The thread size is ____________ and is tapped __________ deep.
14. Describe how the large hole in the holder is to be made.

____

____

____

____

____

15. Is a tolerance indicated for the final diameter? _________ If so, what is it? _________
16. What is the size of the key on the base?

____

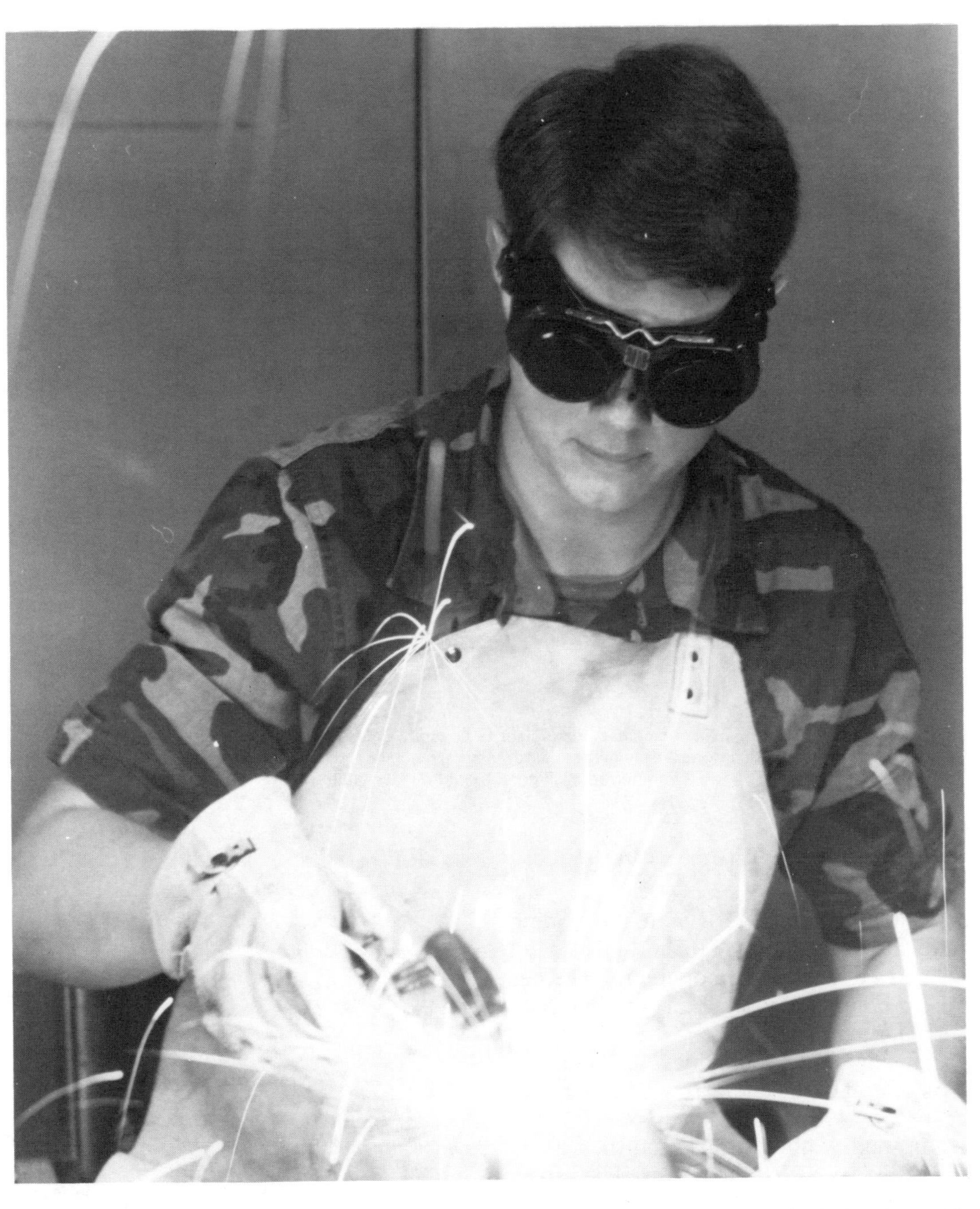

The United States Army and other branches of the Armed Forces offer an excellent opportunity for learning welding skills. Best of all, you get paid while you learn.

Unit 19

SURFACING WELDS

Surfacing welds are frequently employed to extend the service life of parts subjected to constant wear or abrasive conditions. They are also used to build up worn surfaces of shafts and bearing plates, Fig. 19-1.

SURFACING WELD DIMENSIONS

The *buildup thickness dimension* is shown on the SAME SIDE of the reference line as the surfacing weld

Fig. 19-1. Many components of earth moving machines are subject to constant wear. They can be rebuilt with surfacing welds with a considerable savings over replacement with new parts. On modern earth-moving machines, where high strength steels are used, E7018 electrodes will normally give best weld buildup material.

GENERAL USE OF SURFACE WELD SYMBOL

The *surfacing weld symbol* is ADDED to the welding symbol to indicate the surface(s) to be built up by welding, Fig. 19-2. The same symbol applies whether the buildup is to be made by single or multiple pass welds.

Since the surfacing weld symbol does NOT indicate the welding of a joint, it has no arrow or other side significance.

The surfacing weld symbol is shown on the side of the reference line TOWARDS the reader. The arrow points to the surface on which the weld is to be deposited.

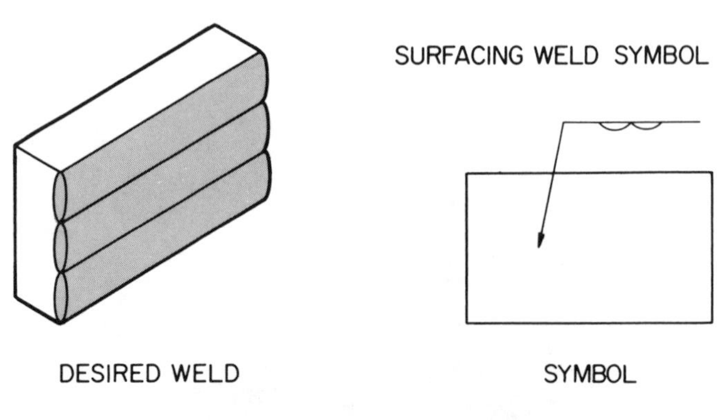

Fig. 19-2. Note how the surfacing weld symbol is used. Since no joint is welded, symbol has no arrow or other side significance. Arrow points towards area to be surfaced.

symbol, and to the LEFT of the symbol. Fig. 19-3 shows an example.

The thickness dimension indicates the MINIMUM height of the buildup. If no dimension is shown, no specific height of weld deposit is required.

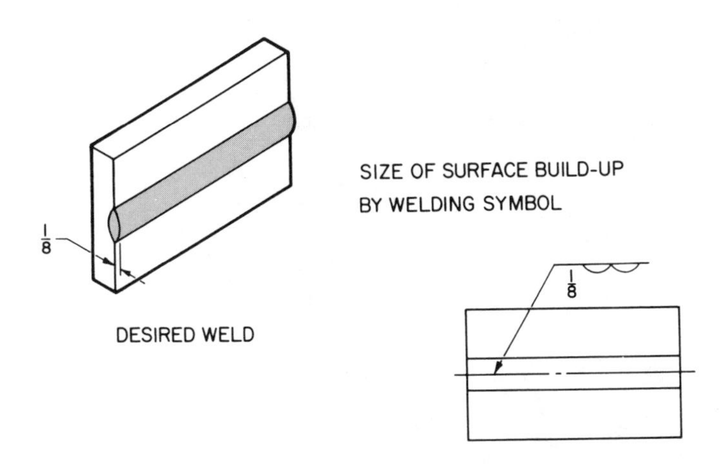

Fig. 19-3. Thickness of weld buildup is denoted by dimension to left of surface weld symbol.

LOCATION, EXTENT, AND ORIENTATION OF SURFACING WELDS

The *location, extent,* and *orientation* of the buildup area will be indicated on the print when only a PORTION of the area is to be surfaced. This is shown in Fig. 19-4.

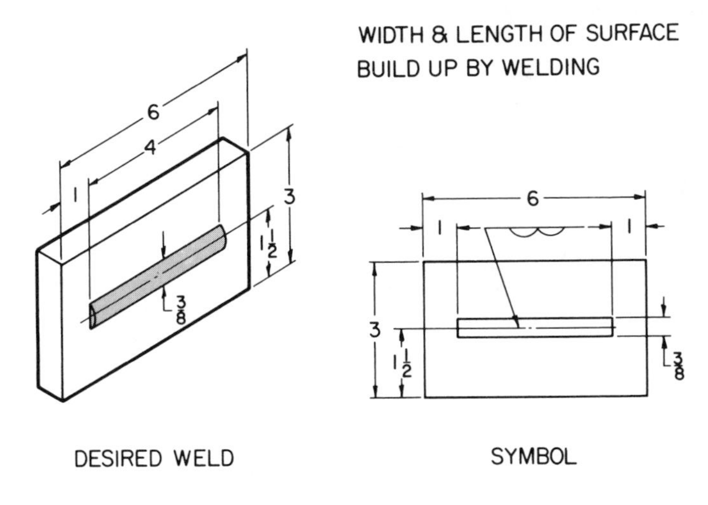

Fig. 19-4. Location, extent, and orientation of buildup area will be given on drawing.

Unless required, no dimension other than the height of the deposit will be shown on the welding symbol when the entire area of a plane or curved surface is to be built up. Refer to Fig. 19-5.

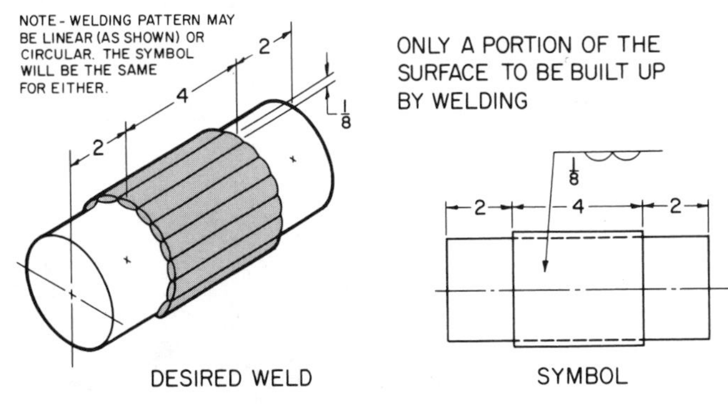

Fig. 19-5. On circular objects, welding pattern may be linear, as shown, or circular. Weld dimensions, other than buildup height, will be shown on print and not on welding symbol.

UNIT 19—TEST YOUR KNOWLEDGE

1. Surfacing welds are frequently used to: (Check the correct answer.)
   a. ____ Extend the service life of parts that are subject to abrasive conditions.
   b. ____ Build up worn shafts and bearing surfaces to reduce or eliminate the need for expensive new replacements.
   c. ____ Add a hardsurfacing material to parts that are subject to hard wear.
   d. ____ All of the above.
   e. ____ None of the above.
2. Draw the welding symbol for a surfacing weld that is 5.0 mm thick.
3. Since the surfacing weld symbol does NOT indicate the welding of a joint, it: (Check the correct answer.)
   a. ____ Has no arrow or other side significance.
   b. ____ Has arrow or other side significance.
   c. ____ It has arrow or other side significance only on special welding applications.
   d. ____ All of the above.
   e. ____ None of the above.
4. The surfacing weld symbol is shown on the side of the reference line ____________________.
   The arrow points ____________________.

5. The thickness dimension on the surfacing welding symbol indicates ____________________ of the buildup.

6. If no buildup thickness dimension is shown on the surfacing welding symbol, ____________________ ____________________.

Robotics are taking over many monotonous and dirty welding jobs. Welding of automotive components is readily adaptable to robot production techniques. (Pontiac)

Unit 20

FLANGE WELDS

Flange welds are used to join the edges of two or more members. They are primarily used with light gage metal. The edges to be welded may be flared or flanged.

Flange welds are indicated by the flange weld symbol. Fig. 20-1 shows several examples. The symbol has NO both sides significance.

An *edge-flange* weld symbol is used when BOTH members are flanged. Both lines that form the symbol curve before they contact the reference line. Refer to A and B of Fig. 20-1.

A *corner-flange* weld symbol is used when only ONE of the members is flanged. Only one of the lines for forming the symbol curves before touching the reference line. See C and D of Fig. 20-1.

DIMENSIONS OF FLANGE WELDS

Flange weld dimensions are shown on the SAME SIDE of the reference line as the weld symbol. This is also shown in Fig. 20-1.

The *radius* and *height* above the points of tangency dimensions are separated by a plus (+) sign. They are located to the LEFT of the weld symbol. This is illustrated in Fig. 20-2.

Flange weld size is indicted by a dimension placed outward of the flange dimensions. Also see Fig. 20-2.

Root openings of flange welds are NOT shown on the welding symbol. This dimension, if required, is specified elsewhere on the print.

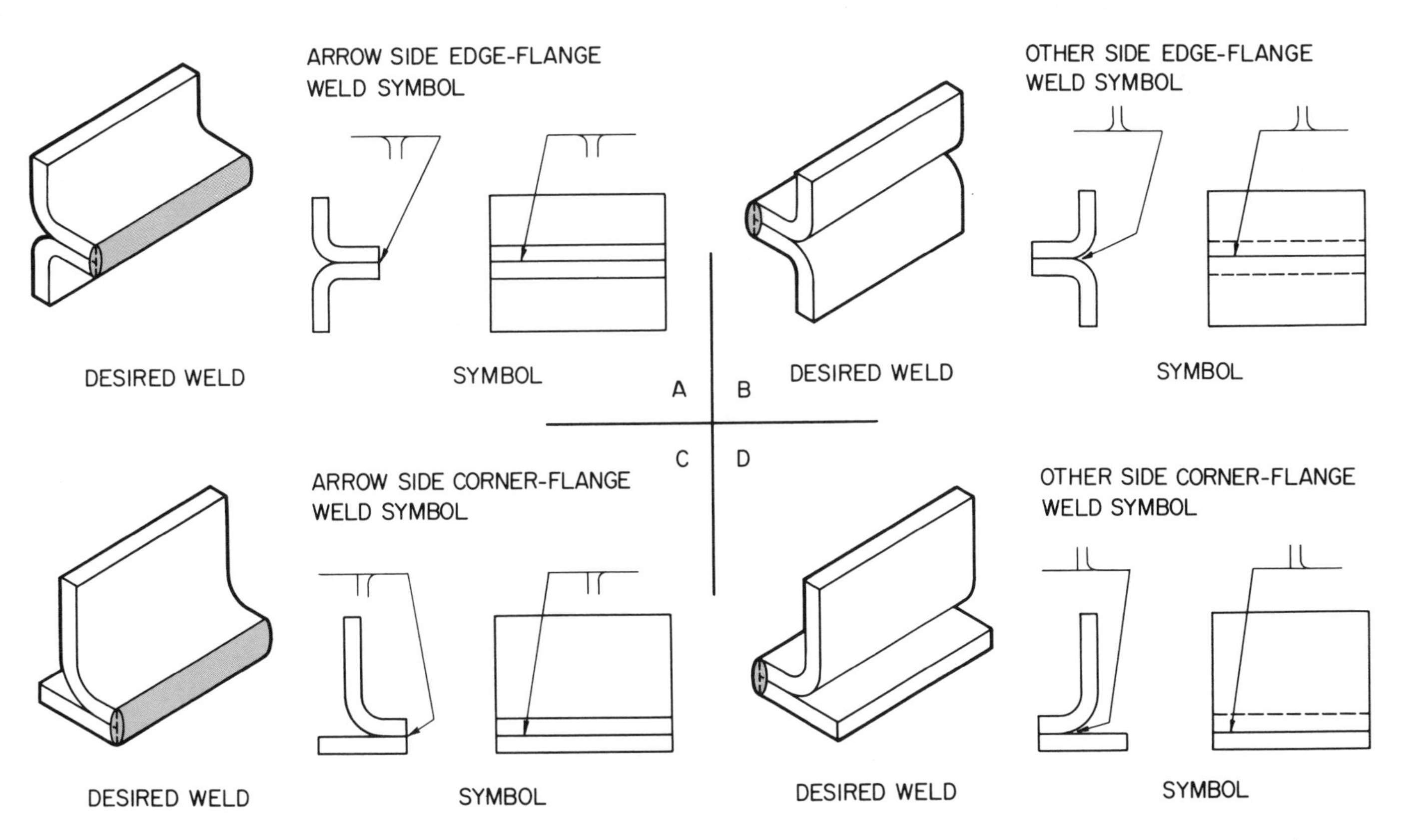

Fig. 20-1. Study how flange weld symbol is used to denote welds. The symbol does not have both sides significance. Note difference between edge and corner-flange symbols.

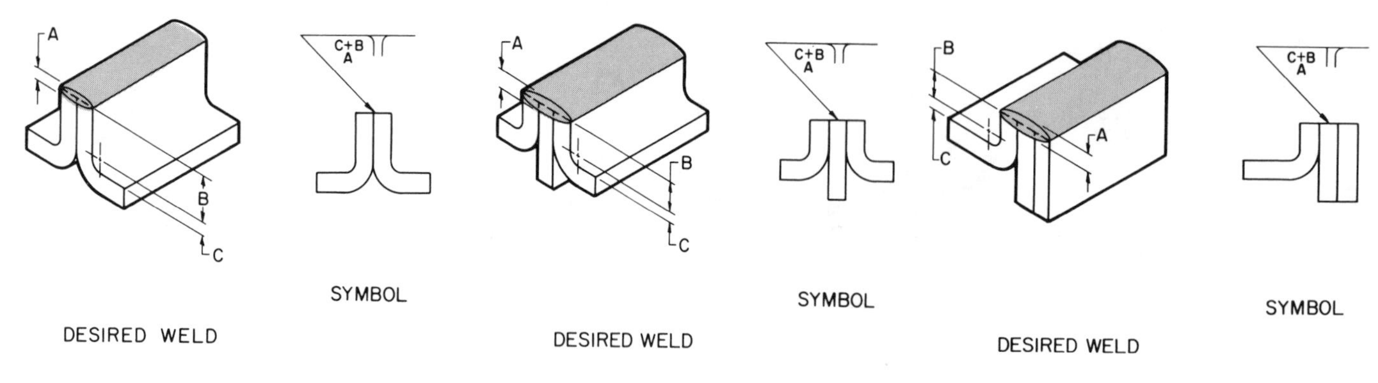

Fig. 20-2. Radius and height above tangency dimensions are shown separated by a plus (+) sign and are given to left of weld symbol. Flange weld size is indicated by a dimension placed outward from flange dimension.

MULTIPLE JOINT FLANGE WELDS

Some assemblies require one or more pieces to be inserted between the two outer members. The SAME SYMBOL used for the two member joints is utilized regardless of the number of pieces inserted.

UNIT 20—TEST YOUR KNOWLEDGE

1. Flange welds are primarily used with ________ ____.
2. Flange weld dimensions are shown on the ________ side of the reference line as the ______________.
3. Draw the flange welding symbol for this weld.

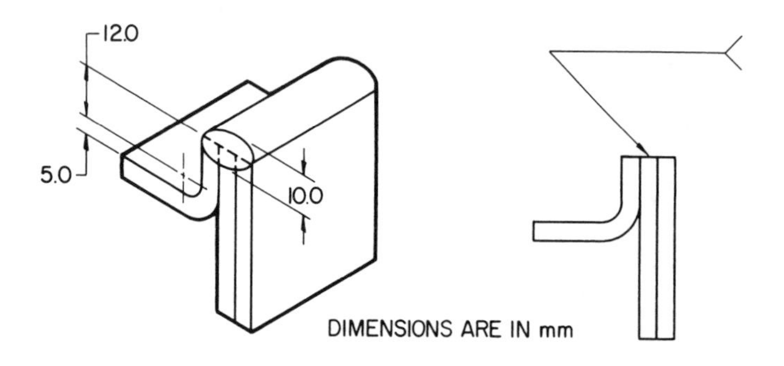

4. Draw the flange welding symbol for a joint with the same dimensions as that in Question 3. However, in this case, there are three pieces of metal between the two outer pieces.

Unit 21

BRAZED JOINTS

Brazing is the term given to the process in which metal pieces are joined by heating them to a suitable temperature above 800°F (428°C) but below their melting point. Nonferrous filler metal, having a melting point below that of the base metal, is added to the weld metal.

In brazing, the strength of the joint depends upon the alloying of the filler metal with the base metal. The filler metal, usually a brass alloy, is distributed between the closely fitted surfaces of the joint by capillary action.

At times, the welder will encounter jobs that specify brazed joints, Fig. 21-1. If such joints require no joint preparation, other than cleaning the members, only the arrow with the brazing process indicated in the tail will be shown on the print.

For other brazing operations, the application of conventional weld symbols, with minor variations, to the brazed joint will be employed. See Fig. 21-2.

NOTE: All ferrous and nonferrous metals, including aluminum and magnesium, can be joined by brazing. When dissimilar metals are brazed and one part fits within the other, it is advisable to check the coefficient of thermal expansion, especially in the brazing temperature range. This will assure that the proper clearance can be provided for the filler metal to enter the joint.

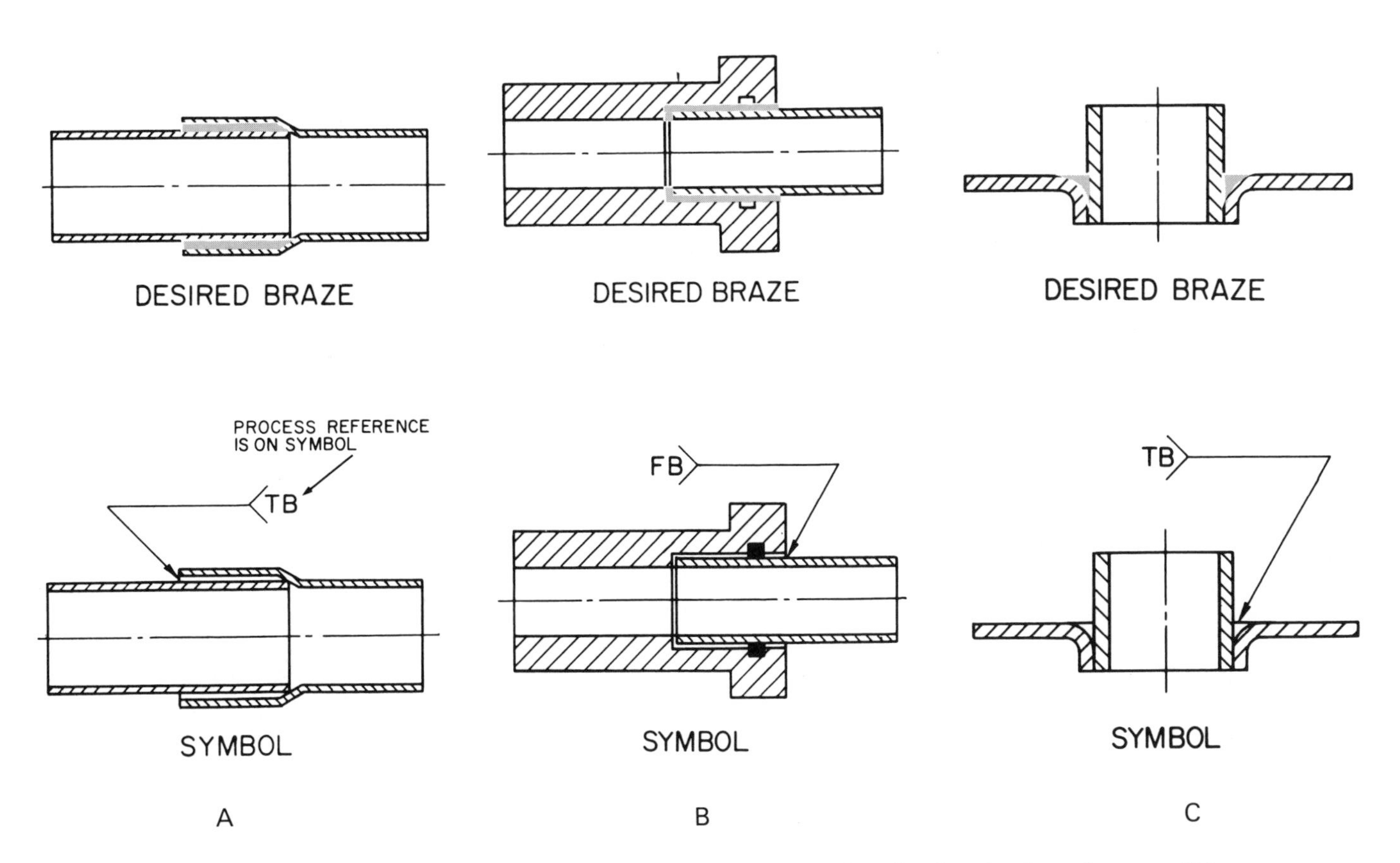

Fig. 21-1. Application of brazing symbol. When joints do not need preparation other than cleaning, only arrow with brazing process in tail will be on print. A—Torch braze. B—Furnace braze. C—Torch braze.

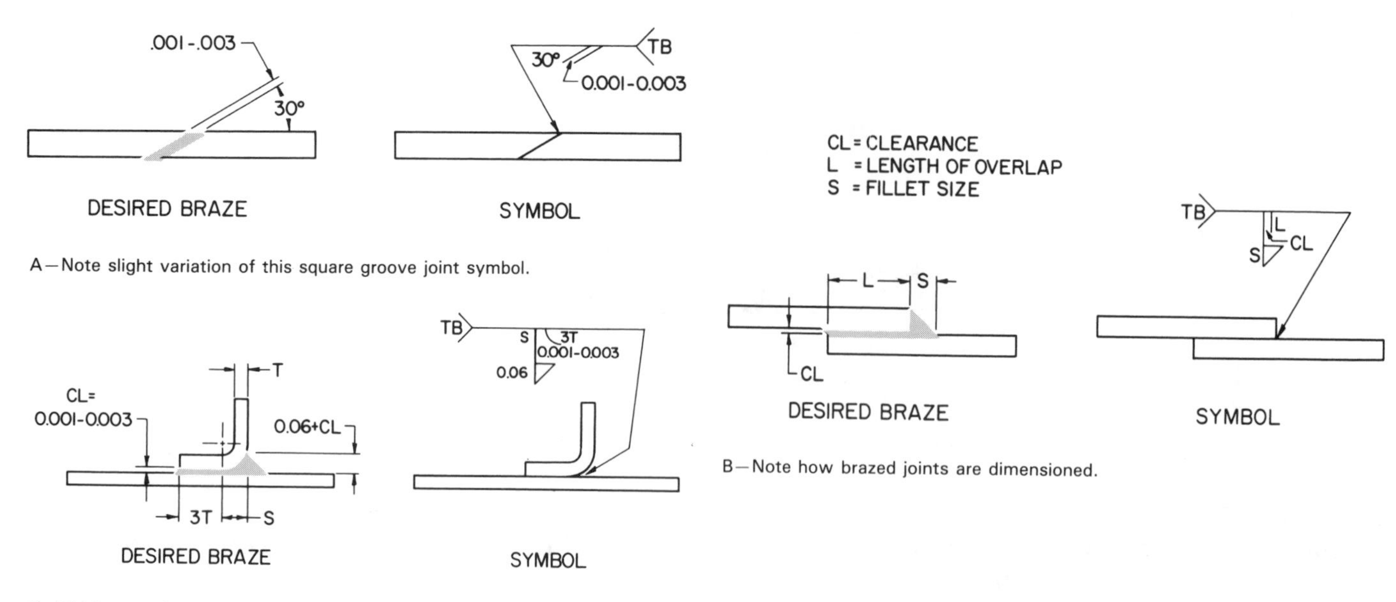

Fig. 21-2. Conventional weld symbols will be used to indicate type of brazed joint.

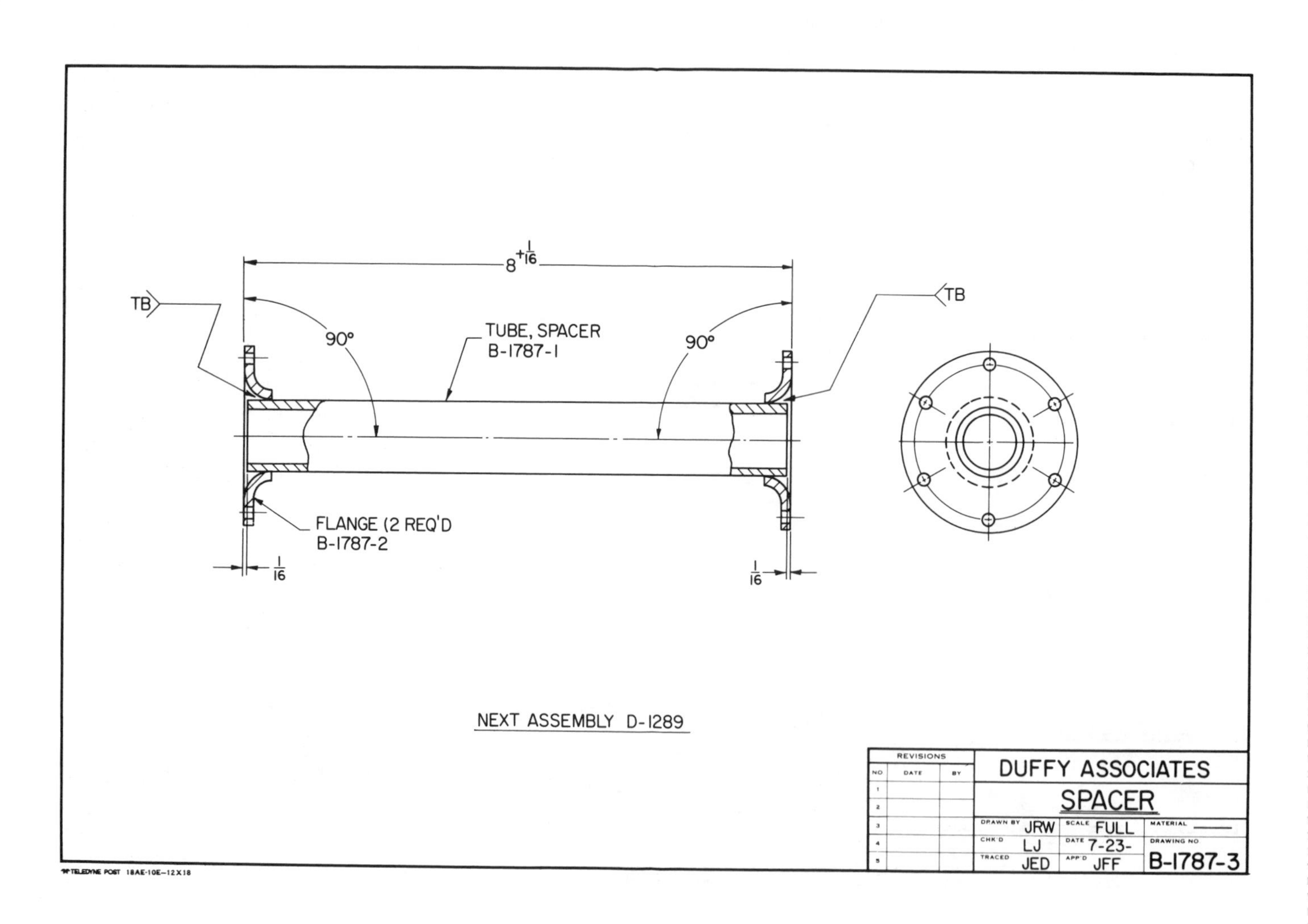

Use this print (B-1787-3) to answer the questions in Unit 21—Test Your Knowledge.

UNIT 21—TEST YOUR KNOWLEDGE

When answering the following questions, refer to the provided print (B-1787-3).

1. What is the name of the part? ____________
2. How many components make up the assembly?

   ____

   Name them!

   a. ____________ c. ____________

   b. ____________ d. ____________
3. What brazing process is specified? ________
4. How many places are brazed? ____________
5. What is the MAXIMUM acceptable length of the unit?________
6. What is the MINIMUM acceptable length of the unit?________
7. What is the next assembly? ____________
8. Must the flanges be square with the centerline of the tube? _____ If so, what indicates this condition? ____________________
9. How far is the tube inset from the face of the flange?________
10. Briefly explain the difference between welding and brazing.

    ____

    ____

    ____

    ____

This oxy-fuel multicutting head is operated from the console in front of the operator. An electronic scanner moves over a line or silhouette tracing, providing torch movement to make the desired cuts. It can also be equipped with preprogrammed shapes for which it is only necessary to key in the dimensions. (Linde Div., Union Carbide)

Unit 22

PIPE WELDING

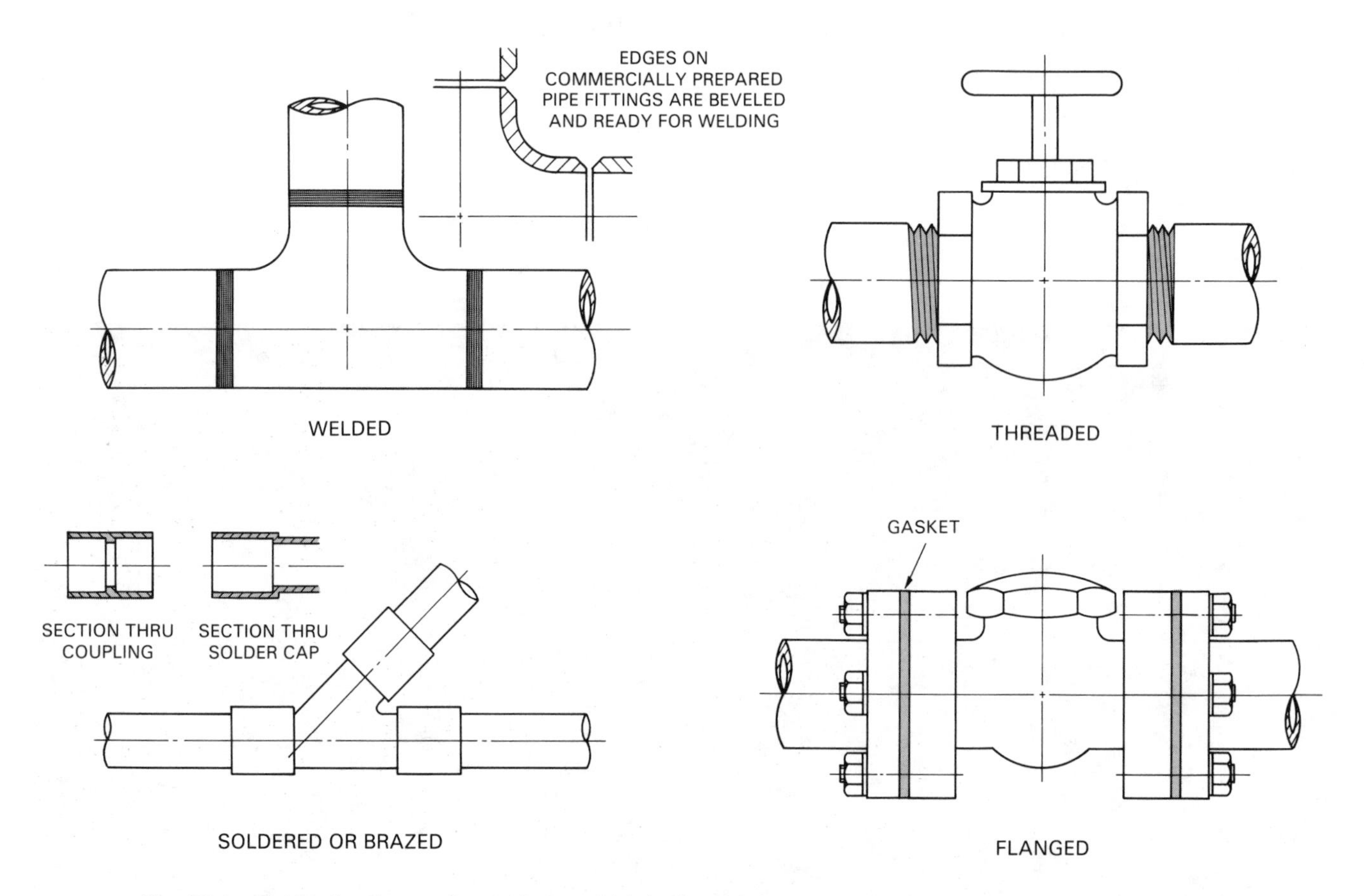

Fig. 22-1. Welding is only one of several ways of fabricating a piping system. Systems can also be joined by soldering, threaded fittings, or by flanged fittings that use gaskets.

A *piping system* is an assembly of pipe or tube (thin wall pipe) sections with the fittings (valves, joints, etc.) necessary to direct and control fluids in either liquid or gaseous form. As shown in Fig. 22-1, welding is one of the many methods employed to fabricate piping system.

Pipe size is determined by its nominal INSIDE DIAMETER (ID) which differs slightly from the actual inside diameter. *Tube* is measured by its OUTSIDE DIAMETER (OD).

NOTE! The welding of critical or high pressure pipe systems (systems where weld defects could be costly and life threatening) require the skills of specially trained and qualified welders.

PIPING DRAWINGS

Piping drawings show the assembly of a pipe system. Fig. 22-2 gives an example.

Pipe drawings are made in either DOUBLE-LINE or SINGLE-LINE drawings. Compare the two types in Fig. 22-3.

Most pipe layouts are single-line drawings in isometric or oblique views because orthographic

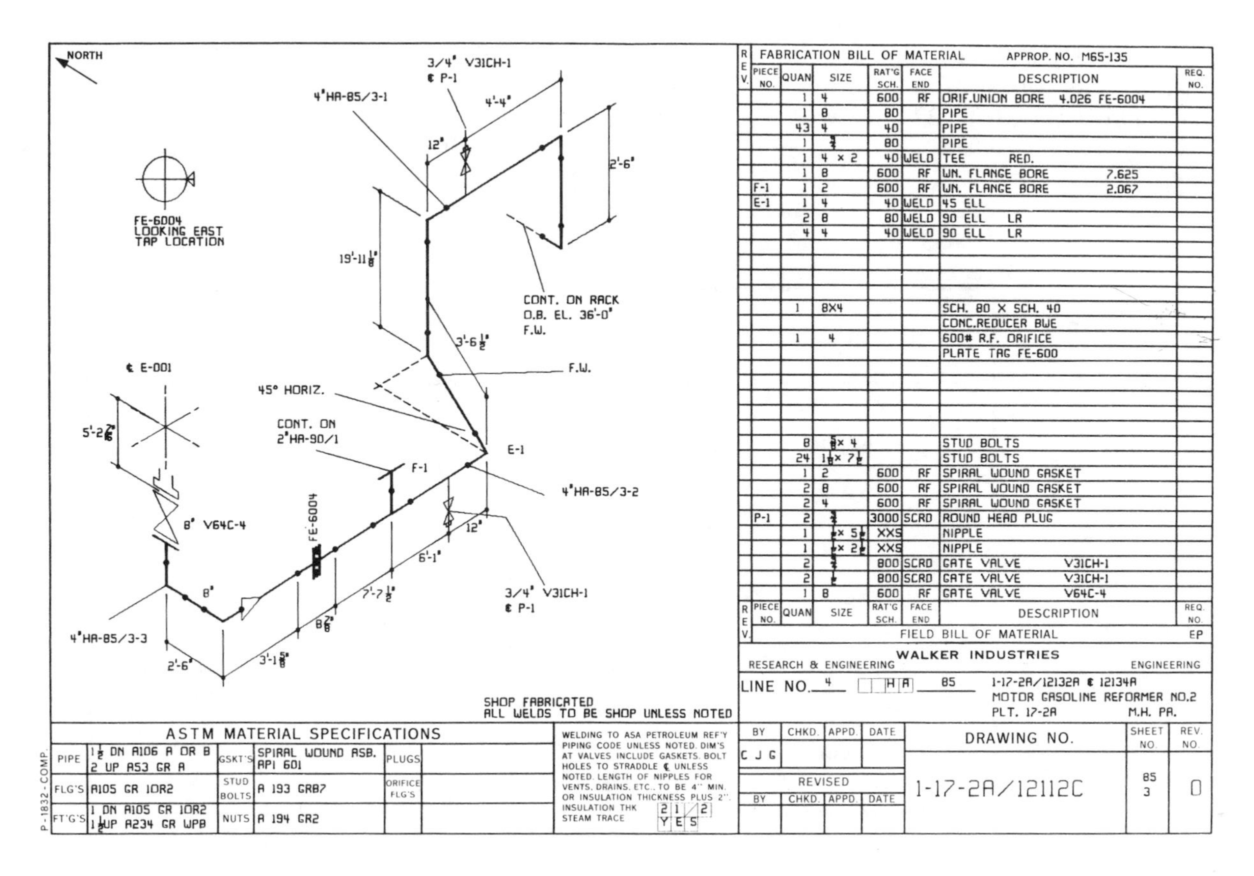

Fig. 22-2. Piping drawing shows how to construct a pipe system. This pipe drawing was prepared by a Computer-Aided Design and Drafting (CADD) process. As you will learn, special symbols are used to denote components.

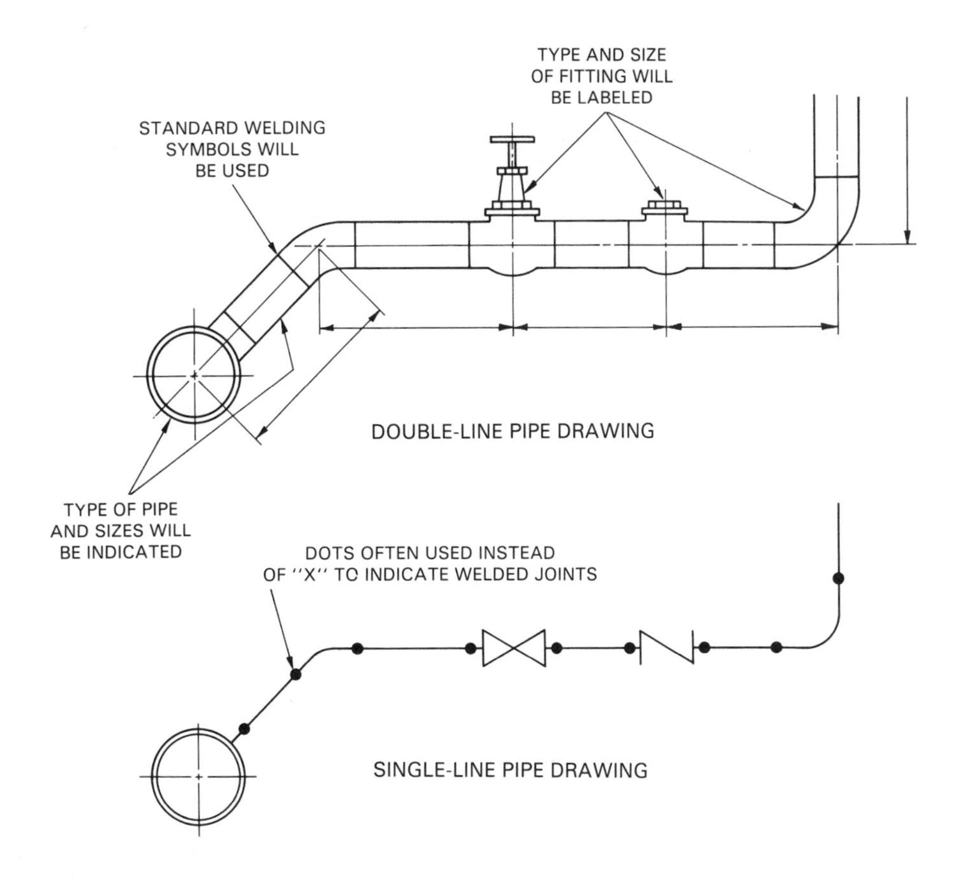

Fig. 22-3. Piping drawings may be presented as either double-line or single-line prints. Double-line drawings are used to make scale layouts where dimensional accuracy is important. Single-line drawings are diagrammatic in nature.

projection views (three-view drawings) are difficult to read. Fittings are normally shown by standard, conventional symbols, Fig. 22-4. However, some firms have developed their own series of pipe symbols that may vary slightly from the ANSI (Ameraican National Standards Institute) symbols.

Double-line pipe drawings are used to make scale layouts where dimensional accuracy is important. Pipe and fittings are drawn to exact scale. Single-line drawings are diagrammatic.

FITTING	FLANGED	SCREWED	WELDED*
BUSHING	NONE		
CAP	NONE		NONE
CROSS, STRAIGHT			
ELBOW, 45°			
ELBOW, 90°			
ELBOW, TURNED DOWN			
ELBOW, TURNED UP			
JOINT, CONNECTING PIPE			
LATERIAL			NONE
PIPE PLUG	NONE		NONE

Fig. 22-4. Note a few of many pipe symbols. While many companies use ANSI symbols, some large companies have devised their own symbols.

FITTING	FLANGED	SCREWED	WELDED*
REDUCER, CONCENTRIC			
JOINT, EXPANSION			
TEE, STRAIGHT			
TEE, OUTLET UP			
TEE, OUTLET DOWN			
SLEEVE			
UNION			
VALVE, CHECK			
VALVE, GATE			
VALVE GLOBE			

*A ● MAY BE USED INSTEAD OF THE "X" TO REPRESENT A BUTT WELDED JOINT

(Fig. 22-4 Continued from Page 180)

Dimensions on piping drawings give the LOCATION of components. They are made to the centerlines of the pipes and fittings.

A BILL OF MATERIALS or FABRICATION SCHEDULE may be found on the drawing or on an attached sheet. One example is shown in Fig. 22-5.

Equipment installations that utilize piping are sometimes shown as a pictorial drawing. See Fig. 22-6.

REV.	FABRICATION BILL OF MATERIAL					APPROP. NO. M65-135	
REV.	PIECE NO.	QUAN	SIZE	RAT'G SCH.	FACE END	DESCRIPTION	REQ. NO.
		1	4	600	RF	ORIF.UNION BORE 4.026 FE-6004	
		1	8	80		PIPE	
		43	4	40		PIPE	
		1	3/4	80		PIPE	
		1	4 × 2	40	WELD	TEE RED.	
		1	8	600	RF	WN. FLANGE BORE 7.625	
	F-1	1	2	600	RF	WN. FLANGE BORE 2.067	
	E-1	1	4	40	WELD	45 ELL	
		2	8	80	WELD	90 ELL LR	
		4	4	40	WELD	90 ELL LR	
		1	8X4			SCH. 80 X SCH. 40	
						CONC.REDUCER BWE	
		1	4			600# R.F. ORIFICE	
						PLATE TAG FE-600	
		8	5/8 × 4			STUD BOLTS	
		24	1 1/8 × 7 1/2			STUD BOLTS	
		1	2	600	RF	SPIRAL WOUND GASKET	
		2	8	600	RF	SPIRAL WOUND GASKET	
		2	4	600	RF	SPIRAL WOUND GASKET	
	P-1	2	3/4	3000	SCRD	ROUND HEAD PLUG	
		1	1/2 × 5 1/2	XXS		NIPPLE	
		1	1/2 × 2 1/2	XXS		NIPPLE	
		2	3/4	800	SCRD	GATE VALVE V31CH-1	
		2	1/2	800	SCRD	GATE VALVE V31CH-1	
		1	8	600	RF	GATE VALVE V64C-4	
REV.	PIECE NO.	QUAN	SIZE	RAT'G SCH.	FACE END	DESCRIPTION	REQ. NO.
	FIELD BILL OF MATERIAL						EP

WALKER INDUSTRIES

RESEARCH & ENGINEERING — ENGINEERING

LINE NO. 4 | | | H | A | 85 — 1-17-2A/12132A & 12134A

MOTOR GASOLINE REFORMER NO.2

PLT. 17-2A — M.H. PA.

...ESS NOTED

...ROLEUM REF'Y
...NOTED. DIM'S
...GASKETS. BOLT
...℄ UNLESS
...NIPPLES FOR
... TO BE 4" MIN.
...CKNESS PLUS 2".

2	1	/	2
Y	E	S	

BY	CHKD.	APPD.	DATE	DRAWING NO.	SHEET NO.	REV. NO.
C J G				1-17-2A/12112C	85 3	0
REVISED						
BY	CHKD.	APPD.	DATE			

Fig. 22-5. A bill of materials or fabrication schedule is commonly included on pipe drawing. However, if too lengthy to fit on drawing, it is presented on a second sheet. Reference to this sheet will be made on pipe drawing.

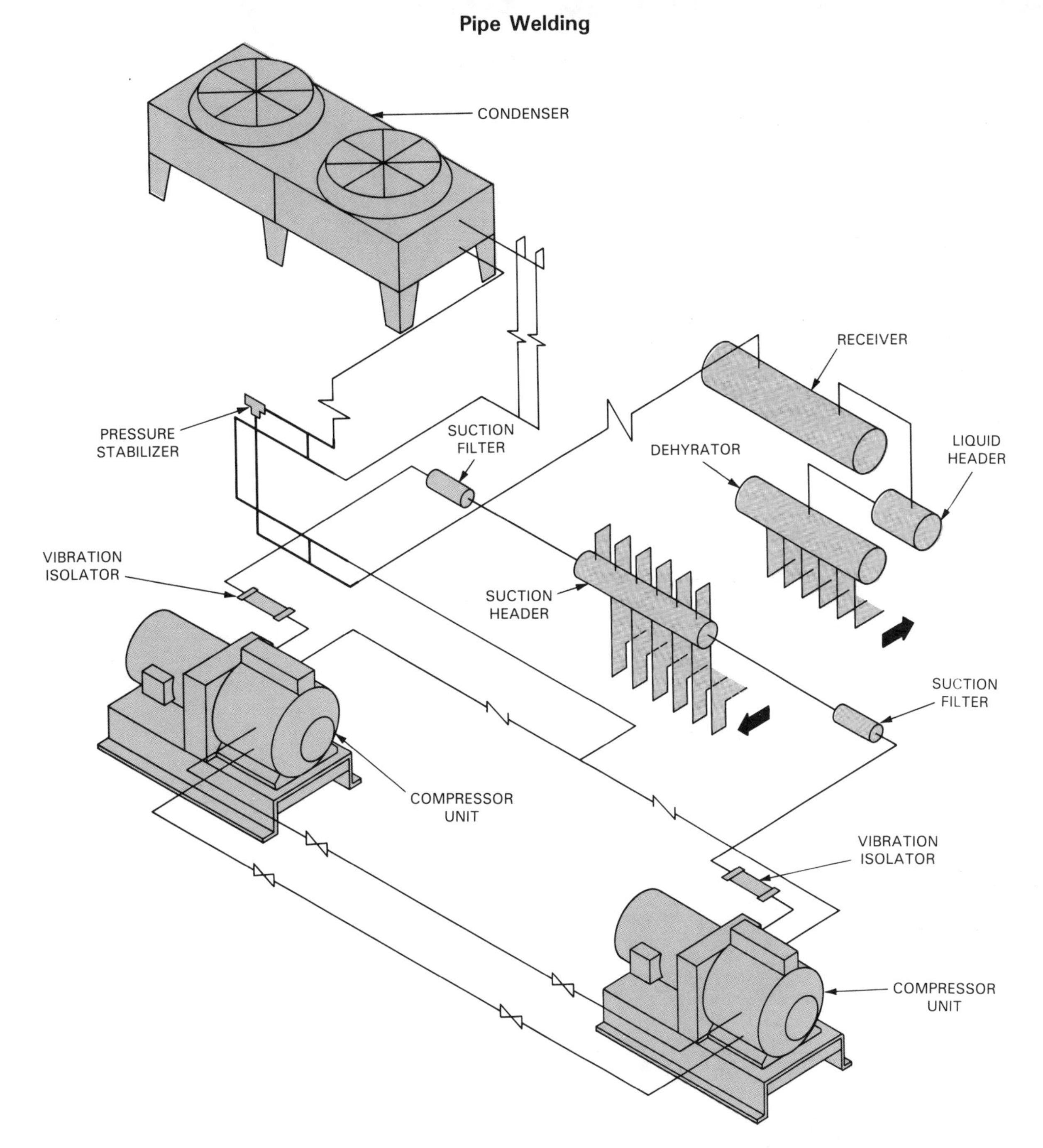

Fig. 22-6. This is a pictorial of a commercial refrigerating system with roof-mounted, air-cooled condensers. There are six refrigerant lines. Equipment installations that use piping or tubing may be shown in this manner.

UNIT 22—TEST YOUR KNOWLEDGE

Part I

Answer the following questions.

1. List at least four methods employed to fabricate pipe systems.
   a. ______________________
   b. ______________________
   c. ______________________
   d. ______________________
2. What is a piping system?
   ____
   ____
   ____
   ____
3. Pipe size is determined by______________________
   ______________________.
4. Tube size is measured by its ______________________.
5. What is a pipe drawing? ______________________
   ____

6. Pipe systems may be made as either __________ or __________ drawings.
7. Dimensions on piping drawing give the __________ of components. They are made to the __________ __________.
8. Fittings are shown on pipe drawings by standardized symbols. Sketch the symbols for the following welded fittings. The "x" or "•" may be used to denote the weld.

a. 45 deg. elbow

b. 90 deg. elbow

c. Reducer

d. Expansion joint

e. Tee

f. Union

g. Check valve

h. Gate valve

Part II

Identify the pipe fittings on the drawing shown on page 185.

a. __________
b. __________
c. __________
d. __________
e. __________
f. __________
g. __________
h. __________

Part III

Refer to the drawing on page 185 and answer the following questions.

1. Where is the pipe system to be fabricated? __________
2. Welds are to be made in the __________.
3. List weld specifications for all 8 in. pipe. How is the weld to be inspected?
   a. __________
   b. __________
4. Prepare a sketch of the weld specifications for all 4 in. dia. pipe. The pipe has a wall thickness of 1/4 in.
5. Is this drawing of a complete pipe system? If not, how did you arrive at your conclusion? __________
6. Are all welds to be made in the shop? If not, how did you arrive at your conclusion? __________

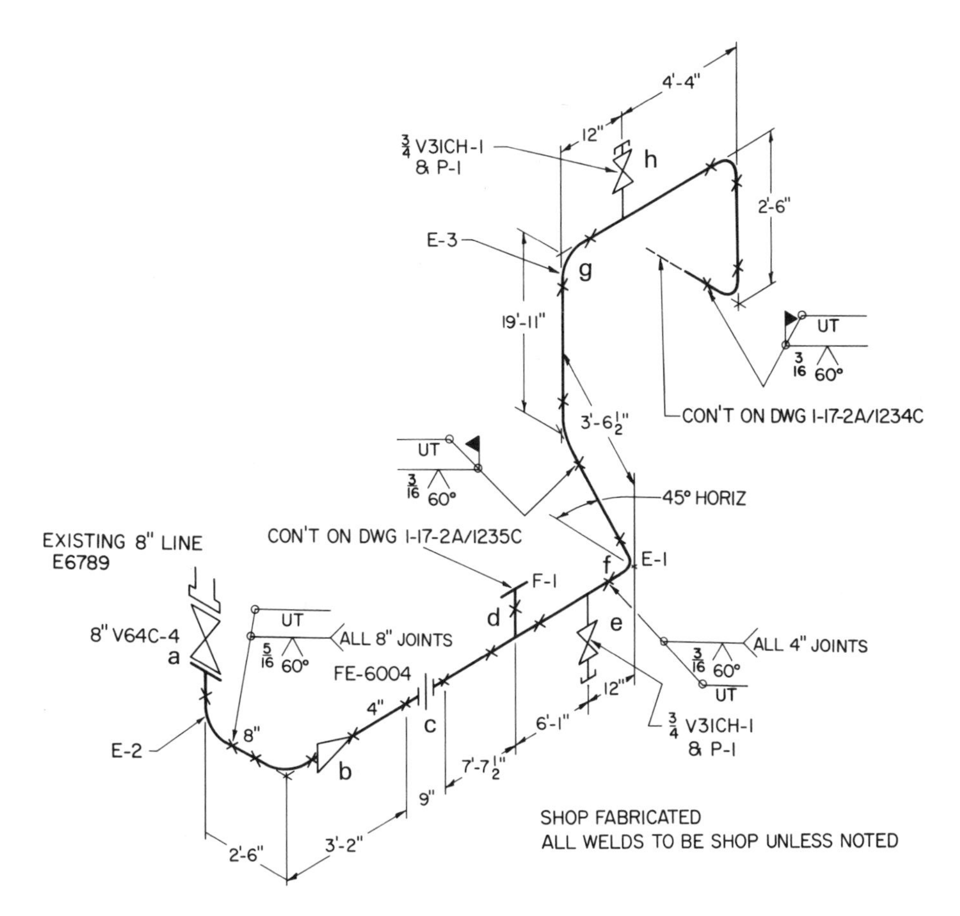

Use this drawing to complete Part II and III of Test Your Knowledge-Unit 22.

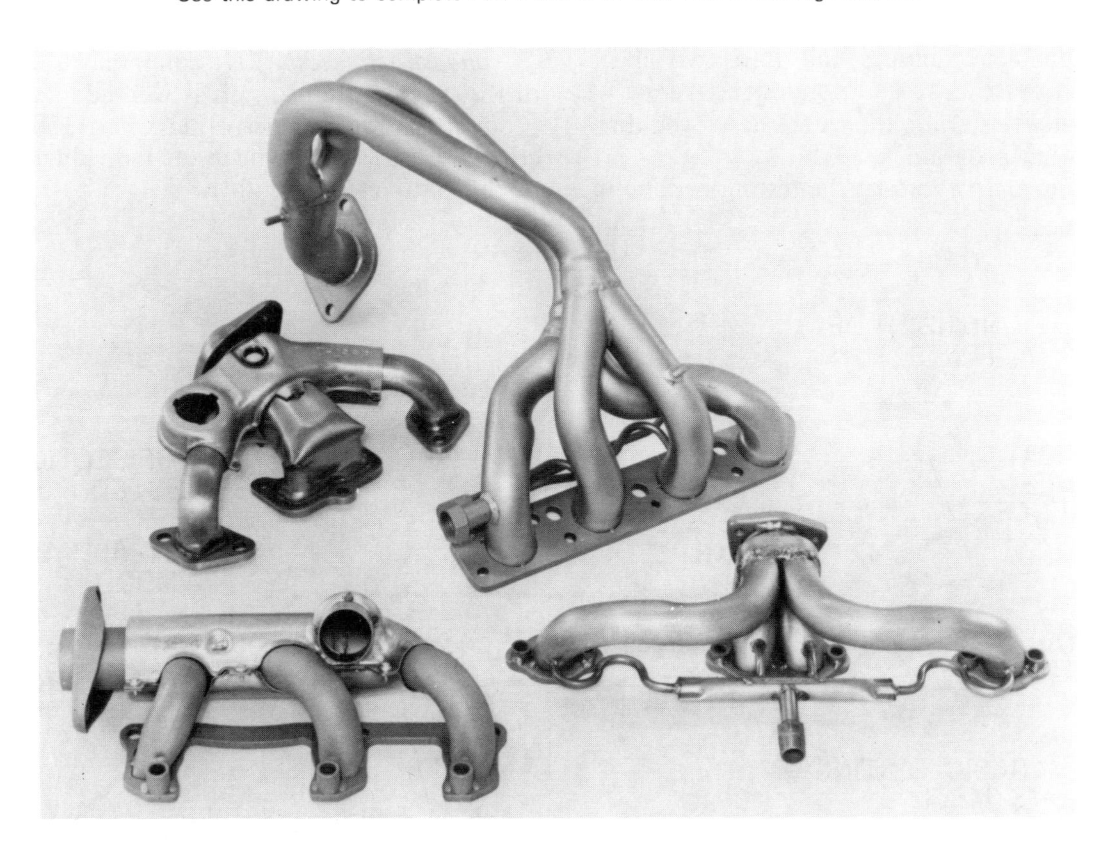

Pipe welding also extends to auto industry. Shown are stainless steel exhaust manifolds fabricated mainly by welding. Each unit is carefully inspected because of danger of exhaust gas leakage to driver and passengers. (American Iron and Steel Institute)

Unit 23

NONDESTRUCTIVE TESTING OF WELDMENTS

Quality control is one of the most important segments of industry. Its purpose is to improve product quality, maintain quality, and help to reduce costs. It is also an important way to maintain product safety.

Quality control can play a vital role in improving the competitive position of the manufacturer. This also translates into continuing employment for the employees.

The days of making weldments "by-guess and by-golly" are long past.

BASIC CLASSIFICATIONS OF QUALITY CONTROL TECHNIQUES

Quality control techniques fall into two basic classifications: destructive and nondestructive.

1. With *destructive testing,* the part is destroyed during the testing program.
2. With *nondestructive testing,* the testing is done in such a manner that the usefulness of the part is NOT impaired.

DESTRUCTIVE TESTING

Destructive testing is costly and time consuming. The specimen is selected at random from a given number of pieces. Statistically, the results indicate the characteristics of the remaining undestroyed, untested pieces. It does not give complete assurance of perfection because a considerable number of defective parts could slip by.

NONDESTRUCTIVE TESTING

Nondestructive testing is a commonly used, basic tool of the welding industry. It is well adapted to testing products where the performance of EACH part is critical. Each weld can be tested individually or as part of the completed assembly.

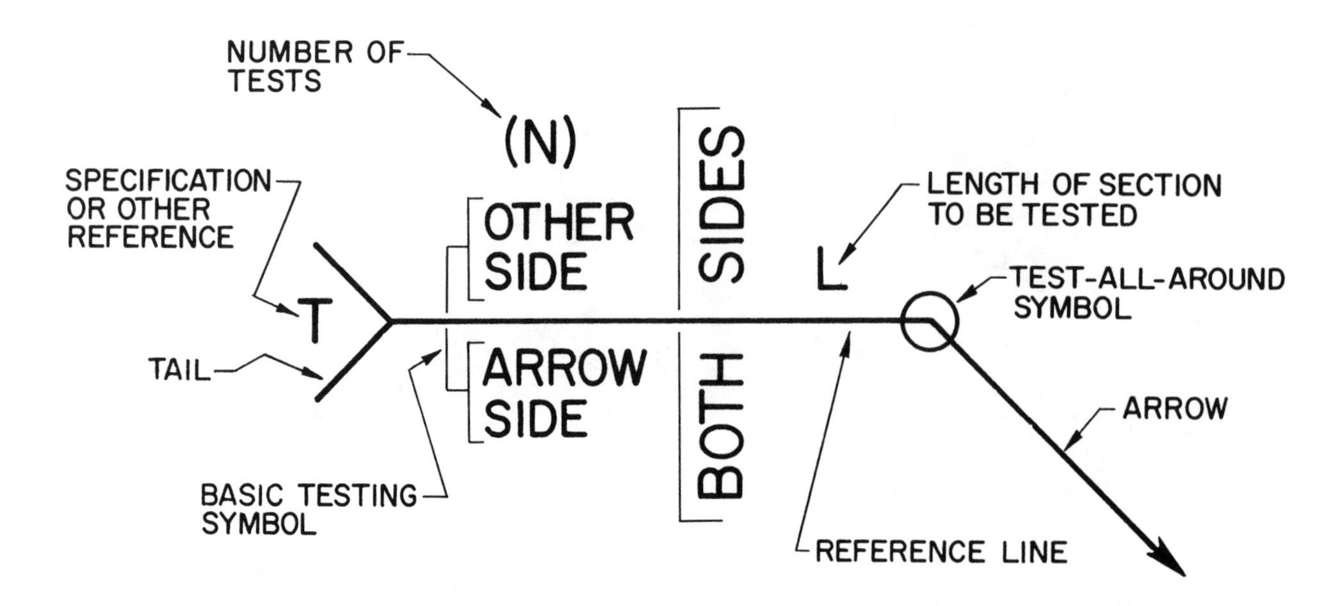

Fig. 23-1. Study testing symbol that denotes required nondestructive tests to be done on a welded joint.

Basic nondestructive testing symbols

Basic *nondestructive testing symbols* that have been approved by the American Welding Society (AWS) are as follows:

TEST	SYMBOL
Acoustic emission	AET
Eddy current	ET
Leak	LT
Magnetic particle	MT
Neutron radiographic	NRT
Penetrant	PT
Dye penetrant	DPT
Fluorescent penetrant	FPT
Proof	PRT
Radiographic	RT
Ultrasonic	UT
Visual	VT

Location significance of arrow

The arrow connects the reference line to the weldment to be tested. The side of the welded joint to which the arrow points is considered the arrow side of the joint. The side opposite the arrow side is considered the other side.

Location of testing symbol

If the test symbol is placed on the arrow side of the reference line (towards reader), the test is to be made on the arrow side of the joint, as shown in Fig. 23-2.

A test to be made on the other side will be indicated by the test symbol being placed on the other side (away from reader) of the reference line. Refer to Fig. 23-3.

Tests to be made on both sides of the welded joint are indicated by test symbols on both sides of the reference line, as in Fig. 23-4.

The test symbol is centered on the reference line when the test has no arrow or other side significance. Fig. 23-5 gives some examples.

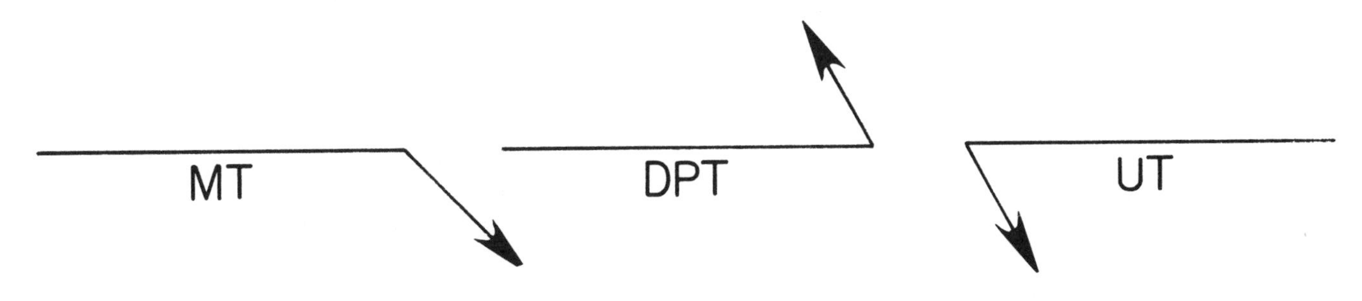

Fig. 23-2. Symbols denote tests to be made on arrow side of each joint.

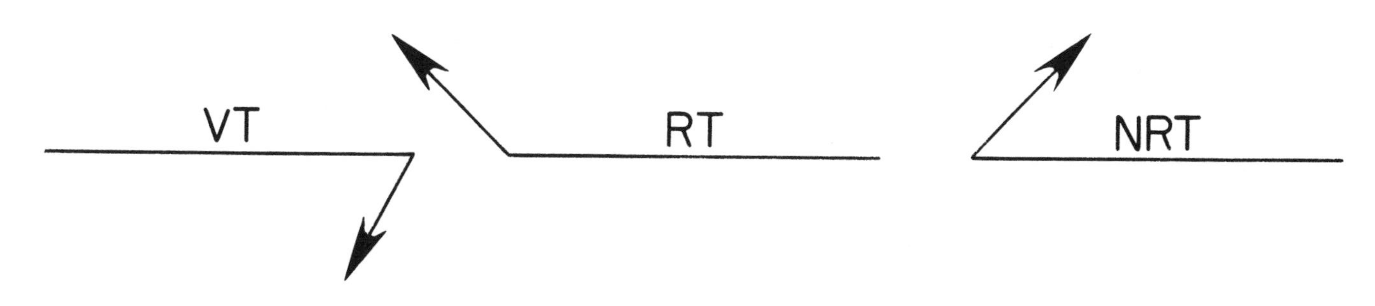

Fig. 23-3. Symbols indicate tests to be made on other side of each joint.

ELEMENTS OF THE TESTING SYMBOL

A *testing symbol,* similar to the welding symbol, is employed to indicate the required nondestructive test to be carried out. Fig. 23-1 summarizes the organization of a testing symbol.

Direction of radiation

The *direction of radiation* testing is specified by a symbol located on the drawing at the desired angle of the test(s). It is in conjunction with the radiographic and neutron radiographic testing symbols. See Fig. 23-6.

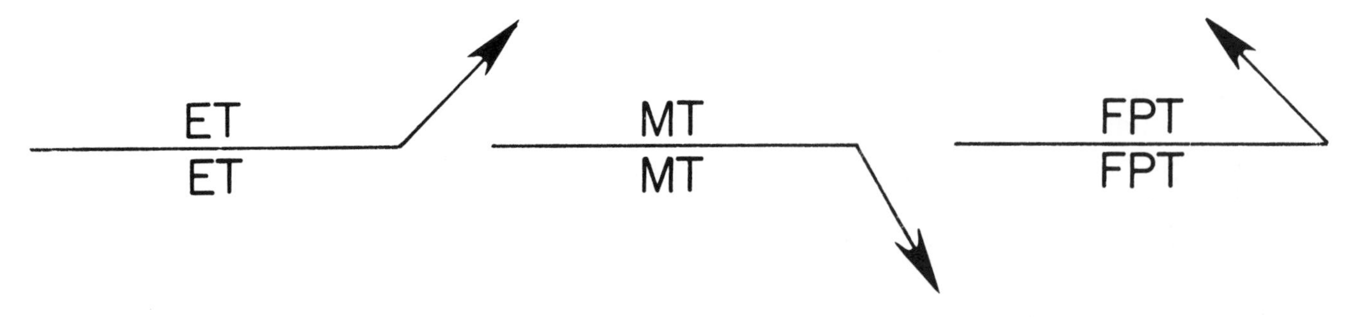

Fig. 23-4. These symbols show that tests are to be made on both sides of joint.

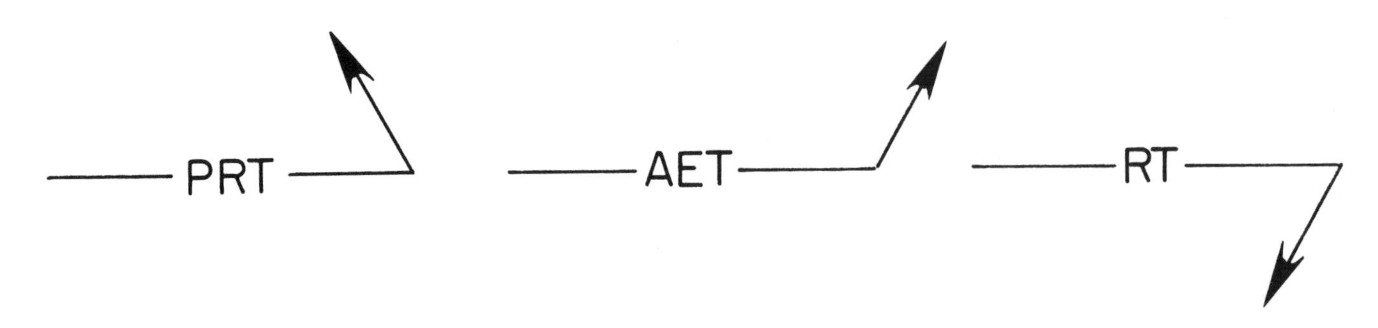

Fig. 23-5. Test has no arrow or other side significance on joint.

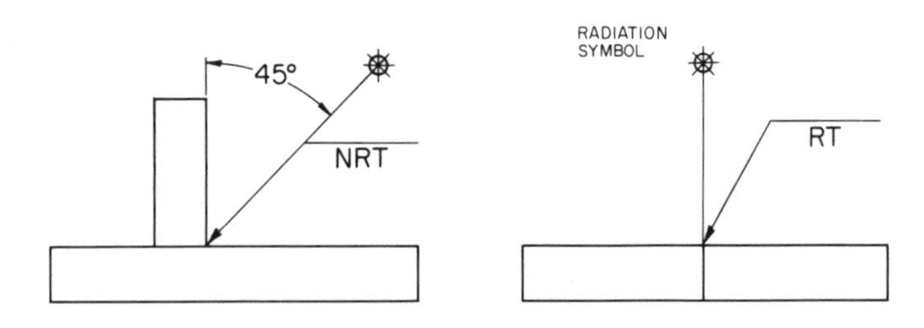

Fig. 23-6. Note radiation symbol (star) and how direction of radiation testing is specified.

Combination of nondestructive testing symbols and welding symbols

Nondestructive testing symbols and welding symbols may be combined on a print. Two simple examples are illustrated in Fig. 23-7.

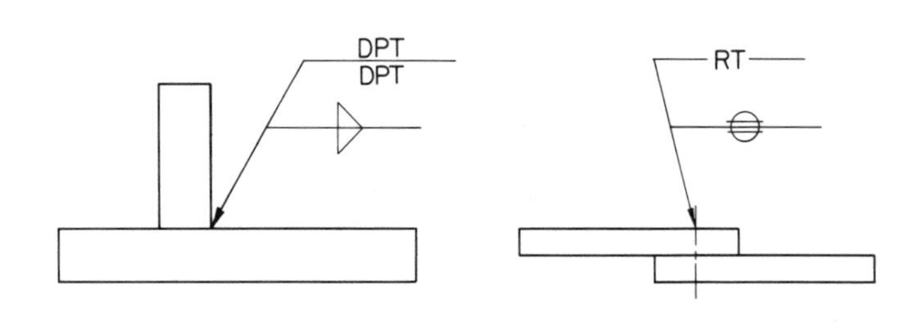

Fig. 23-7. Test symbols are often used in conjunction with weld symbols on multiple reference lines.

Specifying length of section to be tested

At times, a specific section of a welded joint must be tested. The exact location and length of the tested section will be indicated by dimension lines, Fig. 23-8.

When the full length of a welded joint requires testing, no dimension will be shown on the testing symbol.

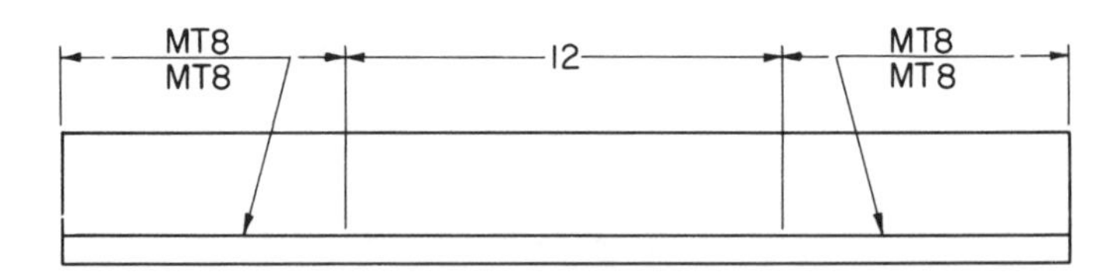

Fig. 23-8. Note how length of weld section is specified for testing. Only ends of weldment are to be tested.

If less than one hundred percent of the weld is to be tested, the percentage of the length to be tested will be indicated as shown in Fig. 23-9. Test locations will be specified by a note elsewhere on the print.

Should tests be required randomly along a weld, the number of tests is shown in parentheses. See Fig. 23-10.

The test-all-around symbol will be used when the test is to be made all around a joint. Appropriate dimensions will locate the test area. Fig. 23-11 gives an example.

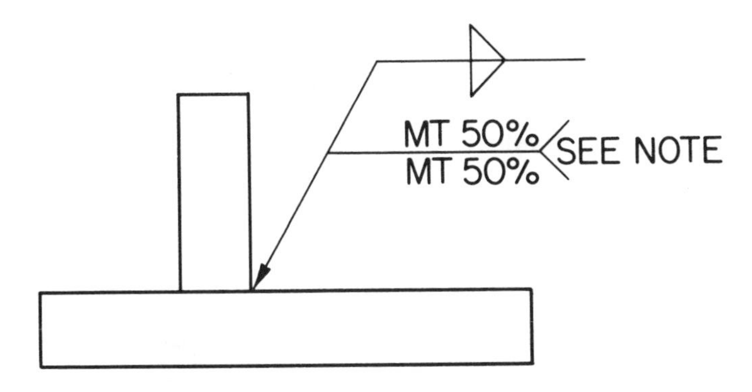

Fig. 23-9. When not all of weld length is to be tested, procedure for selecting test locations will be indicated in tail of testing symbol.

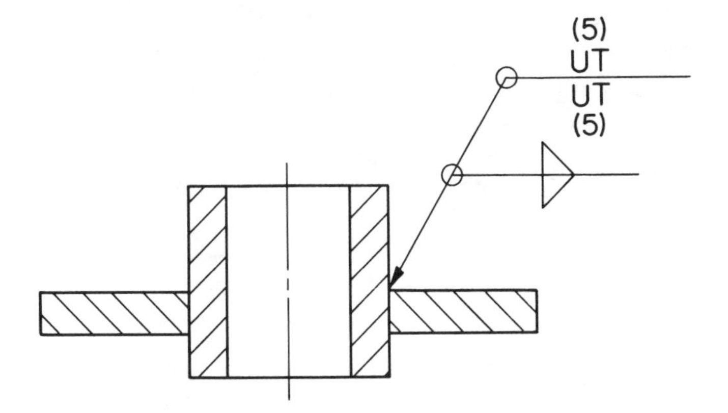

Fig. 23-10. Note how number of random tests, five on each side in this example, to be made on weldment are specified.

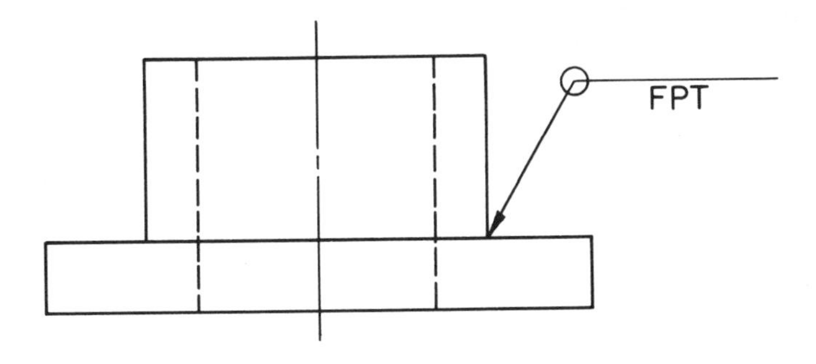

Fig. 23-11. Test-all-around symbol is used when test is required completely around welded joint.

SPECIFYING A SECTION OR PLANE AREA TO BE TESTED

Nondestructive testing of an area or part of a unit, before or after being welded, will be surrounded by a series of straight broken lines having a circle at each change of direction. The symbol specifying the kind of test will also be shown, Fig. 23-12. The test area may or may not be located by dimensions.

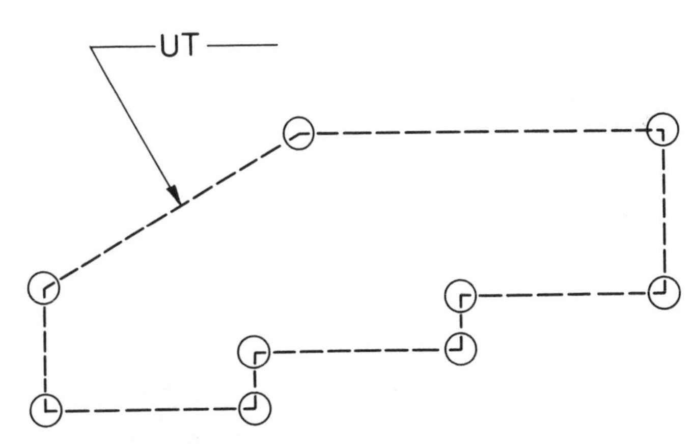

Fig. 23-12. Technique for indicating a specific area for nondestructive testing before or after making weld(s).

UNIT 23—TEST YOUR KNOWLEDGE

1. Quality control is one of the most important segments of industry. Why?

____

____

____

2. There are two basic classifications of quality control techniques. List them and briefly describe each.

   a. ____

   ____

   b. ____

   ____

3. What do the following mean on a testing symbol? Refer to the drawing.

   a. (N) = _______________________
   b. L = _______________________
   c. T = _______________________
   d. [all-around symbol] = _______________________

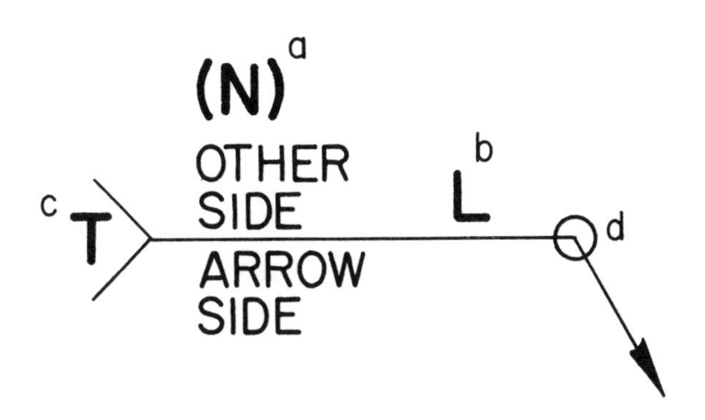

4. If the test symbol is placed on the arrow side of the reference line, the test is to be made on the ______ ______________________.

5. If the test symbol is placed on the other side of the reference line, the test is to be made on the ______________________.

6. Tests to be made on both sides of the welded joint are indicated by ______________________.

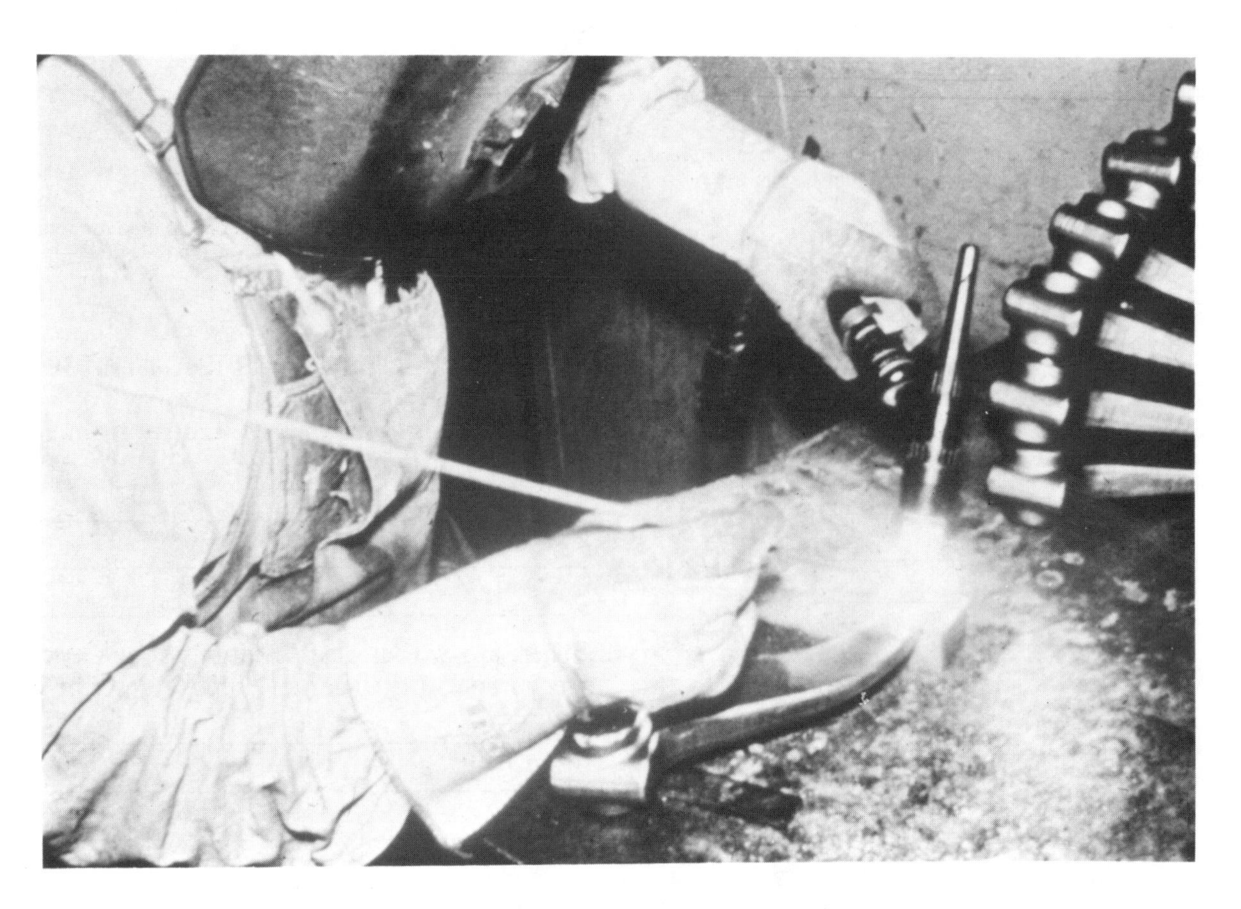

Welders are classified as skilled craftworkers. They must be familiar with metal characteristics and able to read and understand drawings. They work indoors and outdoors, in all kinds of weather. (Lincoln Electric Co.)

REFERENCE SECTION

METRIC SIZE DRAFTING SHEETS

ISO STANDARD

SIZE	MILLIMETERS	INCHES
AO	841 x 1189	33.11 x 46.81
A1	594 x 841	23.39 x 33.11
A2	420 x 594	16.54 x 23.39
A3	297 x 420	11.69 x 16.54
A4	210 x 297	8.27 x 11.69

AMERICAN STANDARD

SIZE	INCHES	MILLIMETERS
E	34 x 44	863.6 x 1117.6
D	22 x 34	558.8 x 863.6
C	17 x 22	431.8 x 558.8
B	11 x 17	279.4 x 431.8
A	8 1/2 x 11	215.9 x 279.4

Left—Metric size drafting paper sheets are exactly proportional in size. Right—Charts show millimeter sizes and customary inch equivalent sizes of ISO papers.

ABBREVIATONS

(SOME FORMER ABBREVIATIONS ARE NOW SYMBOLS)

Term	Abbreviation	Term	Abbreviation
Across flats	ACR FLT	Inside diameter	ID
Centers	CTR	Left hand	LH
Center line	CL	Material	MATL
Centimetre	cm	Meter	m
Chamfer	CHAM	Millimeter	mm
Counterbore	CBORE	Number	NO
Countersink	CSK	Outside diameter	OD
Countersunk head	CSK H	Pitch diameter	PD
Diameter (before dimension)	Ø	Radius	R
Diameter (in a note)	DIA	Right hand	RH
Drawing	DWG	Round	RD
Figure	FIG	Square (before dimension)	□
Hexagon	HEX	Square (in a note)	SQ
Hexagonal head	HEX HD	Thread	THD

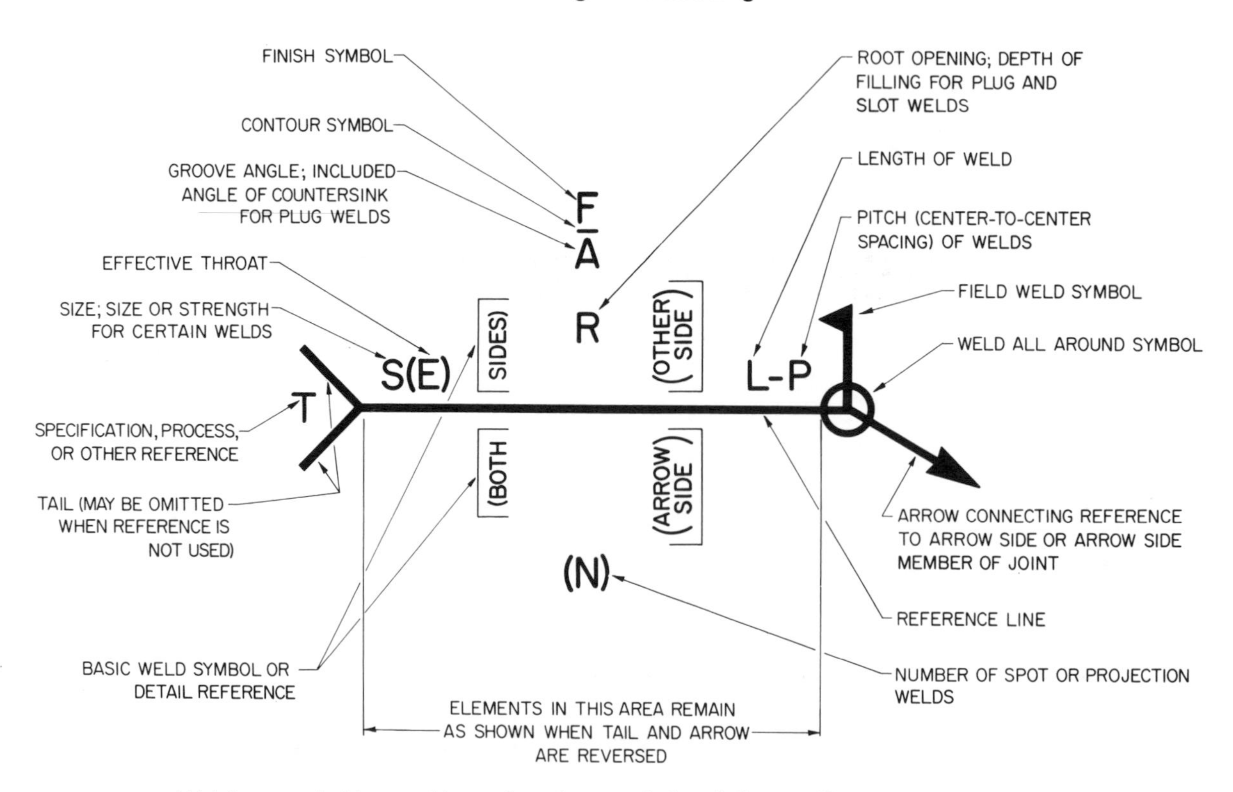

Welding symbol is graphic explanation needed to fully specify weld requirements.

FILLET	PLUG OR SLOT	SPOT PROJECTION	SEAM	GROOVE							BACK OR BACKING	SURFACING	FLANGE	
				SQUARE	"V"	BEVEL	"U"	"J"	FLARE "V"	FLARE BEVEL			EDGE	CORNER

Names and shapes of basic weld symbols.

BASIC WELDING SYMBOLS AND THEIR LOCATION SIGNIFICANCE								SUPPLEMENTARY SYMBOLS		
LOCATION SIGNIFICANCE	SQUARE	V	BEVEL	GROOVE U	J	FLARE-V	FLARE-BEVEL	WELD-ALL-AROUND	FIELD WELD	MELT-THROUGH
ARROW SIDE										
OTHER SIDE									CONTOUR	
BOTH SIDES								FLUSH	CONVEX	CONCAVE
NO ARROW SIDE OR OTHER SIDE SIGNIFICANCE		NOT USED	NOT USED	NOT USED	NOT USED	NOT USED	NOT USED			

BASIC WELDING SYMBOLS AND THEIR LOCATION SIGNIFICANCE								
LOCATION SIGNIFICANCE	FILLET	PLUG OR SLOT	SPOT OR PROJECTION	SEAM	BACK OR BACKING	SURFACING	FLANGE	
							EDGE	CORNER
ARROW SIDE					GROOVE WELD SYMBOL			
OTHER SIDE					GROOVE WELD SYMBOL	NOT USED		
BOTH SIDES		NOT USED	NOT USED	NOT USED	NOT USED	NOT USED	NOT USED	NOT USED
NO ARROW SIDE OR OTHER SIDE SIGNIFICANCE	NOT USED	NOT USED			NOT USED	NOT USED	NOT USED	NOT USED

Basic welding symbols and their location significance.

WELDING SAFETY

You must be familiar with situations that can endanger your safety, and the safety of others, when in a welding facility. The best way to eliminate or control the safety in a welding shop is to eliminate the conditions that could be potentially dangerous.

Listed are a few general safety warnings that pertain to welding:

1. Ultraviolet rays are harmful to eyes. Safety goggles or a helmet with proper lenses will provide protection.
2. Ionizing radiations markedly increase the possibility of eye cataracts. Even small amounts of ultraviolet may cause cataracts. Those operating welding equipment, as well as helpers and other people in the area, must take precautions at all times to prevent eye injury by wearing safety glasses.
3. Always protect eyes, skin, and respiratory system. Provide adequate ventilation during all welding, cutting, brazing, and soldering operations.
4. Carbon dioxide creates a carbon monoxide (CO) problem as it breaks down in the welding arc. Again, keep your work area ventilated.
5. The oxyfuel gas flame is generally safe, but there is still a carbon monoxide (CO) problem in a poorly ventilated area.
6. Coatings on metals can be a problem when the metal is heated. Nitrogen dioxide fumes, coated electrode fumes, and iron oxide and cadmium and zinc coating fumes are examples. Do not breathe these fumes!
7. Red lead paint is sometimes used outdoors for metal finishing and protection. Lead oxide fumes caused by burning lead paint coatings can produce lead poisoning.
8. Cadmium plate is commonly used on small parts. Cadmium oxide fumes, at low levels, can be harmful to lungs and liver.
9. Silver brazing alloy fumes can be quite dangerous.
10. Termeplate is a metallic lead coating and is dangerous when heated.
11. Fluorides in fluxes are common. Fluoride fumes are also harmful.
12. Manganese dioxide is not extremely toxic, but it may cause trouble if ventilation is poor.
13. When gas tungsten arc welding aluminum, there is danger because the ultraviolet frequency is right to form ozone. This gas is most toxic and may cause severe lung and body damage. It is irritating and causes coughing. Keep the arc area well ventilated.
14. Beryllium is toxic in even very small amounts. Therefore, any operation involving beryllium must be contained.
15. Cobalt should be handled similar to beryllium.
16. Thorium is toxic. Therefore, when thoriated electrodes are used, an alpha emission is produced and causes an ionization effect. Ventilate the area.
17. Reducing furnaces may have carbon monoxide (CO) emissions. Such furnaces should be well vented at both charging and discharging ends.
18. When vacuum furnaces or welding chambers are used, the pump exhaust must be vented away from people.
19. Inert gases are dangerous to use in confined spaces. Inert gases displace air and oxygen. Workers entering spaces or tanks that are filled with an inert gas must wear supplied air or self-contained breathing equipment.
20. Oil smoke is a problem. The aromatics produced can be dangerous. Note the table of safe limits for some welding fumes.

MATERIAL	GASES		MILLION PARTS
	ppm	mg/m³	per Cu. Ft.
Acetylene	1000		
Beryllium		.002	
Cadmium Oxide fumes		.1	
Carbon Dioxide	5000		
Copper fumes		.1	
Iron Oxide fumes		10.0	
Lead		.2	
Manganese		5.0	
Nitrogen Dioxide	5.0		
Oil Mist		5.0	
Ozone	.1		
Titanium Oxide		15.0	
Zinc Oxide fumes		5.0	
Silica, crystalline			2.5
Silica, amorphous			20.0
Silicates: Asbestos			5.0
Portland Cement			50.0
Graphite			15.0
Nuisance Dust			50.0

A table of safe limits for some welding fumes. All gases tend to reduce oxygen by replacement. Such gases as argon, helium, carbon dioxide, etc., present this danger.

FORMULAS

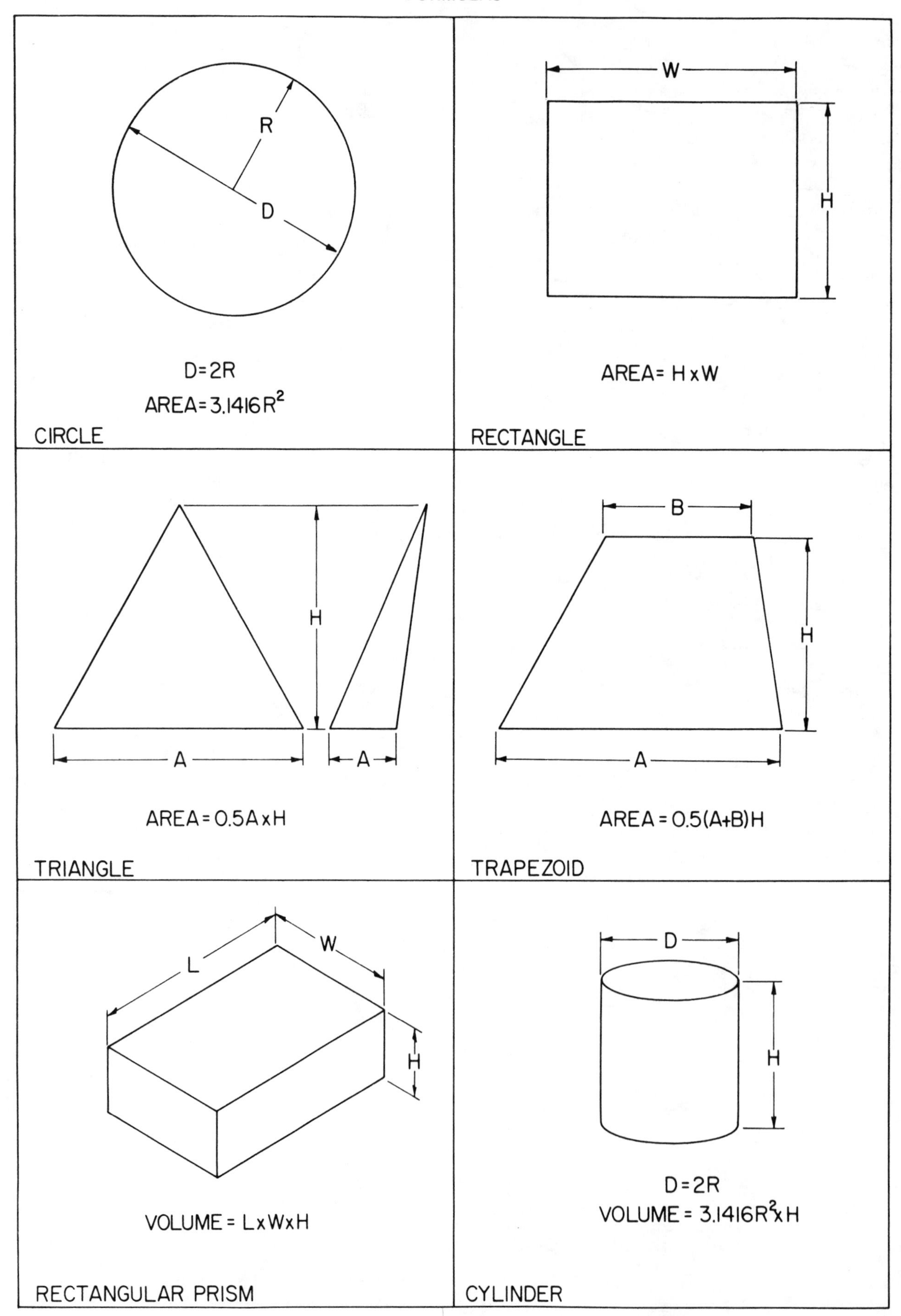
R
D
D=2R
AREA=3.1416R²
CIRCLE
W
H
AREA= H x W
RECTANGLE
H
A
A
AREA=0.5A x H
TRIANGLE
B
H
A
AREA=0.5(A+B)H
TRAPEZOID
L
W
H
VOLUME=L x W x H
RECTANGULAR PRISM
D
H
D=2R
VOLUME=3.1416R² x H
CYLINDER

FORMULAS

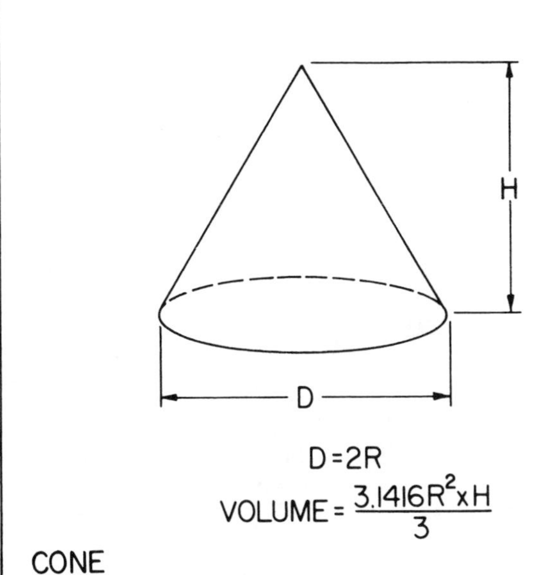

$D = 2R$

$$\text{VOLUME} = \frac{3.1416R^2 \times H}{3}$$

CONE

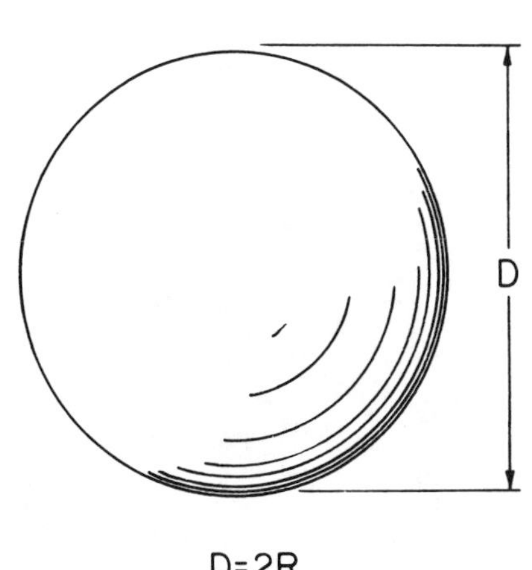

$D = 2R$

$$\text{VOLUME} = \frac{4 \times 3.1416R^3}{3}$$

SPHERE

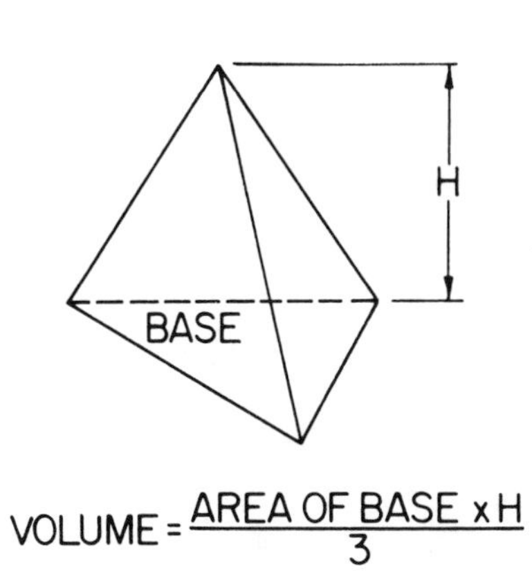

$$\text{VOLUME} = \frac{\text{AREA OF BASE} \times H}{3}$$

TRIANGULAR PYRAMID

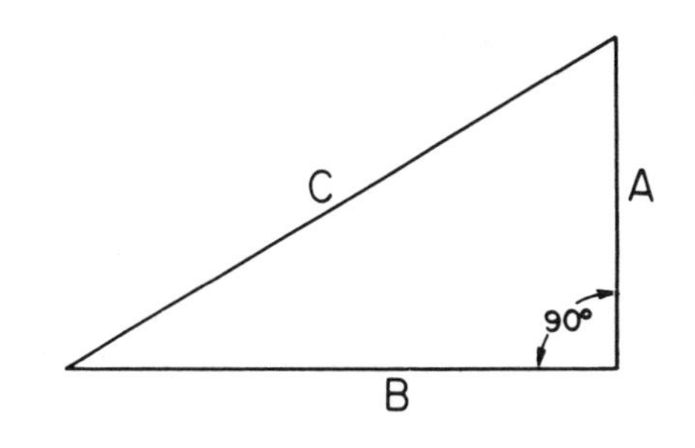

$A = \sqrt{C^2 - B^2}$

$B = \sqrt{C^2 - A^2}$

$C = \sqrt{A^2 + B^2}$

PYTHAGOREAN THEOREM

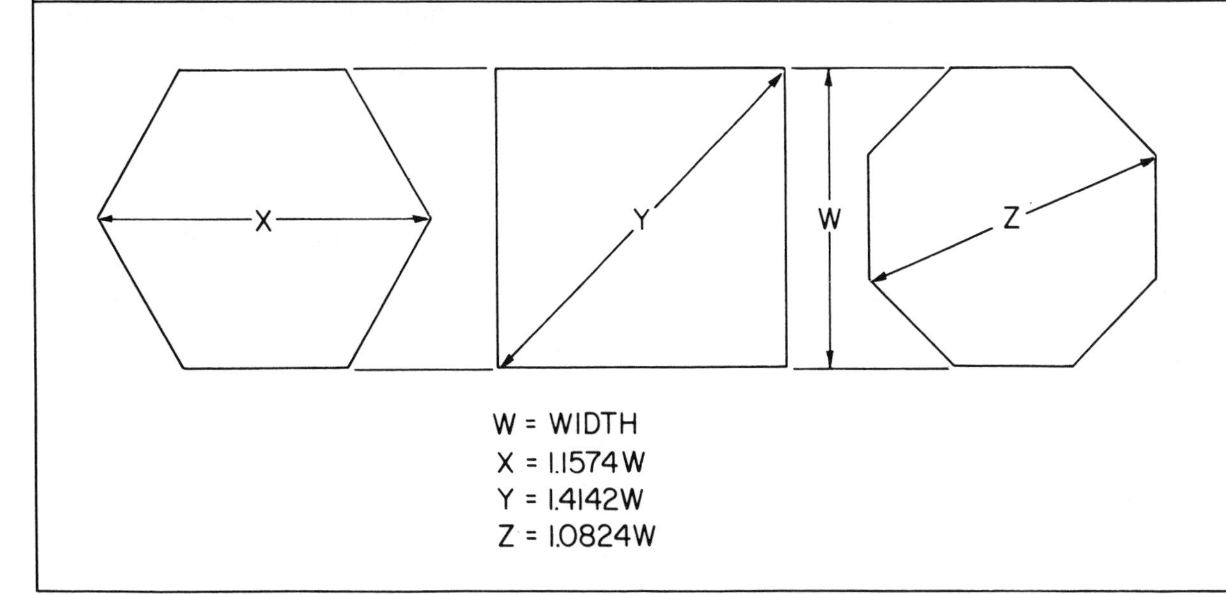

W = WIDTH

X = 1.1574W

Y = 1.4142W

Z = 1.0824W

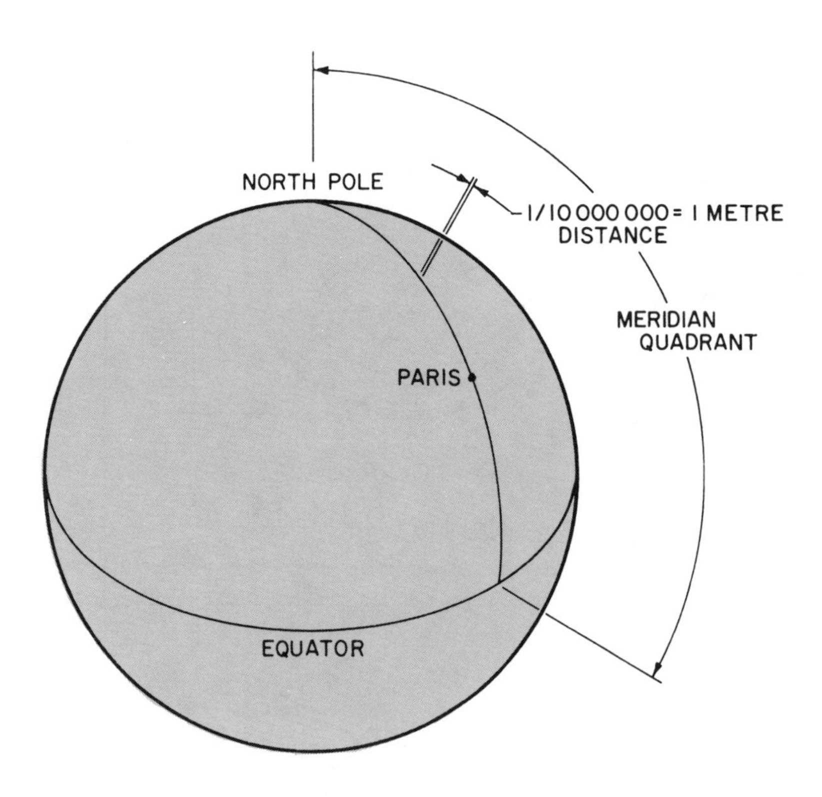

In 1793, the French government adopted a system of standards they called the METRIC SYSTEM. It was based on what they called a METER. The meter was calculated to be one ten-millionth part of the distance from the North Pole to the Equator when measured on a straight line that ran along the surface of the earth through Paris.

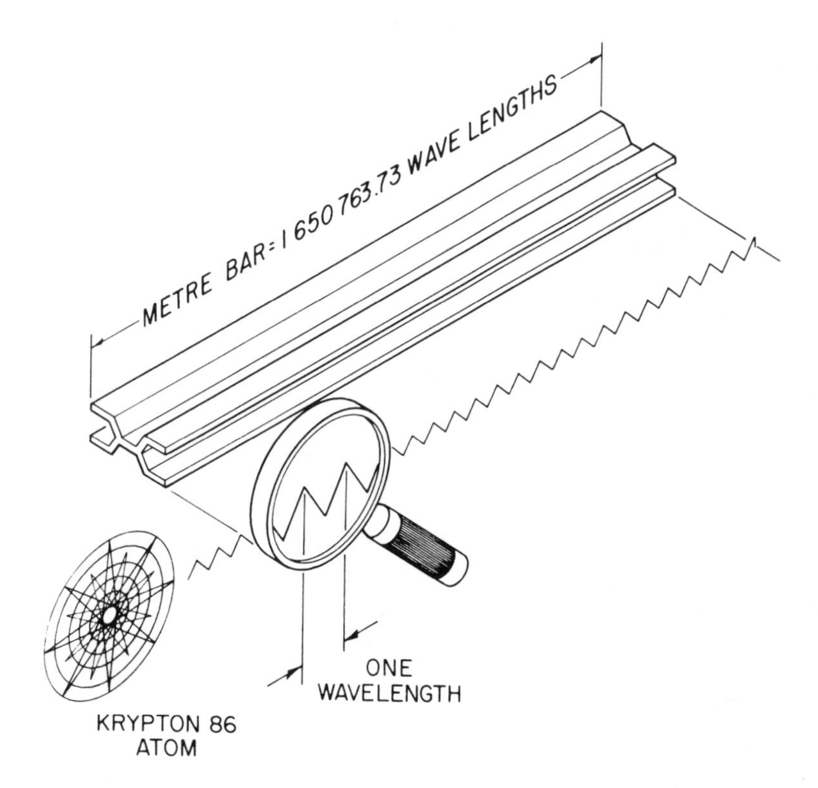

The meter, based on the measurement of the earth, was not accurate for modern use. The modern meter is defined as 1 650 736.73 wavelengths in a vacuum of the orange-red line of the spectrum for krypton-86. An INTERFEROMETER is used to measure this length by means of light waves. The reason the meter is defined in terms of the wavelength of light is so it can be reproduced in any modern scientific laboratory. By using a specific light wave, an accuracy of one part in a hundred million can be maintained.

PREFIXES, EXPONENTS AND SYMBOLS

DECIMAL FORM	EXPONENT OR POWER	PREFIX	PRONUNCIATION	SYMBOL	MEANING
1 000 000 000 000 000 000	$= 10^{18}$	exa	ex'a	E	quintillion
1 000 000 000 000 000	$= 10^{15}$	peta	pet'a	P	quadrillion
1 000 000 000 000	$= 10^{12}$	tera	tĕr'á	T	trillion
1 000 000 000	$= 10^{9}$	giga	ǰi'gá	G	billion
1 000 000	$= 10^{6}$	mega	mĕg'á	M	million
1 000	$= 10^{3}$	kilo	kĭl'ō	k	thousand
100	$= 10^{2}$	hecto	hĕk'to	h	hundred
10	$= 10^{1}$	deka	dĕk'a	da	ten
1					base unit
0.1	$= 10^{-1}$	deci	dĕs'ī	d	tenth
0.01	$= 10^{-2}$	centi	sĕn'tǐ	c	hundredth
0.001	$= 10^{-3}$	milli	mǐl'ǐ	m	thousandths
0.000 001	$= 10^{-6}$	micro	mi'kro	μ	millionth
0.000 000 001	$= 10^{-9}$	nano	năn'o	n	billionth
0.000 000 000 001	$= 10^{-12}$	pico	péc o	p	trillionth
0.000 000 000 000 001	$= 10^{-15}$	femto	fĕm'to	f	quadrillionth
0.000 000 000 000 000 001	$= 10^{-18}$	atto	ăt'to	a	quintillionth

Most commonly used

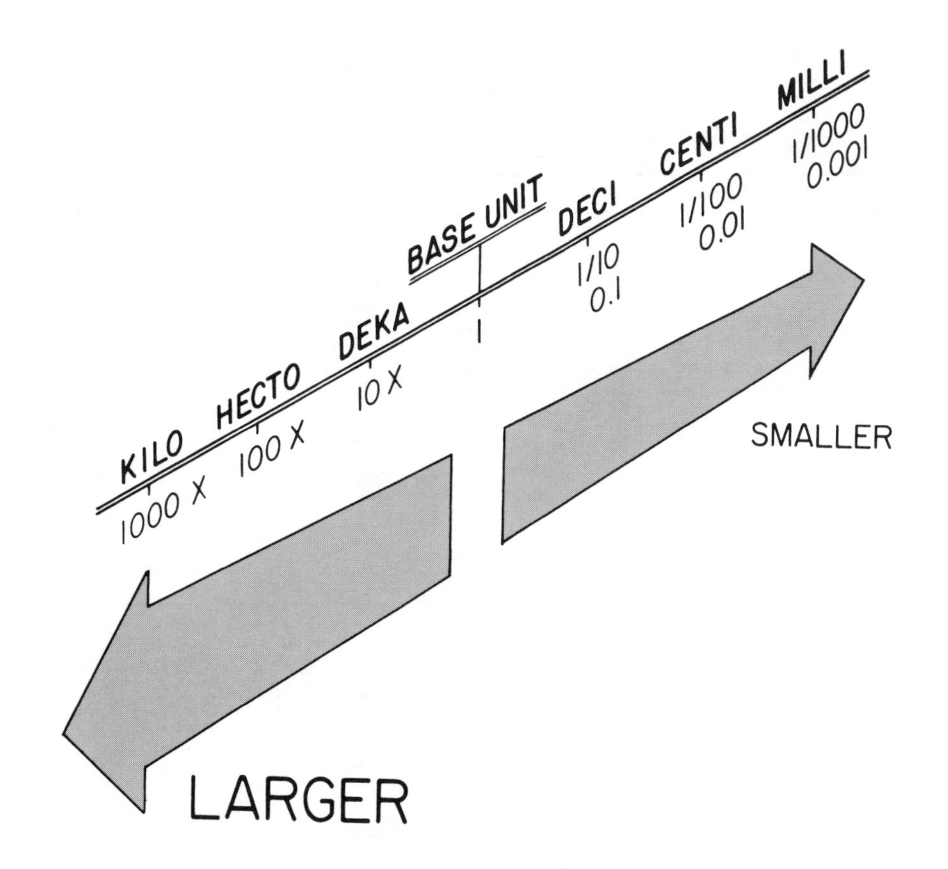

CONVERSION TABLE METRIC TO ENGLISH

WHEN YOU KNOW	MULTIPLY BY: (* = Exact) VERY ACCURATE	APPROXIMATE	TO FIND
LENGTH			
millimeters	0.0393701	0.04	inches
centimeters	0.3937008	0.4	inches
meters	3.280840	3.3	feet
meters	1.093613	1.1	yards
kilometers	0.621371	0.6	miles
WEIGHT			
grains	0.00228571	0.0023	ounces
grams	0.03527396	0.035	ounces
kilograms	2.204623	2.2	pounds
tonnes	1.1023113	1.1	short tons
VOLUME			
milliliters		0.2	teaspoons
milliliters	0.06667	0.067	tablespoons
milliliters	0.03381402	0.03	fluid ounces
liters	61.02374	61.024	cubic inches
liters	2.113376	2.1	pints
liters	1.056688	1.06	quarts
liters	0.26417205	0.26	gallons
liters	0.03531467	0.035	cubic feet
cubic meters	61023.74	61023.7	cubic inches
cubic meters	35.31467	35.0	cubic feet
cubic meters	1.3079506	1.3	cubic yards
cubic meters	264.17205	264.0	gallons
AREA			
square centimeters	0.1550003	0.16	square inches
square centimeters	0.00107639	0.001	square feet
square meters	10.76391	10.8	square feet
square meters	1.195990	1.2	square yards
square kilometers		0.4	square miles
hectares	2.471054	2.5	acres
TEMPERATURE			
Celsius	*9/5 (then add 32)		Fahrenheit

CONVERSION TABLE ENGLISH TO METRIC

WHEN YOU KNOW	MULTIPLY BY: (* = Exact) VERY ACCURATE	APPROXIMATE	TO FIND
LENGTH			
inches	*25.4		millimeters
inches	* 2.54		centimeters
feet	* 0.3048		meters
feet	*30.48		centimeters
yards	* 0.9144	0.9	meters
miles	* 1.609344	1.6	kilometers
WEIGHT			
grains	15.43236	15.4	grams
ounces	*28.349523125	28.0	grams
ounces	* 0.028349523125	.028	kilograms
pounds	* 0.45359237	0.45	kilograms
short ton	* 0.90718474	0.9	tonnes
VOLUME			
teaspoons		5.0	milliliters
tablespoons		15.0	milliliters
fluid ounces	29.57353	30.0	milliliters
cups		0.24	liters
pints	* 0.473176473	0.47	liters
quarts	* 0.946352946	0.95	liters
gallons	* 3.785411784	3.8	liters
cubic inches	* 0.016387064	0.02	liters
cubic feet	* 0.028316846592	0.03	cubic meters
cubic yards	* 0.764554857984	0.76	cubic meters
AREA			
square inches	* 6.4516	6.5	square centimeters
square feet	* 0.09290304	0.09	square meters
square yards	* 0.83612736	0.8	square meters
square miles		2.6	square kilometers
acres	* 0.40468564224	0.4	hectares
TEMPERATURE			
Fahrenheit	*5/9 (after subtracting 32)		Celsius

CONVERSION CHART INCH/mm

Drill No. or Letter		Inch	mm
		.001	0.0254
		.002	0.0508
		.003	0.0762
		.004	0.1016
		.005	0.1270
		.006	0.1524
		.007	0.1778
		.008	0.2032
		.009	0.2286
		.010	0.2540
		.011	0.2794
		.012	0.3048
		.013	0.3302
80	.0135	.014	0.3556
79	.0145	.015	0.3810
	1/64	.0156	0.3969
78		.016	0.4064
		.017	0.4318
77		.018	0.4572
		.019	0.4826
76		.020	0.5080
75		.021	0.5334
		.022	0.5588
74	.0225	.023	0.5842
73		.024	0.6096
72		.025	0.6350
71		.026	0.6604
		.027	0.6858
70		.028	0.7112
		.029	0.7366
69	.0292	.030	0.7620
68		.031	0.7874
	1/32	.0312	0.7937
67		.032	0.8128
66		.033	0.8382
		.034	0.8636
65		.035	0.8890
64		.036	0.9144
63		.037	0.9398
62		.038	0.9652
61		.039	0.9906
		.0394	1.0000
60		.040	1.0160
59		.041	1.0414
58		.042	1.0668
57		.043	1.0922
		.044	1.1176
		.045	1.1430
56	.0465	.046	1.1684
	3/64	.0469	1.1906
		.047	1.1938
		.048	1.2192
		.049	1.2446
		.050	1.2700
		.051	1.2954
55		.052	1.3208
		.053	1.3462
		.054	1.3716
54		.055	1.3970
		.056	1.4224
		.057	1.4478
		.058	1.4732
		.059	1.4986
53	.0595	.060	1.5240
		.061	1.5494
		.062	1.5748
	1/16	.0625	1.5875
		.063	1.6002
52	.0635	.064	1.6256
		.065	1.6510
		.066	1.6764
51		.067	1.7018
		.068	1.7272
		.069	1.7526
50		.070	1.7780
		.071	1.8034
		.072	1.8288
49		.073	1.8542
		.074	1.8796
		.075	1.9050
48		.076	1.9304
		.077	1.9558
47	.0785	.078	1.9812
	5/64	.0781	1.9844
		.0787	2.0000
		.079	2.0066
		.080	2.0320
46		.081	2.0574
45		.082	2.0828
		.083	2.1082
		.084	2.1336
		.085	2.1590
44		.086	2.1844
		.087	2.2098
		.088	2.2352
43		.089	2.2606
		.090	2.2860
		.091	2.3114
		.092	2.3368
42	.0935	.093	2.3622
	3/32	.0937	2.3812
		.094	2.3876
		.095	2.4130
41		.096	2.4384
		.097	2.4638
40		.098	2.4892
		.099	2.5146
39	.0995	.100	2.5400

Drill No. or Letter		Inch	mm
		.101	2.5654
38	.1015	.102	2.5908
		.103	2.6162
37		.104	2.6416
		.105	2.6670
36	.1065	.106	2.6924
		.107	2.7178
		.108	2.7432
		.109	2.7686
	7/64	.1094	2.7781
35		.110	2.7490
34		.111	2.8194
		.112	2.8448
33		.113	2.8702
		.114	2.8956
		.115	2.9210
32		.116	2.9464
		.117	2.9718
		.118	2.9972
		.1181	3.0000
		.119	3.0226
31		.120	3.0480
		.121	3.0734
		.122	3.0988
		.123	3.1242
		.124	3.1496
	1/8	.125	3.1750
		.126	3.2004
		.127	3.2258
		.128	3.2512
30	.1285	.129	3.2766
		.130	3.3020
		.131	3.3274
		.132	3.3528
		.133	3.3782
		.134	3.4036
		.135	3.4290
29		.136	3.4544
		.137	3.4798
		.138	3.5052
		.139	3.5306
28	.1405	.140	3.5560
	9/64	.1406	3.5719
		.141	3.5814
		.142	3.6068
		.143	3.6322
27		.144	3.6576
		.145	3.6830
		.146	3.7084
26		.147	3.7338
		.148	3.7592
		.149	3.7846
25	.1495	.150	3.8100
		.151	3.8354
24		.152	3.8608
		.153	3.8862
23		.154	3.9116
		.155	3.9370
		.156	3.9624
	5/32	.1562	3.9687
22		.157	3.9878
		.1575	4.0000
		.158	4.0132
21		.159	4.0386
		.160	4.0640
20		.161	4.0894
		.162	4.1148
		.163	4.1402
		.164	4.1656
		.165	4.1910
19		.166	4.2164
		.167	4.2418
		.168	4.2672
		.169	4.2926
18	.1635	.170	4.3180
		.171	4.3434
	11/64	.1719	4.3656
		.172	4.3688
17		.173	4.3942
		.174	4.4196
		.175	4.4450
		.176	4.4704
16		.177	4.4958
		.178	4.5212
		.179	4.5466
15		.180	4.5720
		.181	4.5974
14		.182	4.6228
		.183	4.6482
		.184	4.6736
13		.185	4.6990
		.186	4.7244
		.187	4.7498
	3/16	.1875	4.7625
		.188	4.7752
12		.189	4.8006
		.190	4.8260
11		.191	4.8514
		.192	4.8768
		.193	4.9022
10	.1935	.194	4.9276
		.195	4.9530
9		.196	4.9784
		.1969	5.0000
		.197	5.0038
		.198	5.0292
		.199	5.0546
8		.200	5.0800

Drill No. or Letter		Inch	mm
7		.201	5.1054
		.202	5.1308
		.203	5.1562
	13/64	.2031	5.1594
6		.204	5.1816
5	.2055	.205	5.2070
		.206	5.2324
		.207	5.2578
		.208	5.2832
4		.209	5.3086
		.210	5.3340
		.211	5.3594
		.212	5.3848
3		.213	5.4102
		.214	5.4356
		.215	5.4610
		.216	5.4864
		.217	5.5118
		.218	5.5372
	7/32	.2187	5.5562
		.219	5.5626
		.220	5.5880
2		.221	5.6134
		.222	5.6388
		.223	5.6642
		.224	5.6896
		.225	5.7150
		.226	5.7404
		.227	5.7658
1		.228	5.7912
		.229	5.8166
		.230	5.8410
		.231	5.8674
		.232	5.8928
		.233	5.9182
A		.234	5.9436
	15/64	.2344	5.9531
		.235	5.9690
		.236	5.9944
		.2362	6.0000
		.237	6.0198
B		.238	6.0452
		.239	6.0706
		.240	6.0960
		.241	6.1214
C		.242	6.1468
		.243	6.1722
		.244	6.1976
		.245	6.2230
D		.246	6.2484
		.247	6.2738
		.248	6.2992
		.249	6.3246
E	1/4	.250	6.3500
		.251	6.3754
		.252	6.4008
		.253	6.4262
		.254	6.4516
		.255	6.4770
		.256	6.5024
F		.257	6.5278
		.258	6.5532
		.259	6.5786
		.260	6.6040
G		.261	6.6294
		.262	6.6548
		.263	6.6802
		.264	6.7056
		.265	6.7310
	17/64	.2656	6.7469
H		.266	6.7564
		.267	6.7818
		.268	6.8072
		.269	6.8326
		.270	6.8580
		.271	6.8834
I		.272	6.9088
		.273	6.9342
		.274	6.9596
		.275	6.9850
		.2756	7.0000
		.276	7.0104
J		.277	7.0358
		.278	7.0612
		.279	7.0866
		.280	7.1120
K		.281	7.1374
	9/32	.2812	7.1437
		.282	7.1628
		.283	7.1882
		.284	7.2136
		.285	7.2390
		.286	7.2644
		.287	7.2898
		.288	7.3152
		.289	7.3406
L		.290	7.3660
		.291	7.3914
		.292	7.4168
		.293	7.4422
		.294	7.4676
M		.295	7.4930
		.296	7.5184
	19/64	.2969	7.5406
		.297	7.5438
		.298	7.5692
		.299	7.5946
		.300	7.6200

Drill No. or Letter		Inch	mm
		.301	7.6454
N		.302	7.6708
		.303	7.6962
		.304	7.7216
		.305	7.7470
		.306	7.7724
		.307	7.7978
		.308	7.8232
		.309	7.8486
		.310	7.8740
		.311	7.8994
		.312	7.9248
	5/16	.3125	7.9375
		.313	7.9502
		.314	7.9756
		.3150	8.0000
		.315	8.0010
O		.316	8.0264
		.317	8.0518
		.318	8.0772
		.319	8.1026
		.320	8.1280
		.321	8.1534
		.322	8.1788
P		.323	8.2042
		.324	8.2296
		.325	8.2550
		.326	8.2804
		.327	8.3058
		.328	8.3312
	21/64	.3281	8.3344
		.329	8.3566
		.330	8.3820
		.331	8.4074
Q		.332	8.4328
		.333	8.4582
		.334	8.4836
		.335	8.5090
		.336	8.5344
		.337	8.5598
		.338	8.5852
R		.339	8.6106
		.340	8.6360
		.341	8.6614
		.342	8.6868
		.343	8.7122
	11/32	.3437	8.7312
		.344	8.7376
		.345	8.7630
		.346	8.7884
		.347	8.8138
S		.348	8.8392
		.349	8.8646
		.350	8.8900
		.351	8.9154
		.352	8.9408
		.353	8.9662
		.354	8.9916
		.3543	9.0000
		.355	9.0170
		.356	9.0424
		.357	9.0678
T		.358	9.0932
		.359	9.1186
	23/64	.3594	9.1281
		.360	9.1440
		.361	9.1694
		.362	9.1948
		.363	9.2202
		.364	9.2456
		.365	9.2710
		.366	9.2964
		.367	9.3218
U		.368	9.3472
		.369	9.3726
		.370	9.3980
		.371	9.4234
		.372	9.4488
		.373	9.4742
		.374	9.4996
	3/8	.375	9.5250
		.376	9.5504
V		.377	9.5758
		.378	9.6012
		.379	9.6266
		.380	9.6520
		.381	9.6774
		.382	9.7028
		.383	9.7282
		.384	9.7536
		.385	9.7790
W		.386	9.8044
		.387	9.8298
		.388	9.8552
		.389	9.8806
		.390	9.9060
	25/64	.3906	9.9219
		.391	9.9314
		.392	9.9568
		.393	9.9822
		.3937	10.0000
		.394	10.0076
		.395	10.0330
		.396	10.0584
X		.397	10.0838
		.398	10.1092
		.399	10.1346
		.400	10.1600

Drill No. or Letter		Inch	mm
		.401	10.1854
		.402	10.2108
		.403	10.2362
Y		.404	10.2616
		.405	10.2870
		.406	10.3124
	13/32	.4062	10.3187
		.407	10.3378
		.408	10.3632
		.409	10.3886
		.410	10.4140
		.411	10.4394
		.412	10.4648
Z		.413	10.4902
		.414	10.5156
		.415	10.5410
		.416	10.5664
		.417	10.5918
		.418	10.6172
		.419	10.6426
		.420	10.6680
		.421	10.6934
	27/64	.4219	10.7156
		.422	10.7188
		.423	10.7442
		.424	10.7696
		.425	10.7950
		.426	10.8204
		.427	10.8458
		.428	10.8712
		.429	10.8966
		.430	10.9220
		.431	10.9474
		.432	10.9728
		.433	10.9982
		.4331	11.0000
		.434	11.0236
		.435	11.0490
		.436	11.0744
		.437	11.0998
	7/16	.4375	11.1125
		.438	11.1252
		.439	11.1506
		.440	11.1760
		.441	11.2014
		.442	11.2268
		.443	11.2522
		.444	11.2776
		.445	11.3030
		.446	11.3284
		.447	11.3538
		.448	11.3792
		.449	11.4046
		.450	11.4300
		.451	11.4554
		.452	11.4808
		.453	11.5062
	29/64	.4531	11.5094
		.454	11.5316
		.455	11.5570
		.456	11.5824
		.457	11.6078
		.458	11.6332
		.459	11.6586
		.460	11.6840
		.461	11.7094
		.462	11.7348
		.463	11.7602
		.464	11.7856
		.465	11.8110
		.466	11.8364
		.467	11.8618
		.468	11.8872
	15/32	.4687	11.9062
		.469	11.9126
		.470	11.9380
		.471	11.9634
		.472	11.9888
		.4724	12.0000
		.473	12.0142
		.474	12.0396
		.475	12.0650
		.476	12.0904
		.477	12.1158
		.478	12.1412
		.479	12.1666
		.480	12.1920
		.481	12.2174
		.482	12.2428
		.483	12.2682
		.484	12.2936
	31/64	.4844	12.3031
		.485	12.3190
		.486	12.3444
		.487	12.3698
		.488	12.3952
		.489	12.4206
		.490	12.4460
		.491	12.4714
		.492	12.4968
		.493	12.5222
		.494	12.5476
		.495	12.5730
		.496	12.5984
		.497	12.6238
		.498	12.6492
		.499	12.6746
	1/2	.500	12.7000

ALUMINUM ASSOCIATION DESIGNATION SYSTEM

NUMBER GROUP	PRINCIPLE ALLOYING ELEMENT
1XXX	Aluminum—100% purity or greater
2XXX	Copper
3XXX	Manganese
4XXX	Silicon
5XXX	Magnesium
6XXX	Magnesium and silicon
7XXX	Zinc
8XXX	An element other than mentioned above
9XXX	Unassigned

The last two digits in this system indicate similar alloys before the present identification was adopted. For example, the alloy 5052 was formerly 52S, 7075 was known as 75S.

The letter "H," when used to designate temper, is followed by two numbers. For example, 3003-H14. The first digit following the "H" denotes the process used to produce the temper. The second number indicates the actual temper (degree of hardness):

2 1/4 hard (2/8)
4 1/2 hard (4/8)
6 3/4 hard (6/8)
8 Full hard (8/8)

STEEL CLASSIFICATION TABLES

TABLE I CARBON STEELS

Type	Carbon Content
Low	0.05 to approximately 0.30 percent carbon
Medium	0.30 to approximately 0.60 percent carbon
High	0.60 to approximately 0.95 percent carbon

TABLE II SAE-AISI CODE CLASSIFICATION (First digit)

The first number of the SAE-AISI Code Classification System frequently, but not always, indicates the basic type of steel. When carbon or alloy steel contains the letter "L" in the code, it contains from 0.15 to 0.35 percent lead to improve machinability. These steels are also known as free-machining steel. The prefix "E" before the alloy steel designation indicates it is made only by electric furnace.

1 — Carbon
2 — Nickel
3 — Nickel-chrome
4 — Molybdenum
5 — Chromium
6 — Chromium-vanadium
7 — Tungsten
8 — Nickel-chromium-molybdenum
9 — Silicomanganese

COLOR CODES FOR MARKING STEELS

S.A.E. Number	Code Color	S.A.E. Number	Code Color	S.A.E. Number	Code Color	S.A.E. Number	Code Color
	CARBON STEELS	2115	Red and bronze	T1340	Orange and green	3450	Black and bronze
1010	White	2315	Red and blue	T1345	Orange and red	4820	Green and purple
1015	White	2320	Red and blue	T1350	Orange and red		CHROMIUM STEELS
X1015	White	2330	Red and white		NICKEL-CHROMIUM STEELS	5120	Black
1020	Brown	2335	Red and white	3115	Blue and black	5140	Black and white
X1020	Brown	2340	Red and green	3120	Blue and black	5150	Black and white
1025	Red	2345	Red and green	3125	Pink	52100	Black and brown
X1025	Red	2350	Red and aluminum	3130	Blue and green		CHROMIUM-VANADIUM STEELS
1030	Blue	2515	Red and black	3135	Blue and green	6115	White and brown
1035	Blue		MOLYBDENUM STEELS	3140	Blue and white	6120	White and brown
1040	Green	4130	Green and white	X3140	Blue and white	6125	White and aluminum
X1040	Green	X4130	Green and bronze	3145	Blue and white	6130	White and yellow
1045	Orange	4135	Green and yellow	3150	Blue and brown	6135	White and yellow
X1045	Orange	4140	Green and brown	3215	Blue and purple	6140	White and bronze
1050	Bronze	4150	Green and brown	3220	Blue and purple	6145	White and orange
1095	Aluminum	4340	Green and aluminum	3230	Blue and purple	6150	White and orange
	FREE CUTTING STEELS	4345	Green and aluminum	3240	Blue and aluminum	6195	White and purple
1112	Yellow	4615	Green and black	3245	Blue and aluminum		TUNGSTEN STEELS
X1112	Yellow	4620	Green and black	3250	Blue and bronze	71360	Brown and orange
1120	Yellow and brown	4640	Green and pink	3312	Orange and black	71660	Brown and bronze
X1314	Yellow and blue	4815	Green and purple	3325	Orange and black	7260	Brown and aluminum
X1315	Yellow and red	X1340	Yellow and black	3335	Blue and orange		SILICON-MANGANESE STEELS
X1335	Yellow and black		MANGANESE STEELS	3340	Blue and orange	9255	Bronze and aluminum
	NICKEL STEELS	T1330	Orange and green	3415	Blue and pink	9260	Bronze and aluminum
2015	Red and brown	T1335	Orange and green	3435	Orange and aluminum		

STRUCTURAL METALS

SHAPES	LENGTH	HOW MEASURED	* HOW PURCHASED
Sheet less than 1/4 in. thick	to 144 in.	Thickness x width, widths to 72 in.	Weight, foot or piece
Plate more than 1/4 in. thick	to 20 ft.	Thickness x width	Weight, foot or piece
Band	to 20 ft.	Thickness x width	Weight, or piece
Rod	12 to 20 ft.	Diameter	Weight, foot or piece
Square	12 to 20 ft.	Width	Weight, foot or piece
Flats	Hot rolled 20–22 ft. Cold finished	Thickness x width	Weight, foot or piece
Hexagon	12 to 20 ft.	Distance across flats	Weight, foot or piece
Octagon	12 to 20 ft.	Distance across flats	Weight, foot or piece
Angle	Lengths to 40 ft.	Leg length x leg length x thickness of legs	Weight, foot or piece
Channel	Lengths to 60 ft.	Depth x web thickness x flange width	Weight, foot or piece
I–beam	Lengths to 60 ft.	Height x web thickness x flange width	Weight, foot or piece

* Charge made for cutting to other than standard lengths.

Physical Properties of Metals

METAL	SYMBOL	SPECIFIC GRAVITY	SPECIFIC HEAT	MELTING POINT*		LBS. PER CUBIC INCH
				DEG. C	DEG. F.	
Aluminum (Cast)	Al	2.56	.2185	658	1217	.0924
Aluminum (Rolled)	Al	2.71	–	–	–	.0978
Antimony	Sb	6.71	.051	630	1166	.2424
Bismuth	Bi	9.80	.031	271	520	.3540
Boron	B	2.30	.3091	2300	4172	.0831
Brass	–	8.51	.094	–	–	.3075
Cadmium	Cd	8.60	.057	321	610	.3107
Calcium	Ca	1.57	.170	810	1490	.0567
Carbon	C	2.22	.165	–	–	.0802
Chromium	Cr	6.80	.120	1510	2750	.2457
Cobalt	Co	8.50	.110	1490	2714	.3071
Copper	Cu	8.89	.094	1083	1982	.3212
Columbium	Cb	8.57	–	1950	3542	.3096
Gold	Au	19.32	.032	1063	1945	.6979
Iridium	Ir	22.42	.033	2300	4170	.8099
Iron	Fe	7.86	.110	1520	2768	.2634
Iron (Cast)	Fe	7.218	.1298	1375	2507	.2605
Iron (Wrought)	Fe	7.70	.1138	1500–1600	2732–2912	.2779
Lead	Pb	11.37	.031	327	621	.4108
Lithium	Li	.057	.941	186	367	.0213
Magnesium	Mg	1.74	.250	651	1204	.0629
Manganese	Mn	8.00	.120	1225	2237	.2890
Mercury	Hg	13.59	.032	38.7	37.7	.4909
Molybdenum	Mo	10.2	.0647	2620	4748	.368
Monel Metal	–	8.87	.127	1360	2480	.320
Nickel	Ni	8.80	.130	1452	2646	.319
Phosphorus	P	1.82	.177	43	111.4	.0657
Platinum	Pt	21.50	.033	1755	3191	.7767
Potassium	K	0.87	.170	62	144	.0314
Selenium	Se	4.81	.084	220	428	.174
Silicon	Si	2.40	.1762	1427	2600	.087
Silver	Ag	10.53	.056	961	1761	.3805
Sodium	Na	0.97	.290	97	207	.0350
Steel	–	7.858	.1175	1330–1378	2372–2532	.2839
Strontium	Sr	2.54	.074	–	–	.0918
Sulphur	S	2.07	.175	115	235.4	.075
Tantalum	Ta	10.80	–	2850	5160	.3902
Tin	Sn	7.29	.056	232	450	.2634
Titanium	Ti	5.3	.130	1900	3450	.1915
Tungsten	W	19.10	.033	3000	5432	.6900
Uranium	U	18.70	–	–	–	.6755
Vanadium	V	5.50	–	1730	3146	.1987
Zinc	Zn	7.19	.094	419	786	.2598

* Circular of the Bureau of Standards No. 35, Department of Commerce and Labor.

INDEX

U

V

W

X

Z